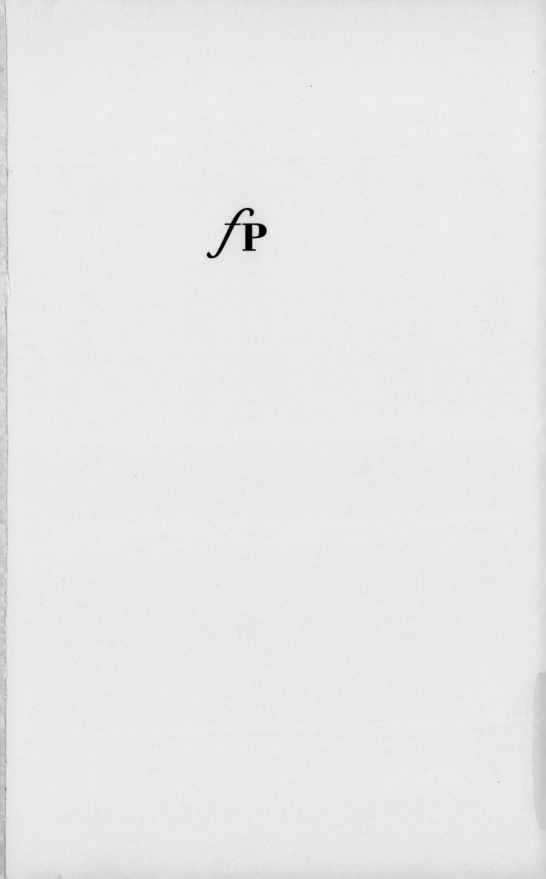

# GERALD L. SCHROEDER

## HOW SCIENCE

## REVEALS

## THE ULTIMATE

## TRUTH

THE FREE PRESS
NEW YORK
LONDON
TORONTO
SYDNEY
SINGAPORE

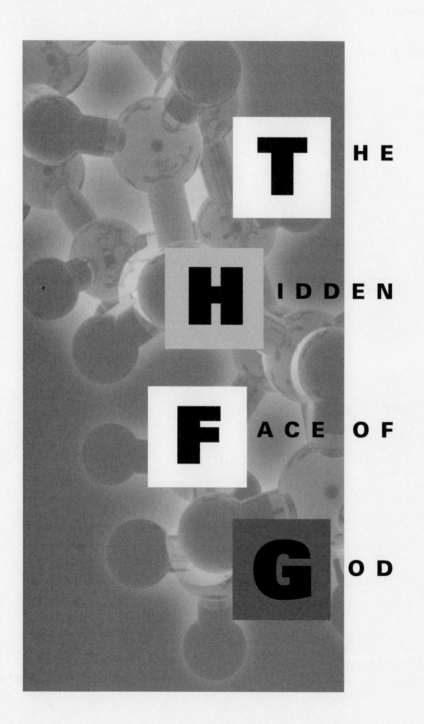

THE HIDDEN FACE OF GOD

THE FREE PRESS
A Division of Simon & Schuster, Inc.
1230 Avenue of the Americas
New York, NY 10020

Copyright © 2001 by Gerald L. Schroeder

THE FREE PRESS and colophon are trademarks
of Simon & Schuster, Inc.

Designed by Karolina Harris

Manufactured in the United States of America

10   9   8   7   6   5   4   3   2   1

Library of Congress Cataloging-in-Publication Data
Schroeder, Gerald L.
The hidden face of God: how science reveals the ultimate truth/Gerald L. Schroeder.
p. cm.
Includes index.
1. Religion and science. 2. God—Proof, Teleological. I. Title.
BL240.2 .S3226 2001
215—dc21          00-050363
ISBN 0-684-87059-2

*To my wife and our children—Avi, Josh, Hadas, Yael, Hanna—*
*who continually inspire me with the magnificence of creation.*

"…[W]hat a wonderful world," as sung by the late, great
Louis Armstrong, a valued associate of my parents,
Dorothy and Herman Schroeder of blessed memory.

# CONTENTS

# PROLOGUE: WE ARE THE

# UNIVERSE COME ALIVE

A single consciousness, an all-encompassing wisdom, pervades
the universe. The discoveries of science, those that search the
quantum nature of subatomic matter, those that explore the molecular complexity of biology, and those that probe the
brain/mind interface, have moved us to the brink of a startling
realization: all existence is the expression of this wisdom. In the
laboratories we experience it as information first physically articulated as energy and then condensed into the form of matter.
Every particle, every being, from atom to human, appears to
have within it a level of information, of conscious wisdom. The
puzzle I confront in this book is this: where does this arise?
There is no hint of it in the laws of nature that govern the interactions among the basic particles that compose all matter. The
information just appears as a given, with no causal agent evident,
as if it were an intrinsic facet of nature.

The concept that there might be an attribute as nonphysical
as information or wisdom at the heart of existence in no way
denigrates the physical aspects of our lives. Denial of the plea-
sures and wonder of our bodies would be a sad misreading of
the nature of existence. The accomplishments of a science based
on materialism have given us physical comforts, invented life-
saving medicines, sent people to the moon. The oft-quoted
statement, "not by bread alone does a human live" (Deut. 8:3),
lets us know that there are *two* crucial aspects to our lives, one of

which is bread, physical satisfaction. The other parameter is an underlying universal wisdom. There's no competition here between the spiritual and the material. The two are complementary, as in the root "to complete."

When we see through the camouflage haze that at times convinces us that only the material exists, when we touch that consciousness, we know it. A joyful rush of emotion sweeps over the entire self. This emotional response—some might call it a religious experience—is reported in every culture, from every period. It tells us that we've come home. We've discovered the essence of being. Everyone has felt it at some time or other. Perhaps at a brilliant sunrise, in a work of art, the words of a loved one. The physical and the metaphysical have joined.

If we dared, we'd call the experience spiritual, even Godly. But there's a reluctance to use the "G" word. "Listen to the Force" is acceptable on the great silver screen. If the *Star Wars* scriptwriter had used "Listen to God," the theater would have emptied in a flash. The reluctance is not surprising, considering the bizarre claims erroneously attributed to God through the ages and especially in our age. A bit of scrutiny reveals that most of those claims are based on the expectations for the putative (and generally misunderstood) God of the Bible that we learned as children. Obviously, when our child-learned wisdom is evaluated by the sophistication of our adult minds, that wisdom is bound to seem naive.

The age-old theological view of the universe is that all existence is the manifestation of a transcendent wisdom, with a universal consciousness being its manifestation. If I substitute the word information for wisdom, theology begins to sound like quantum physics. Science itself has rediscovered the confluence between the physical and the spiritual.

If a spiritual unity does underlie physical reality, it would be natural for people to search for that unity. Regrettably in the rush of our daily obligations we often become disconnected, losing the realization that such a unity might actually exist. Our private worlds today seem to expand almost as rapidly as the universe has been physically expanding since its creation. The scientific discoveries facilitating this nomadic mobility of the mind come at a rate that far exceeds the ability of our cultures to adapt. New technologies simply displace old cultural ties, and in doing so jettison traditions that formerly stabilized society.

In the developing world, those referred to as the poorest of the poor are the landless. In a sense we have become landless nomads, cut off from our roots, even in the midst of wealth. We deal in tokens. Other than artists and the one percent of the population that works on a farm, most of us have no relation to the final product of our labors. We buy and sell stocks of companies making products we barely understand. We deal in the ultimate of tokens, money. Money has no intrinsic value. It may promise security, pleasure, even freedom, but it doesn't provide those insatiable and all too often elusive goals. The resulting angst is almost palpable. Divorce rates exceed 50 percent. Violence in homes crosses socioeconomic divisions. Histories have been exchanged for gossamer hopes of a freedom untethered to tradition.

Accessing the consciousness within which we are embedded requires skills that go beyond our intuitions. The amazing, even startlingly illogical, discoveries in physics and biology during the past few decades have given us the tools to gain scientific insight into the metaphysical underpinnings of our world and, in return, acquire spiritual insight into scientific, empirical fact. Understanding nature's wonders need in no way detract from its majesty. By realizing the interwoven complexity of existence, we experience the oneness both by revelation and by reason.

No monk's life of isolated contemplation is being proposed here, no excluding of oneself from the world. The upsurge of interest in meditation, Eastern religions, and kabala reflects an almost desperate search to rediscover our spiritual roots. Those roots are best found while fulfilling the usual responsibilities of adult life, not within some cloister. Exposing the awe of existence *within* the reality of daily life is what this book is about.

We are, each of us, a part of the universe seeking itself. We struggle between a world that seems totally material and the emotional, even spiritual, pull we all feel at times. To relegate, a priori, those feelings of love and joy and spirituality to some assumed function of our ancestors' evolutionary drive for survival masks the greatest pleasure in life, the experiential realization of the metaphysical.

In the following pages, as we journey through the newly discovered marvels of the cosmos, of life, and finally of the brain/mind interface, I ask only that, as you read, you use these facts to reexamine your opinions concerning the origins, evolution, and essence of this wonderful world in which we live.

# THE PUZZLE OF

# EXISTENCE:

# AN OVERVIEW

*Theology is not simply a matter of interpreting scriptures, be it the Bible, the Koran, or the Tao Te Ching. Theology brings us the amazing concept of a metaphysical Force that brought the physical universe into being. The universe is the physical expression of the metaphysical. All that we know of the putative Creator is found within the physical creation. With this in mind, it is incongruous to describe a theology without the insights of science.*

When I picture the earth and solar system hanging in the vastness of space, I feel an anxious need to grab hold of something stable. We're a minuscule speck, somehow floating in the seemingly endless dark of the night sky. It's at those times that I realize all existence hangs by a thread, the breadth of a hair. But the puzzle of our tenuous place in space is secondary to the most baffling riddle of all: that of existence itself. It's a question we might prefer to ignore.

Why is there an "is"? Why is there something, anything, rather than nothing? In our fascination with life's origin and evolution, we bypass this most basic of conundrums. The very fact and nature of existence, the finite aspects of the physical world we view about us, the limited nature of time, space, and matter from which we and all the universe are constructed, force upon us the unsettling reality that at some level there is the metaphysical. Some undefined whatever, transcendent of the physical, produced the physical? Atheist, agnostic, skeptic, and "believer" all share the understanding that some metaphysical non-thing, metaphysical in the sense of being above or outside of the physical, must have preceded our universe or have our universe imbedded in it. That much is a certainty.

But is this metaphysical force that produced our universe Godly? That is not an easy question. It is, however, probably the most important of questions. And the answer really depends on one's definition of God. Just what are the expectations of how an infinite, incorporeal metaphysical creating force would be made manifest within the finite corporeality of the universe? It may be definitions more than reality that separate skeptic from believer, especially if one's expectations of the metaphysical or of the physical are what was learned as a child.

To have a prayer of probing intelligently the subtleties of the metaphysics of existence, we first had better understand the essence of existence at the physical level. In the coming chapters we'll get into the molecular workings that underlie life and discover a complexity so extreme, so overwhelmingly elaborate, that it outdoes science fiction by a league. And then we'll confront a most intriguing enigma: what allows life to conceive a thought, to imagine a rainbow, a symphony? Is the mind totally the neurological workings of the brain or is there a brain/mind interface where the physical mingles with the nonphysical? That would make the brain an instrument, a sort of antenna that taps into the consciousness of the universe. But I'm getting ahead of my story.

Discoveries in the fields of physics, cosmology, and molecular biology during the past few decades have moved us ever closer to what might be termed a metaphysical truth. First of all, there is the creation itself: how did the universe come into existence? And then there is the orderly yet phenomenally complex information we find encoded time and again in all aspects of nature, and most especially in life. The wisdom contained therein is not at all evident in the physical building blocks from which life, even in its simplest of forms, is composed. And finally, there is the puzzle of consciousness and the brain/mind interface. Is mind more that the sum of the physical parts of the brain? The possibility that mind is an emergent property of brain echoes the emergence of wisdom in nature. Of course, the suggestion of a ubiquitous wisdom raises the problem of *imperfect design*. If life is indeed the result of an intelligent force, why are there so many less-than-optimum aspects in its design? I will briefly discuss these topics in this chapter and then pursue the details in the rest of the book. As the saying goes, God is in the details.

## *Creation and a First Cause*

Is there a universal consciousness, an entanglement that at some subtle level ties all existence together? At the moment of the big bang everything, the entire universe, you included, was part of a homogeneous speck—no divisions, no separations—an undifferentiated iota filled with exquisitely powerful energy, a speck no larger than the black pupil of an eye. And before this there was neither time nor space nor matter. That speck was the entire universe. Not a speck within some vacuous space. A vacuum is space. The speck was the entire universe. There was no other space. No outside to the inside of the creation. Creation was everywhere at once. And then the space and the energy stretched out to form all that exists today in the heavens and on the earth. You (and I too) were once part of a total unity, a single burst of radiation. It formed our bodies and led to our thoughts. As incomprehensible as it may seem, we, the essence that was eventually to form our beings, were present at the creation.

When we leave the workings of the macro-world and enter the world of the atom, we are confronted with an even more surprising yet scientifically confirmed picture of our universe.

First we discover that solid matter, the floor upon which we stand and the foundation that bears the weight of a skyscraper, is actually empty space. If we could scale the center of an atom, the nucleus, up to four inches, the surrounding electron cloud would extend to four miles away and essentially all the breach between would be marvelously empty. The solidity of iron is actually 99.9999999999999 percent startlingly vacuous space made to feel solid by ethereal fields of force having no material reality at all. Hollywood would have rejected such a script out of hand and yet it is the proven reality. But don't knock your head against that space. Force fields can feel very solid.

To experience a force field, hold a heavy object out at arm's length. Feel the downward pull. But what is doing the pulling? Gravity, you say. But what produces the gravity? Gravitons of course. But what are gravitons? For that question there is no answer. Where are the strings that tug down toward the earth? They are there, but other than the pull, they are totally imperceptible.

And then quantum physics came onto the scene, with its micro, micro world, only to prove to us that there is no reality. Not even the one part in a million billion that seemed to be solid. Ask a scientist, a physicist, what an electron or the quarks of a proton are made of. She or he will have no answer. Ask the composition of photons—the wave/particle packets of energy that underlie all those micro-particles of matter. The reply will be along the lines of "Huh?"

That all existence may be the expression of information, an idea, a quantum wave function, is not fantasy and it is not some flaky idea. It's mainstream science coming from such universities as Princeton and M.I.T. There is the growing possibility that for all existence, we humans included, there's nothing, nothing as in "no thing," there. In the following chapters we'll discover that the world is more a thought than a thing, more intangible than real. It's essential to see the magnificent subtlety and bounds of how the physical world works if we are going to speculate concerning the metaphysical.

In science, there are easy questions and hard questions. The easy questions, though in practice hard to solve rigorously, all have material solutions. Examples include Newton's phenomenal insight into the laws of motion and gravity; Einstein's revolutionary concepts of relativity. The hard questions relate to what seem to be the intrinsic qualities of nature. Why is there a force of gravity and why does matter generate such a force? Of what are the fundamental particles of matter, electrons, quarks, composed? What is energy?

And then there is what Professor David Chalmers of the department of psychology at the University of Arizona describes as the very hard question. It is that at which I am aiming. What is the mind?

## *Mind/Matter*

The easy part of that question deals with the brain. Wiggle your toes. Feel them? But where do you feel them? Not in your toes. Toes feel nothing. You feel them in your brain. Anyone who has had the misfortune of having a limb amputated can tell you how the missing limb continues to be felt—in the brain. The brain has within it maps of the body that record every sensation and then project that sensation onto the mental image of the relevant body part. But it certainly feels like I'm feeling my toes in my toes. And it is not just the toes. The entire reality, what we see and what we feel, what we smell and what we hear, is mapped in the brain and then those recorded sensations reach out to our consciousness from within the two-to-four-millimeter (about one-eighth-inch) thin wrinkled gray layer, the cerebral cortex, that rests at the top of each of our brains. There is a reality out there in the world, but what we experience—every touch and every sound, every sight, smell and taste—arises in our heads. All our mental images, fantasy or factual, are built on our life's experiences and these are totally physically based. Pure abstraction, even in mathematics, is more than difficult. It is impossible. Even numbers and symbols of calculus take on mental forms.

Through the dedicated and precise work of extraordinarily skilled scientists we now know how and where in the brain each of our sensations is processed and stored. That has been mapped to near perfection. And I explore those maps in the coming chapters. The processes are nothing less that mind-boggling.

And then comes the hard part of the hard question: the sound of music. The waves of sound impinge upon my eardrum and in a beautifully complex path become converted to bio-electrical pulses that are chemically stored in the cortex of my brain. (We'll look at that path in detail.) But how do I hear the sound? Up to and including the storage of the data in the brain, it's all biochemistry. But I don't hear biochemistry. I hear sound. Where's the sound generated in my head? Or the vision; or the smell? Where's the consciousness? Just which of those formerly inert atoms of carbon, hydrogen, nitrogen, oxygen, and on and on, in my head have become so clever that they can produce a thought or reconstitute an image? How those stored biochemical data points are recalled and replayed into sentience remains an enigmatic mystery.

Organization separates living matter from what we perceive as inert, lifeless matter. But there is no innate physical difference between the atoms before and after they organize, learn to take energy from their surrounding environment, and become, as a group, alive. However there is what appears to be a qualitative transition between the awesome biochemistry by which the brain physically records the incoming data and the consciousness by which we become aware of that stored information. In that passage from brain to mind we may be looking for a physical link that does not exist.

Could the consciousness we perceive as the mind be as fundamental as, let's say, the phenomena of gravity generated by mass, or the electrical charge generated by a proton? Gravity and charge are emergent properties. They emanate from substrate particles, as for example gravity from mass, and extend throughout the universe. And though each particle continually emits its field, the emission does not "use up" the particle. For all the force of the field, it seems the force emerges from, but is not made of, the particle from which it arises. Though it originates with the substrate particle, the field is not part of the particle.

Time provides an additional clue in the brain/mind puzzle. We drift in a river of time. There's no possibility of swimming upstream, of going back in time. Destroy every clock, every item that feels the passage of time. The flow of time continues unabated. Time is an intrinsic, ubiquitous quality of our universe, irrespective of whether or not we measure its passage. Might consciousness also be an intrinsic, all-present part of nature, of the universe? In that case every particle would have some aspect of consciousness within. The more complex the entity, the greater would be its awareness of the consciousness housed within.

If quantum physics is correct, and it has an excellent track record of being on the mark, then this suggestion is not so very far from mainstream thought. Every particle, every body, each aspect of existence appears to be an expression of information, information that via our brains or our minds, we interpret as the physical world. Physicist Freeman Dyson, upon his acceptance of the Templeton Prize, stated the idea this way: "Atoms are weird stuff, behaving like active agents rather than inert substances. They make unpredictable choices between alternative possibilities according to the laws of quantum mechanics. It appears that mind, as manifested by the capacity to make choices, is to some extent inherent in every atom. The universe is also weird, with its laws of nature that make it hospitable to the growth of mind. I do not make any clear distinction between mind and God. God is what mind becomes when it has passed beyond the scale of our comprehension." These are the words of one of today's leaders in physics!

From the tiniest grain of sand to the brain of an Einstein, all existence, animate and inanimate, is the product of the same ninety-two elements that are themselves the harvest of the energy of the creation. At every turn an underlying commonality, a unity, emerges from within the diversity.

Now quantum physics has moved us to a more uncanny realization: that at some level this commonality extends to what can be described as an instantaneous mutual awareness among all particles, regardless of the distance of separation. What occurs on one side of the universe seems to affect events on the other side and all points in between. If indeed there is a universal consciousness, this could explain the interrelatedness of particles even when separated by large distances. The most famous demonstration of this is presented in the persistently arcane observations obtained in the "double-slit" experiment. Particles, be they ethereal photons of energy, featherweight electrons or massive atoms, passing through a slit, call it "A," somehow know, and react accordingly, to conditions at far-off slit "B." Among those variable conditions at B is whether or not a conscious observer is present. These data carry a hint that perhaps—just perhaps—consciousness affects the physical manifestation of existence.

It took humanity millennia before an Einstein discovered that, as bizarre as it may seem, the basis of matter is energy, that matter is actually condensed energy. It may take a while longer for us to discover that there is some non-thing even more fundamental than energy that forms the basis of energy, which in turn forms the basis of matter. The renowned former president of the American Physical Society and professor of physics at Princeton University, recipient of the Einstein Award and member of the National Academy of Sciences, John Archibald Wheeler, likened what underlies all existence to an idea, the "bit" (the binary digit) of information that gives rise to the "it," the substance of matter. If we can discover that underlying idea, we will have ascertained not only the basis for the unity that underlies all existence, but most important, the source of that unity. We will have encountered the hidden face of God.

In his book *The Medium Is the Massage* [sic], Marshall McLuhan wrote: "We're not sure who discovered water, but we're pretty sure it wasn't the fish." When one is totally and continually immersed in a single medium, awareness of that unique medium fades.

All we know of ourselves and of the universe, all we know of existence, is what the consciousness of the mind tells us. The one constant aspect of my life is that I am myself. Even in my most fantastic dreams I remain conscious of being me. We swim in a stream of consciousness and accept it without even noticing it.

## (Imperfect) Design

When searching for hints of the metaphysical within the realm of the physical, we first must realize that scientists learn to investigate *how* the world works, not *why* it exists. I earned my three academic degrees at the Massachusetts Institute of Technology. My third degree was a Ph.D. in earth sciences and nuclear physics. Ph.D. as in doctor of philosophy. But while with my doctorate of philosophy I learned a great deal in how to probe the functioning of the world, I learned precious little about the philosophy of the world, the study of the why of existence. Until recently science has dealt almost exclusively with a cause-and-effect, materialistic view of reality. The transition from classical physics to quantum physics is only gradually forcing questions of why into the curriculum. The "how" it works is physics. The "why" of existence is already metaphysics.

Another mental hurdle to traverse, and perhaps the most common error in the ongoing, but unnecessary, dispute between science and theology, is the assumption that intelligent design means perfect design. From a human perspective, it's obviously wrong. The *Titanic* in boats, the *Electra* in planes, Three Mile Island and Chernobyl in power generation, all make the point fairly clearly. From a theological stance, the story is the same. Judaism, Christianity, and Islam all draw their understanding of creation from the opening chapters of the Book of Genesis. Genesis claims that a single, eternal, omnipotent and incorporeal God created the universe. This would imply intelligent design on the level of the Divine, and yet from the Bible's view, the universe is far from perfect.

The Bible has broken this sad feature of reality to us by the fourth chapter of Genesis, with Abel's murder by Cain. And remember, according to the text, Abel was the good guy. By chapter four, the good guy is dead. By chapter six, God brings the famous Flood at the time of Noah. From a biblical perspective, very early in the game—in fact, after only six chapters—things had gone from bad to worse and God let them all happen. Society had become so degenerate that God lamented: "I regretted that I made them [all humanity]" (Gen. 6:7). It's at that point that God, in a sense, pressed the reset button and brought on the Flood. The conditions of the world changed and the nine-hundred-year human life-spans that preceded the Flood gradually decreased to the ninety or so years we still experience. Apparently nine-hundred-year longevity was not the optimum design for humans. For the moment, it's irrelevant whether or not you believe that the Flood and all its details actually occurred, or whether or not nine-hundred-year life-spans were possible. What is important here is the biblical message being presented: that the original Divine design of the world was flawed, required Divine retuning, and the Tuner acknowledged this need. According to the Bible, intelligent design, even at the level of the Divine, is not necessrily perfect design.

Biblically, debacle follows upon debacle. Psalm 23 begins "The Lord is my shepherd, I shall not want." This certainly sounds like good news. But just before, we read "My God, my God, why have you forsaken me?" That's Psalm 22.

Opening our eyes to the possibility of imperfect design even at the level of the Divine makes the theological claim of design having been included in the creation much more difficult to dismiss. In fact, biblically speaking, the Designer seems not even to be always visibly active in the accomplishment of the design. As theologically nonkosher as it may sound, at times it seems as if the biblically described Creator is actually feeling Its way along in the unfolding drama of existence. The opening chapter of the Bible, in a brief thirty-one sentences, describes the development of our universe from chaos at its creation to the culminating symphony of life. Seven times in those few sentences we're told "God saw it is good." At the end of the process God was so pleased that the creation was described as "very good." Sounds as if at times it might not have been so good.

What we learn from these biblical events is that intelligent design can be complex design, but it is not necessarily flawless design, even when that design is the work of the Creator. If your image of God is based on a simplistic model of the Divine, don't expect that image to rest easily with the Bible's concept of God or with the real world. Out there on the street, the innocent are often the victims and the guilty at times merely walk away.

There is an ancient source of biblical interpretation, known as kabala, that can help us understand the paradox of what seems to be an occasionally absent God. Based on nuances in the biblical text, the kabala teases out meanings not always apparent in a casual reading. A point of clarification is essential. The popular conception that kabala is mysticism is in error. Kabala is logic, but such deep logic that discovering it can lead to a mystical experience. The kabalistic approach is in essence mathematical. There are two sides to the equation of existence. One side deals with the material world, the other, the spiritual. Any activity to the one brings a parallel activity in the other. Kabala is not the study of God. It would be hubris to think that the finite can comprehend the infinite. Kabala is the study of how the Infinite interacts with the finite creation, what might be called the spiritual physics of the universe.

Two basic concepts govern this mode of study. First, that a

unity pervades and underlies all existence. This is the meaning of "the Eternal is One" (Deut. 6:4).* And then the concept of *tzimtzum*, the Hebrew word for contraction, a spiritual pulling back by the Eternal to yield spiritual space for existence. It is this *tzimtzum* that masks the unity of the creation and allows the slack or leeway in the system. At times the universe seems grandly divine; at times totally natural. Like an Escher painting, with its figure/ground reversals popping back and forth even as we look at it. We struggle to keep one dimension in focus. God, having introduced this level of indeterminacy into the system, allows the world to operate within that range. With humans, it shows itself as our ability to exercise our free will choices. According to the Bible, only when events get way off course does God manifestly step in and redirect the flow. Other than that it all looks natural.

When we piece together these aspects of what the metaphysical might be, we find that the common ground held both by skeptic and believer is actually quite broad. All hold that an eternal non-thing preceded our universe; that our universe had a beginning; that intelligent design, even Divine intelligent design, is not necessarily perfect design. All agree that our universe operates according to laws of great power. One and only one facet of the metaphysical separates the skeptic from the believer. That is whether or not that which preceded creation is immanent and active in the creation. Believer says yes. Skeptic says no. But even on this point, the putative presence of the biblical God can be quite tenuous, and often not evident to the untrained eye.

WE humans like to label things, to wrap our minds around a concept, to define and package it; in essence to limit it so that the concept finds harmony within our human definition of logic.

---

*Biblical quotations are based on translations from the original Hebrew.

But how does someone label or even think about that which is not part of our physical world? Confining the metaphysical to a physical description totally misses the "meta" aspect.

> And Jacob asked him, and said "Tell me your name." And he said "Wherefore is it that you ask after my name?" (Genesis 32:30)
>
> And [the people] will say to me "What is His name?" What shall I say to them? And God said to Moses, "I will be that which I will be." And He said thus shall you say to the children of Israel "I will be" sent me to you. . . . This is My name *le-olam*. (Exod. 3:13–15)

*Le-olam*—a Hebrew word with three root meanings: forever, and also hidden, and also in the world. This is my name forever hidden in the world. So how to recognize the presence of the metaphysical?

> On that day [the Eternal] shall be One and Its name One. (Zech. 14:9)
>
> The Eternal is One. (Deut. 6:4)

That is to say, the Eternal *is* One.

But don't think that this is the kind of *one* after which might come the quantities two, three, and four. Nothing as superficial as a number is being revealed in these statements. Rather, the infinite metaphysical as perceived by the physical is an all-encompassing, universal unity. A total oneness is as close as the several trillion neural connections in our brains can come in our quest to discern the infinite.

> You shall know this day and place it in your heart that the Eternal is God in heaven above and on earth below; *ain od*. (Deut. 4:39)

*ain od*—a Hebrew expression in this verse meaning there is nothing else (compare Deut. 4:39 with Deut. 4:35).

That is to say, there is *nothing else*. Nothing other than this singular totality.

Everything, everything with no exception, is a manifestation of an eternal unity, a transcending ubiquitous consciousness, which many label as God. When you touch that unity, you perceive and also experience the wonder within which you and all the rest of creation are embedded. As the rush of emotion sweeps through your body, your level of consciousness moves from the personal aspect of being self-aware and closes the gulf between the local physical and the universal metaphysical.

This transition in awareness finds parallel in the Bible through our changing relationship with God. At Eden there's a parent-child bond, with the parent (God) setting arbitrary rules—such as don't eat from the tree of knowledge. Later, following the covenant between God and Abraham, a partnership emerges as Abraham and then Moses argue with God over how God should manage the world. And finally, in the Song of Songs, the ultimate relationship is reached, that of a lover (we humans) seeking the loved one (God), and in turn, the loved one seeking the lover. When the encounter finally occurs, we touch and blend with the unity that is all existence. The hidden face of God is to be discovered in that unity.

"The king has brought me into his chambers; we will be glad and rejoice in thee" (Song of Songs 1:4). "All the writings of the Bible are holy," Rabbi Akiva declared, "but the Song of Songs is the holiest of all" (Talmud).

God is not mentioned in the Song of Songs. Even at the ultimate encounter, the face of God remains hidden.

# PHYSICS

# AND METAPHYSICS

*If we could know all the properties of all the elements, and every nuance of the laws of nature, we could predict that sodium and chlorine can combine to form sodium chloride, "common" table salt. But could we predict that a collection of atoms can join together to form a brain and then a thought?*

*System A can produce system B where B is more complex than A. But system A cannot produce a new system, call it B, whose basic parameters are of a type that are totally different from those of A. Try for example to imagine a new type of universe, the form of which has neither time nor space nor matter. Or picture a boundary such as a wall that has an inside edge but no outside edge. Or an existence in which the only dimension is time—no space and no physical items showing the effects of time's passage. All of these systems are outside our range of imagination because all our imaginings are built of the parameters in which we live.*

*And yet when we observe the sentient wisdom of life emerging from the physical structures of life we are witnessing the emergence of a parameter not evident in those structures.*

The classrooms in which I teach introductory and second-level sciences are located in different campus buildings. The five-minute walk from one to the other winds through the alleys of the Old City of Jerusalem, known in Arabic as Al Kuds. It is a spectacular passage. With the turn of a head, you can see two thousand years of history and toil. Here three cultures, Jewish, Christian, and Moslem (chronologically listed), are learning slowly and at times painfully to live together. A while ago, I shared that walk with a colleague, the academic dean of a well-known American university.

After a few comments that skirted the issue, he broached what was uppermost in his mind, though he did it in the third person. "The head of our physics department complains that he wears two hats. During the week it's that of his lab. Comes the weekend, it's the garb of religion. His beliefs have him trapped in a dichotomy. How can one believe in a Creator that is immanently interested in the creation and also believe in the efficacy of scientific investigation, a system based on predicable laws of nature?"

How indeed? In Newton's time it was acceptable to mix physics and metaphysics. Newton himself, though he was among the first to discern the universality of the laws of nature, found no conflict with his firm belief in the God of creation. There was enough mystery left to be explained in the physical world that one could imagine God being active in almost every event.

Scientific progress has changed much of that. During the past three centuries, the mechanisms of many of those "mysterious" phenomena have been fully explained in natural terms. Awe in the finely tuned workings of nature may remain, however much of the mystery is gone. A sunset has lost none of its magnificent beauty, but the ambiguity of what paints the sky red each evening has departed from the experience. We now know it is not the sun or God saying good night to the mountain as the Ulm Uncle told Heidi. It is the preferential scattering of the blue end of the visible light spectrum and the greater penetrating power of the longer red wavelengths that pushes the red photons of light through the atmosphere to the cones of our eyes. (Cones are the receptors in our eye's retina that react chemically and then electrically to the different wavelengths of photons we refer to as light and color.)

Sounds dull, perhaps even confusing to the scientifically uninitiated. Yet knowing the physics steals none of the wonder as the sky turns from blue to crimson, then deep purple, and finally embraces the black of night as Earth's rotation from west to east leaves the sun ever further in the west. While my wife and I are kissing our kids good night in Jerusalem, my nephew is having lunch in Florida. The mystery that remains in the sunset is the riddle of why and how a mixture of seemingly inert, unthinking atoms of carbon, hydrogen, oxygen, and several other varieties can produce humans capable of having the subjective experience we refer to as beauty, or the love that would have us kiss our kids good night. Science is no closer to answering those questions today than it was a century ago.

And if we are watching that sunset from the hills of Jerusalem, we find an even more difficult question: why does the name of this city which has no natural resources, no major industry, no particular strategic importance, not even an indigenous water supply, appear on the front pages of newspapers worldwide more frequently than any city other than Washington, D.C.?

Both theology and science probe the nature of reality. Both seek to find an order in the workings of the world. For theology, it is a given that all reality is imbued with a transcendent spirituality. The surprise of science is that discoveries starting in the early 1900s have moved ever closer to the implication that the world we see about us, the objects in our daily lives that we take for granted as being solid, our bodies included, are expressions or manifestations of something as ethereal as energy. And that below the energy lies information, a totally nonmaterial basis for existence. While not calling this information spiritual, science has significantly closed the gap between the material and the spiritual.

Debating whether or not there is a "ghost in the system," to use the phrase of my colleague Dennis Turner, that gives meaning to sunsets is futile without understanding the system. In the chapters that follow, we'll explore the wonders of the physical world. But beyond the awesome phenomena that we'll discuss lies the most profound puzzle of all. Why is there existence, even of empty space? Even of time? The basic enigma is not whether or not we evolved from an ape, or the ages of rocks. The real question is why is there "being"? The existence of existence is amazing, awesome. We are so much a part of existence that we take it for granted—as a "given" in scientific terms. But step back from the subjectivity and think about it. What caused the big bang? What caused existence? Can we even expect an answer to these questions?

In our arrogance, we humans at times forget we have limits that ultimately constrain what we can know. Far beyond the uncertainty of measurements taught first by the Heisenberg principle of uncertainty, and later amplified to the limits of probability brought by quantum mechanics, there exists a twofold bound to our knowledge.

First the neurological (nerve-based) makeup of our brains may be large, counted even in the trillions of synaptic connections among the individual nerves, with each connection storing a bit of information, but the number is finite and so is our ability to comprehend. A chimpanzee may look at a computer and wonder what it is (if chimps have the ability to wonder). They may realize (if chimps realize) that they are mentally limited; that they do not have sufficient and proper neural connections in their brains to design and build a computer; that there are beings wiser than they and those beings who build the computers are visibly present. To the chimp, there is no question of the existence of the intellectually superior humans that are able to make computers. The chimp knows there is a limit to that which a chimp can comprehend.

We humans don't have that visible cap to our knowledge. There's no life form above us intellectually; nothing to show us that we have cerebral bounds. In theory everything we need to know we can learn. In theory even the sky is not the limit. But that theory, in actuality, cannot be true. Far beyond what the wonders of science may or may not be capable of discovering in the workings of our magnificent universe, there is an unbridgeable limit to our knowledge. That is simply because the amount of information stored in the universe is finite, not infinite. To retain a piece of knowledge, an electron in a computer or in a brain must be elevated to a slightly higher potential. That is, some amount of energy must be added to it. By agreement, that elevation can be indicative of some bit of information. What that particular datum might describe is not important for the present. But what is important is that information storage requires energy, tiny though that energy may be. (Amazingly, our brain's system of information storage uses close to the minimum required energy.) All indications are that our universe is finite. There may be other universes, but ours has its physical limits.

A finite universe has only a finite amount of energy within it. And that means it can store only a finite amount of information, a finite amount of knowledge. To be sure, we can never know all the information therein. It is spread throughout the vast reaches of space. But accessing the totality is not the point so much as the very fact that the totality is limited. It is no wonder then that even as we may write the symbol for infinity within our mathematical equations, we cannot subjectively comprehend—that is, internalize—its meaning. The sky of wisdom has a limit.

Realizing this was for me an epiphany, a qualitative change in my concept of the world's potential. Surround every star in the universe with a host of planets and populate all the planets with billions of Einsteins and Newtons. Let them study and theorize for all eternity and still what they can know will have a predictable limit. A similar limit came with the collapse of determinism. For 150 years the scientific community had a love affair with Laplace's theory of determinism—his erroneous theory that given total knowledge of the entire universe at any given moment one could, using the laws of cause and effect, predict all future events. Then came quantum mechanics, uncertainty, and the fuzzy indefinite boundaries of reality. With those discoveries, determinism evaporated. The cosmos became illogical. Our inability to completely understand our world had been revealed.

Experiments in physics laboratories regularly produce results for which there is no known explanation. *A* proceeds to *C* without passing through *B*. Of course, the operative word here is "known." With undying faith in our powers of reason and the calculating powers of computers, we may in the future discover the mechanism that allows what we might refer to as situation *A* to leap to situation *C* without passing through what now appears to be the essential stage of *B*. There is, however, a growing certainty that dimensions exist in our universe that we cannot sense, regardless of how clever and precise our instrumentation may eventually be. And that the physics of the universe operates within these insensible dimensions as well as within the four spatial measurements of length, width, height, and time that we do experience. Events occurring in those insensible dimensions affect us. In laboratories, we can see the results of experiments but we can't follow the reactions that lead to those results. The paths of those reactions may reside outside the physical measurements of length, width, height, and time. Physics has entered the metaphysical, the realm beyond the physically perceivable, in the fullest sense of that word.

Infinity is not within our reach, neither through the ponderings of philosophy nor the research of the laboratory. But unity is, a unity that encompasses and binds together all existence. The universe is an expression of this oneness.

For many, especially among persons alienated by superficial material aspects of Western society and drawn by default to monochromatic, top-down versions of spirituality, it comes as a surprise that an all-encompassing unity is the core concept not only of science but also of biblical religion. Experiencing that unity is the goal of both disciplines. Oneness is, in fact, the biblical definition of God. Both Judaism and Christianity have discovered the essential theological statement to be "Hear Israel the Eternal our God the Eternal is One" (Deut. 6:4; Jesus as quoted in Mark 12:29). Simply read, the meaning of this biblical verse is that there is one God. But, as we have seen, the oneness stated here is not the integer one after which follow two, three, and four. The Eternal is One teaches a truth far more profound than that there is only one eternal Creator. The One of this verse reveals an Infinity as perceivable by the noninfinite; our experience of an all-encompassing unity.

Monotheism does not limit its claim to there being only one God. Biblical monotheism teaches that everything is an expression of this Unity. "There is nothing else" (Deut. 4:39 as read in juxtaposition to Deut. 4:35). We are intimately a part of the whole. Therefore studying how that whole works must add to the depth of life's experience, much as knowing the politics of the times of Shakespeare adds to our understanding of *Hamlet*. One might conceive of a science without religion, but it is an oxymoron to conceive of religion without science. Revelation and nature are the two aspects of one creation. Theology and science present two versions of that one reality, each version seen from its own unique perspective.

The three religions of Jerusalem claim that humankind is created in the image of God, but they give no description of what that image entails. They even insist that God has no perceptible image. But that means our image of God must not be a thought, since our thoughts are built of images stored within our brains. How does one imagine, or even relate to, an imageless "absolute"?

Perhaps experiencing the Absolute is more a mode of thinking rather than imaging any specific thought. Finding joy in the simple fact of existence could do for starters in that way of thinking. The Bible suggests that we "Worship the Eternal with joy" (Ps. 100). Joy is the biblical sign of reverence. Associating one's self-image and interests with the holistic totality of existence might be the next step. Like it says, "Love your neighbor as yourself" (Lev. 19:18; Matt. 22:37; Mark 12:29). The difficulty is that our world may or may not have a transcendent underlying spirit. But it certainly has the material needs of daily life, and they seem to take priority over our best desires. Spirituality not rooted in the reality of life's material needs rarely effects a meaningful gain.

Professor of kabala Nadine Shenkar describes this bringing of the metaphysical into the material with an exquisite analogy. A tree lifts its leafy branches toward heaven and embraces the blessing of rain. If we didn't know better, we'd expect that the rain enters the tree's leaves and fruit directly from above. It would seem that the growth was from the top down. But not so. The rains must first enter the ground, and only then can they carry nutrients to the leaves and fruits. In this world, reality is rooted in the material. It flows from the bottom up. To reap the beauty of this world, we too must work from the bottom up. The spiritual must first enter the physical before life can flourish. It is no wonder, for example, that the biblical Sabbath is the most physically satisfying day of the week and simultaneously the most spiritual.

There is a range of scientific tools that impact all aspects of life, both physical and spiritual. I'll start with physics, lead into biology, and end with neuroscience, the brain/mind interface, specifically the puzzle of consciousness. Discipline is required at each step, but the reward will be, here and now, the satisfaction of understanding the wisdom that underlies reality. No need to wait for a hoped-for next world or some hypothetical reincarnation for the pleasure of these insights. If there is a Divine plan, reincarnation may or may not be part of that plan. That's up for debate. But for certain, physicality is in the scheme. We have a body. Linking the physical with the spiritual is the goal, and the lesson of the tree.

The pleasureful rush of emotion we experience at the beauty of a sunset touches that link. But something far more grand than the color of the sky ignites our emotions. Buying a few cans of paint at the local hardware store and mixing them to match the red we saw streaming across the evening sky somehow does not give us the same feeling of ecstasy. Viewing a work of art such as Auguste Rodin's *The Burghers of Calais* might also give that pleasure. Even such a totally intellectual experience as discovering a piece of deep wisdom can incite this feeling of awesome wonder. Some common aspect within these diverse experiences transcends the details, resides in the eternal.

In a world made up of such divergent and differentiated entities as stars and pencils and galaxies and people, all of us have felt those rare moments of joy at discovering the undifferentiated whole that lies beneath that complexity. In that moment our sense of individual self dissolves into the immense fabric of the universe. I find a touch of irony in our emotional perception of an all-encompassing unity having a material counterpart in the impersonal laws of physics. Science has revealed that the totality of physical existence is the expression of a single base reality, variegated fields of force, or material-less ringlets of energy each expressing itself in the material variety we see about us. Scientific inquiry of nature has exposed a metaphysical unity. The transcendent beauty we find in a sunset resonates with those physical cosmic roots.

In his famous book, *The Guide of the Perplexed*, the philosopher Moses Maimonides over eight hundred years ago succinctly stated this position: "We must form a conception of the existence of the Creator according to our capacities; that is, we must have a knowledge of metaphysics (the science of God), which can only be acquired after the study of physics (the science of nature); for the science of physics is closely connected with metaphysics and must even precede it in the course of studies. Therefore the Almighty commenced the Bible with the description of the creation, that is with physical science." In Maimonides' time, the idea that science might have something to add to our understanding of spirituality was such an anathema to the religious establishments that his book was burned by Jews and Christians alike!

The light of what we label as Divine is split into two parts. One part is revealed directly. That is the prophetic experience. The other part is hidden in the wisdoms of nature. The era has come when those hidden wisdoms are being discovered, exposing a new and undreamed-of synthesis in what is superficially perceived as a multifaceted and divergent universe.

The two hats worn by my colleague's friend are made of one fabric. They represent one reality but seen from two perspectives. In the following pages, I'd like to investigate both views of our world. I think we will find that Maimonides' claim was well founded. In fact, the Eternal is One.

# THE WORKINGS

# OF THE UNIVERSE

## THE PHYSICS OF METAPHYSICS

*As the "clay" of matter is energy, so the building block of energy is information, wisdom. The universe is the expression of this wisdom. The universe is the expression of an idea.*

One hundred years ago, a physics professor would have lost tenure on the spot if caught teaching the concept that matter in all its forms of solids, liquids, and gases was actually condensed energy. What hokum it would have seemed. There is conservation of energy and conservation of matter, or so it was believed. Energy may change form, perhaps from radiant to thermal, or from kinetic to potential, and so might matter change, from solid to liquid or gas, but in a closed system the sum total of energy and the sum total of matter remained fixed. That each was distinct from the other seemed too obvious to be questioned.

Then came Einstein, relativity, and $E = mc^2$. Einstein hypothesized, and it has since been confirmed, that matter, $m$, intrinsically represents a specific amount of energy, $E$. And the type of matter was immaterial. As bizarre as it seems, a gram of rose petals and a gram of uranium contain identical amounts of energy. The constant, $c^2$, in the equation is the speed of light squared or multiplied by itself. That $c^2$ is a massive value tells us that even a tiny amount of matter contains a huge quantity of latent energy. Having personally witnessed the detonation of six nuclear weapons, I suggest we all work and pray for peace. The few grams of mass converted into energy during those tests turned the mountain on which I stood into a quivering Jello-like fluid.

Einstein's discovery of the energy/matter relationship is far greater than merely stating that we can get $X$ kilocalories or $Y$ kilojoules from $Z$ grams of matter. Einstein's insight taught the world that every item, every plant and person and star in a galaxy, is a form of condensed energy, energy in its tangible form. If you had lived all your life in Antarctica and seen only ice and snow, and then were shown a kettle out of which billowed a cloud of steam, could you have believed that the hot vapor was made of the same stuff as the frigid ice, that both were water, but water at different energy levels? We and all we see are frozen energy. If you heat any item far beyond the temperatures that break apart molecules and atoms, ultimately it will revert to pure energy, blending with the radiance of all existence.

Science had discovered the metaphysics within physics, the nonphysical nature of the physical world. In 1900 the German physicist Max Planck had proposed an explanation for the surprising discovery that heated objects radiate light at discrete and fixed energies, rather than as the more logical continuous smear of energies. Aspects of Planck's work were to be essential for the development of Einstein's special laws of relativity, published just five years later. Planck suggested that radiant energy exists only in discrete packages that he called quanta. These quanta of energy, also known as photons, are emitted from a heated object when energized electrons fall from higher to lower orbits.

The implications of this rash idea were immense. If electrons orbiting a nucleus can reside only at specific levels of energy, with no intermediate stages allowed, then how does the electron change from one energy level to another? Not by gradually or even rapidly moving across the divide between orbits. Such a transition would imply that for a finite time, no matter how brief, the electron had an energy intermediate between the higher and the lower orbits. But observations showed that such gradualism was forbidden. The electron simply leaps from one orbit to another in zero time. This step-like transition in energy states is reminiscent of the step-like punctuations in morphology found in the fossil record. Totally illogical as these leaps in nature may be, those are the observed facts. Sometimes nature moves in leaps.

Planck, in short, discovered that the expression of energy at the atomic level is both intricate and nonobvious, even illogical. With Planck's insight, the way was now prepared for quantum mechanics. Building on the concept of the quanta, in 1923, almost a decade after Einstein published his general relativity theory (no longer a theory; now it is a law), the French physicist Louis de Broglie introduced an idea that was even more bizarre in its assertions than Einstein's claim that matter really was a form of energy. De Broglie claimed that "as a result of a great law of nature every bit of energy of proper mass is intrinsically related to a periodic phenomenon of frequency."

In simpler language, all matter had related to it a wavelength and a particular frequency, that is, a certain number of wave cycles per second. Not only had we learned that matter was energy and not "really" matter, we now had to believe that all matter is both particle-like and wave-like. Everything—you and I included—has a wave function. Seventy years of experiments have sustained both Einstein's and de Broglie's preposterous, counterintuitive claims. Absurd though these principles seem to the human mind (which works strongly by deduction—what we have seen to be true in the past should be a good indication of what we will see to be true in the future), this wisdom has made possible transistors and lasers and cellular phones, and even aspects of microbiology. Every piece of electronics that fills our homes, from TVs to microwave ovens, is based on these conceptual, counterintuitive breakthroughs. The universe we have discovered behaves in a manner most illogical. It does not comply with human reason.

The "quantum weirdness" of nature has profound implications. Most significantly it tells us that the world simply is not as it seems. A superficial reading of nature finds differentiation; disparate entities: stars and stones and bottled water and even life and death. At a deeper level that same nature reveals unity. I'm on our balcony. The afternoon Jerusalem sun is filtering through the yellow-green finger leaves on a row of eucalyptus trees planted a century ago to mark the property line. ("So the field of Ephron, . . . and the cave which was therein *and all the trees that were in the field, that were in all the border round about* were made over to Abraham," Genesis 23: 17, 18). De Broglie tells me the leaves and the light are one. Not poetically, though that also, but physically they are one. The fact that he has been proven correct fills me with joy. The universe quietly reveals its unity. God is polite, knocking only gently. We have to listen carefully if we are to hear the report.

THE matter/energy relationship, the quantum wave functions, have profound meaning. Science may be approaching the realization that the entire universe is an expression of information, wisdom, an idea, just as atoms are tangible expressions of something as ethereal as energy. Four basic forces govern the physical interactions of all matter. The first of the four to have been quantified as a force was gravity. On the 28th of April in 1686, Isaac Newton presented his *Principia Mathematica* to the Royal Society of London. The third book contained a description of the universal laws of gravitation, deducing that gravity was intrinsic to all matter. Mass attracts mass throughout the universe in a manner identical to its attraction here on earth, with the force of that attraction being proportional to the masses of matter involved, irrespective of the type. A kilogram of air has the same gravitational effect as a kilogram of steel. Newton's concept of a universal law established a new paradigm, that the universe was really a uni-verse. Uni as in unified.

But what *is* it that produces the force of gravity, that pulls an apple to the ground, or holds the distant moon in orbit? For a force to act at a distance, for the earth to reach out to the moon and hold it, there must be something making contact between the two bodies, otherwise how can the force be felt at a distance? We call that force gravity without really knowing what it is. Anyone who predicts that we are approaching the end of science is fooling himself. The secrets of gravity may one day unlock more human potential than the sum total of the technology we possess today.

For centuries the effects of electrostatic attraction had been observed. Rubbing dry wool did something to the wool (we now know it was scraping off electrons) that drew lint to it, even lint several inches away. Sparks of light darted in the cloth as PJ's were being pulled on, and for some reason the force pulled on our hair. Some strange power was at work.

Magnets presented the same enigma. The magnet is here and the iron nail is there. Empty space lies between. Slowly at first and then with increasing speed, the nail moves toward the magnet. No problem. Everybody knows magnets attract iron. But how? What reaches out from the magnet, beckoning the iron to come hither? It's a lesson in the magic of our universe. Take two horseshoe magnets. Try to push north onto north and south onto south. Look closely at the space between the opposing poles. Nothing is seen, but the force is mightily felt as the poles constantly slip aside, avoiding one another.

By the early 1800s Michael Faraday (1791–1867), the English chemist and physicist, had become intrigued by the seeming impossibility of action at a distance. Something must be carrying the force of the magnet's pull on the nail. Newton, a century and a half earlier, had suffered the same quandary in his work when he postulated the laws of gravity, describing what gravity did, without knowing how it did it or what it was.

The second of the four basic forces was "discovered" two hundred years after Newton. The breakthrough came in 1864, when James Clerk Maxwell was able to integrate the electric forces with the pulling power of a magnet, reducing (or increasing) the two seemingly separate phenomena into one: the electromagnetic force. Maxwell's accomplishment in integrating the two forces was prophetic. It was the first step in a drive to unify all forces—each a separate manifestation of a single unified field.

Both the magnetic and the electrostatic forces are theoretically carried by a single type of entity, photons, totally invisible and massless, observed only in their effects, as iron is drawn to a magnet and lint to wool. But what actually produces the photon in the magnet that reaches out to the iron, or in the nucleus of an atom that sends it hurtling off toward the orbiting cloud of electrons? And what is energy in the first place? We talk of energy so often that we've convinced ourselves it is a real and tangible entity. But that simply is not the case. On close inspection, energy is only a convenient concept for quantifying observed effects. We can't handle a piece of energy. We can store energy in a battery, but that's chemistry. The actual essence of energy remains elusive. Does that move energy closer to the information it may represent than to the matter it can form?

It's the electrostatic force, a force mediated by unseen theoretical photons, that holds you together, just as it does the floor on which you stand. It also keeps you from sliding through the openings in the floor like spaghetti through a colander with oversized holes. The world we see as solid is made solid not by matter, but by ethereal forces carried in photons (themselves a theoretical construct) traveling immense distances between the nuclei and surrounding electron clouds.

I press my hand on a table top. I can't penetrate the wood. But a hammered nail pierces it easily, because the invisible binding forces of the molecules of iron are far stronger than those invisible forces binding the molecules of wood. The blow of the hammer allows the nail to break the bonds created by the invisible photons that produce the very real molecular bonding.

The world of atoms and molecules consists of wavelike particles separated from each other by voids, held in place by never seen, massless photons, traveling at the speed of light among particles that are not only particles but also waves. If you can conceptualize this melee in an intelligible way, I have an urgent suggestion: publish.

If these hypothetical photons traveling between protons in the nucleus and orbiting electrons are what keep the electron cloud on its stable course about the proton-rich nucleus, just how does the photon travel? Travel takes time, yet the photon must work its magic instantly. As the photon leaves the nucleus on its instant journey toward an orbiting electron cloud, how does it know what trajectory to follow? By some estimates, the electron is flying at one tenth of the speed of light. That is quite a clip. Thus the photon must traverse a curved path or somehow "anticipate" where the electron will be so it can make its binding contact. Of course the theoretical physicist might brush these questions aside as being irrelevant, even naive, telling us the electron is a cloud and not a particle. So what holds which part of the cloud in place? The question is relevant and perplexing. To accomplish their tasks, these photons seem to have a great deal of wisdom built into them.

We take these phenomena of gravity and magnetism and electricity for granted. But attempts to understand what produces them has opened a world as bizarre as anything Alice encountered in Wonderland. And this is not fiction.

The four basic forces—gravity, the electrostatic force, and the so-called strong and weak nuclear bonds—have no logical explanation for their existence, but because of them we have a user-friendly universe filled with order and stability, and in at least one location, life conscious of its own existence. As the renowned physicist Freeman Dyson stated so eloquently, "Nature has been kinder to us than we had any right to expect." Consider how the four forces fit together.

1. The strongest of the forces, or the fields these forces produce, is appropriately termed the strong nuclear force. It's what

lets atoms form and keeps them and you from disintegrating into a jumble of protons, neutrons, and electrons—the sub-atomic particles of which all atoms are composed.

2. The weak nuclear force, some one thousand times weaker than its strong nuclear partner, in a sense works against the binding force within the nucleus and allows for certain types of nuclear disintegration. Sometimes a nucleus expels a particle to increase its overall stability. In doing so it often changes into another element. In some cases an orbital electron is drawn into the nucleus in a process known as electron capture. A reaction takes place between the captured electron (negative charge) and a proton (positive charge) of the nucleus, in which both disappear and a neutron (no charge) appears in their place. Some radioactive transitions are amazing. For example, when an atom of the metal radium expels an alpha particle (two protons and two neutrons), it undergoes a metamorphosis, changing from a metal to a gas known as radon. From solid metal to fluid gas in one step—something to be wary of when you purchase a home. Radon gas can emanate from radium in the structure or from the ground below the house, creating a serious health hazard. Radon, in its decay, pulls the same stunt as did its parent radium. It expels another alpha particle and changes from a gas back into a metal. Both the gas and the metal are composed of the same basic building blocks, protons and neutrons. Such is the wonder that underlies the physics of our universe.

3 and 4. The two remaining forces are the electromagnetic force and gravity. While the former is some one hundred times weaker than the strong nuclear force, gravity is a phenomenal $10^{42}$ times weaker. That's the strong force divided by a one with forty-two zeros after it. Yet for all its weakness, gravity was the first of the forces to be "discovered" (by Newton) and quantified. That is because of two of its characteristics. Gravity is al-

ways attractive, and it is the only force of the four that acts at large distances. Once beyond the size of an atom, $10^{-10}$ meters, neither of the nuclear forces is significant. And beyond a few centimeters the same is true for the electromagnetic force. That leaves gravity in charge of most of the space in the universe. It's the force that shaped the structure of the universe, pulling the gases of the early universe into huge galactic clouds, and then squeezing those gases into swirling masses of stars and planets. The sun, the moon, and the earth, when seen from space, are all perfect circles. Gravity did that, pulling the mass of each equally in from all sides, shaping the celestial bodies into spheres.

ONE hundred and three years after Maxwell's brilliant insight that unified the electrostatic and magnetic forces, Abdus Salam and Steven Weinberg proposed a theory unifying the weak and electromagnetic forces. What appear to be disparate particles carrying these fields are actually aspects of the same entity, photons, made manifest at different energies. Once again, beneath the seemingly fractionated nature of existence lies a deeper, powerful idea: that of a unified order.

There is a nuance in the Bible that may portend these discoveries or may just be a coincidence but is interesting nonetheless. We read at the closing of the six days of Genesis, "And God saw all that had been done and behold it was very good" (Gen. 1:31). In the nineteen-hundred-year-old translation of Genesis into Aramaic by the sage Onkelos, the verse is read not as "and behold it was very good," but as "and behold it was a unified order." Unity and order were the stamp of completion. Indeed, unification of all four forces is theoretically possible. Unfortunately, to test any such "final" theory, we would require energies comparable to those present just following the big bang, far beyond the reach of foreseeable technology.

Still, the current Holy Grail of physicists teaches one clear lesson: What superficially appears as diversity is, upon deeper scrutiny, unity. For several years I taught physics for second and third graders in my home laboratory. One of the experiments we did together was to tape strips of colored paper to the head of a centrifuge and then watch them spin. As the speed of the centrifuge increased, we first would see a blur and then, in a snap, the blur turned white as the wavelengths of the colors mixed in our brains. Who would believe that a mixture of every color could possibly result in white? Or conversely, if one had never seen a rainbow or sunlight passing through the prism of a drop of water, could it possibly be that red and green and blue and yellow in all their shades were tucked within the white light of the sun? We live downstream of the prism. We see the world through the prism of creation. Looking back through that prism toward creation yields a most amazing vision, the unification not only of the basic forces of nature, but also of the particles that give rise to those forces. Physics has discovered the oneness of existence.

All our activities must comply with these universal forces of nature. They set limits upon the extent to which we can manipulate matter. The universal forces are themselves confined to act within the parameters of two natural constraints. The Pauli exclusion principle, formulated by Wolfgang Pauli (1900–1958) in 1928, forbids two electrons in an atom or molecule from occupying an exactly equivalent energy state. If this were not the case, electrons in orbit around the nucleus would all soon fall to the lowest level, the orbit closest to the nucleus. Ultradense particles would form with no possibility for the variety of chemical reactions we take for granted.

The second constraint demands that electrons envelop an atomic nucleus at discrete and quantized orbits. These permissible orbits are fixed and constant throughout all nature, each being separated from adjacent orbits by a fixed, given amount of energy. Why this should be an inherent part of our universe is anyone's guess. But if electrons were not confined to specific quanta, if they could assume any and all energy levels, then chemical stability would be a dream unknown, as electrons constantly changed orbits under the influence of the slightest wisp of incoming energy. Molecules would form and disintegrate in a totally unpredictable manner. There would be no chemistry.

Every schoolchild learns the simple model of an atom in which the nucleus is like a sun, with electrons orbiting about it like planets. These two principles of atomic structure combine to make that model possible, with a nucleus composed of protons and neutrons and with electrons in stable orbits about that nucleus. Carbon on earth has identical properties to carbon in the farthest galactic cloud, and all indications are that those properties are extant throughout the universe. This uniformity is what makes chemistry possible, the predictable behavior of atoms reacting with other atoms to form molecules and molecules grouping together repeatedly forming copies of the most amazingly complex products. The DNA of genetic material found in every one of the trillion cells of a newborn baby is but one example of the faithful reproducibility of nature.

The physical world is a wonder-filled phenomenon of unity. The same laws that govern the ten thousand billion billion stars distributed among the hundred billion galaxies of our universe, stretching over some 15 billion light years of space, also govern chemical reactions within the 0.001 centimeter of a cell. From the $10^{-5}$ meters of an organic cell to the $10^{26}$ meters of the universe, from the mass of an atom, $10^{-26}$ kilograms, to the mass of the sun, $10^{30}$ kg, it's one set of laws. Why? It didn't have to be this way. Why is the universe so intelligible, so consistent? Science alone cannot say. Perhaps we are encountering a hint of the metaphysical held within the physical.

Consider the discovery made by Antoine-Laurent Lavoisier (1743–1794). The French chemist did for chemistry what the English mathematician and physicist, Newton, a century earlier had done for motion. Both appreciated that a single unified set of laws must operate everywhere. Lavoisier realized that all

chemical reactions follow paths that are quantitatively repro-
ducible, and which are fixed by the composition of the molecules involved. With this brilliant insight, he laid the basis for
all future chemical engineering. After Lavoisier, we could believe that two atoms of hydrogen and one atom of oxygen can
combine into $H_2O$, water, anywhere in the universe.

Unfortunately, the extent of Lavoisier's successes in his
younger years was matched by the tragedy of his death. The
episode is a lesson in the futility of unbridled spirituality. The
French Revolution was in full swing, its leaders filled with the
desire to remake the world into a place good for all. It was a sad
example of the failure of one-sided spirituality, of a theory run
amok. In a bizarre denial of history, E. O. Wilson in *Consilience*
tells us that "the Enlightenment thinkers . . . got it mostly right
the first time. The assumptions they made [were] of a lawful material world, the intrinsic unity of knowledge, and the potential
of indefinite human progress . . . a dream of a world made orderly and fulfilling by free intellect." Yet Wilson writes that the
same Enlightenment prepared the ground for the French Revolution's reign of terror in which the state cannibalized its own
founders along with France's leading intellectuals. Lavoisier was
among them. In 1794, Lavoisier was guillotined. The universe
may be intelligible, but we cannot remake it in *our* own image.

Spinoza's ideal in which we might find our way to perfection
by rational processes alone is an age-old dream. The nature of
the human psyche makes it an unrealistic goal. Both science and
religion acting alone have produced irrational and horrific behavior. Like rain in the growth of a tree, the blessing of insight
from above must mesh with the roots of material reality from
below. Neither by itself can achieve the goal of peace on earth
and goodwill toward all.

WE take as givens, as axioms with no apparent explanation,
the laws that run the physical universe. They seem completely

logical once we know them. But that is because we live with them constantly. Mass attracts mass in direct proportion to the amount of mass present. A kilogram of iron and a kilogram of air have the same gravitational pull. Logical. What else would one expect? It's the mass that engenders the gravitational field.

But must it have been so? Think about the equal but opposite electrical charges of protons and electrons. The former are called, by convention, positive; the latter negative. No problem. That is, no problem until we think a bit deeper. The rest mass of a proton, $1.673 \times 10^{-24}$ grams, is 1,836 times heavier than the rest mass of an electron. If an electric charge followed the same "logical" law of nature patterned by gravity, then the charge of a proton would be 1,836 times greater than that of an electron. Had that been the case our universe would be a very different place. Huge clouds of electrons, 1,836 times more tightly packed, would circle each atom's nucleus to maintain electrical neutrality. Chemical bonding would be hopelessly weak. Alternatively, electrons might have had the same mass as protons and so the same charge as per the gravity/mass relationship. But heavy electrons would require vastly heavier doses of energy to move them in atomic orbit and to compel the electrons to take part in chemical reactions (chemical reactions are in effect exchanges of electrons among atoms). None of these scenarios makes stable chemistry possible.

What good fortune dictated that charge relationships are not proportional to mass relationships? And if mass, the amount of stuff making up the subatomic particle, does not induce electrical charge, what does? A good question. Perhaps charge is proportional to surface area. Perhaps the particle aspects of protons and electrons have similar surface areas but vastly dissimilar densities. Since we do not have a clue as to their basic composition, anything is possible, even if none of it is logical.

Physicists probe the structure of matter by colliding subatomic particles such as protons and electrons at very high velocities. Analyzing the products of the collisions as the fractured pieces careen off in varying directions reveals hints of their constituents. This has opened a Pandora's box. The more deeply matter is probed, the more bizarre it seems. Particles are composed of lesser particles which in turn are composed of still lesser particles, in what has been termed a particle zoo (see Figure 1). The strong possibility exists that at the bottom of the pile we will discover that all the particles are varied manifestations of an underlying energy, which in turn may be the manifestation of something even more ethereal. Call it wisdom, or an idea, information. The Hebrew word integrating all these would be *emet*, reality. It would be the interface between physics and metaphysics. In that case the divide between the physical and metaphysical would be no divide at all. It would be a continuum in which one leads smoothly into the other.

The hint of a flow between the physical and the metaphysical, between the roots of the tree and its branches rich with leaves and fruit, emerges from de Broglie's equating waves and particles,

$$hv = mc^2$$

where $h$ is a constant, known as Planck's constant; $v$, pronounced *nu*, is the wave frequency (the number of cycles per second) associated with the mass, $m$, and $c^2$ is the speed of light squared or multiplied by itself.

Consider the significance of this statement. The left side of the equation describes a wave; the right side a particle. But waves are extended expressions of energy, while particles are discrete entities, having edges and ends, or so we thought in the good old days of logical, classical mechanics. Then came de Broglie and quantum mechanics, and the clear water of reason, the humanly logical universe, became blurred. The wave/particle duality that Einstein had discovered in light, Louis de Broglie extended to matter. There were to be "matter waves." Energy and matter, waves and particles, are all expressions of some deeper reality in which particles and fields of energy and even time blend. If beneath all the weirdness there is logic, a

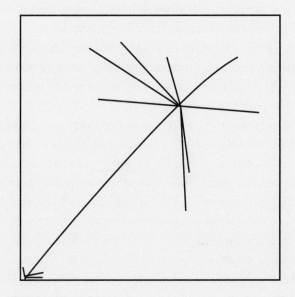

Figure 1

The Particle Zoo

*A cloud chamber showing tracks of a single incoming subatomic particle "exploding" into a starlike configuration of nine less massive particles, one of which then "decays" into still lesser particles (lower left corner). Cloud chambers provide visual representation of the actual paths traveled by charged particles as they careen through a chamber in which ultrapure gas saturated with a vapor, such as water vapor, is suddenly cooled to produce supersaturation. As the charged particle passes, it strips electrons from surrounding water atoms, causing the water vapor to condense along its path. In doing so, it produces the track. We have here the opposite of the statement on United States coins: e pluribus unum—one out of many. With subatomic particles, we see e unum pluribus—many out of one!* Webster's New Collegiate Dictionary *gives an alternative translation as "one composed of many." This seems to be the reality not only of the U.S.A., but of nature as well.* (Figure after J. Hornbostel and Brookhaven National Laboratory)

thought, a preexisting law, we will have discovered the continuum that links the metaphysical with the physical.

From particle to wave to energy to *idea*. That's the pattern from our perspective, from the inside of creation looking out. The flow is exactly in the opposite direction if we attempt to visualize the path from creation to us.

We experience the physical world of time, space, and matter. That is the information our five senses access. But biblical religion and now physics claim there is more to nature than meets the eye (the ear, nose, etc.). Physics says there are possibly ten, eleven, or twenty-six dimensions. We cannot access all of them, but our inability to do so is not because they are mystical. A bit mysterious they may be, but mystical has nothing to do with the physics that led to these suppositions. One hundred years ago, no J. A. Wheeler would have dared to suggest that all we see about us is actually the expression of condensed information. He'd have been dismissed as a mystic. But then one hundred years ago who would have dreamt that the solid world is really 99.9999999999999 percent empty space made solid by hypothetical, force-carrying, massless particles? And that even that minuscule fraction of matter that *is* matter may not actually *be* matter, but wavelets of energy that we material beings sense as matter?

Metaphysics has entered mainstream, peer-reviewed, university-approved physics, though of course not by that name. In academia it's called quantum mechanics. Modern science has broken the barrier that once separated the illogical from the logical. After 2,350 years, modern physics has lifted itself from the erroneous quagmire of materialist Greek philosophy which promulgated the concept that if you can't touch it or see it, it isn't there. In a sense "heaven" and earth have been joined.

There is even a physics for free will. Quantum mechanics teaches that while the general path of a reaction may be predictable, the exact path is not. There is a probabilistic spread in the path that connects cause with effect. That divergence opens a window of opportunity for what might be choice. It's a topic I'll hold till we get to the functioning of the mind.

PHYSICS has demonstrated that a single substrate underlies all existence. One might ponder: why then is there the diversity seen about us? Differentiation, not simplicity, is the mark of our universe. Solids, liquids, gases, stars, stones, people, trees. According to science, that's simply the way an initial unity played itself out on the stage of the universe. In brief, the scientific scenario reads like this:

The big bang produced, from nothing, a universe composed of photons, energy-packed radiations, unimaginably hot and compressed beyond description, a soup of energy, nearly homogeneous throughout. (What produced the big bang remains a question.) The universe was born as an undifferentiated unity, the oneness science struggles so to rediscover today. But that unity was locked into the ultimate of black holes, a highly compressed mass, so dense with such a huge pull of gravity that nothing, not even something as ethereal as light, could escape. The name black hole arises from the fact that any light passing within the horizon of a black hole is sucked in by the black hole's gravity. With the big bang, the entire universe was such an entity, a near infinite gravitational pull housing and holding everything. That might have been both the beginning and the end of the story. The universe might have remained locked literally within itself. But something happened. We call it inflation, a theory first described by the Russian astrophysicist Alex Starbolinski and a year later by the American Alan Guth.

The universe began to evolve, space stretching, expanding out from a scale (size) that may have been no bigger than a grain of mustard or the black pupil of an eye. What caused this "explosion"? Science cannot say for certain. It was a onetime phenomenon that set the course of history. The map was not exact, but the general direction was set. The laws of nature ensured that, with expansion, galaxies, stars, and planets would emerge. They were, in a sense, preprogrammed into the scheme of things. With each doubling in the scale of the universe, the temperature halved as the waves of radiation were stretched. As the temperature plummeted from an unknown immense value to the relative chill of $10^{13}$ degrees Kelvin (that's ten trillion degrees!), a small fraction of the energy condensed into the first forms of stable matter, the protons and neutrons that were to eventually compose the entire material world.

Now the nuclear forces took over. Through the alchemy of nature two protons could combine with two neutrons to produce helium, the second lightest of all elements, hydrogen being the lightest element, consisting of a single proton. But that transition to helium requires several steps, as the protons and neutrons join in Lego-like fashion. Fortunately (or as expected?), the joining of a proton and a neutron to form a heavy isotope of hydrogen known as deuterium is sufficiently unstable that it can form only at temperatures below $10^9$ degrees Kelvin. That threshold occurred when the universe was a bit over three minutes old.

Alchemy, even universal alchemy, requires heat. By about three minutes the universe had stretched and cooled to a level at which nucleosynthesis could no longer continue. The deuterium instability as well as other instabilities "froze" the universe into a composition having approximately 23 percent helium (by weight) and 77 percent hydrogen, or approximately 7 percent helium (by atoms) and 93 percent hydrogen, with a tiny amount of lithium, the third lightest element, also present. But the major component of the universe, by far, was the photons of radiation. That is very much the universe's composition even today. We live in an expanding mass of energy peppered with a small "contamination" of nuclear particles, about one particle for each ten billion photon rays. We and all we see about us are made of those relatively few, contaminating particles.

After three minutes, the force of gravity became the dominant factor. Building on minuscule inhomogeneities in the expanding universe, differences in density as small as one part in ten thousand, gravity pulled together galactic-sized clouds of hydrogen, shaping them into swirling arrays that often mirrored the exponentially shaped spirals on the shells of snails today, or of soap bubbles as water spins down the drain of a sink.

The material of the universe consisted of hydrogen and helium. But neither hydrogen nor helium is very useful for building something as complex as life. The path taken toward life was elegant. While not great for making life, adequately compressed hydrogen is superb for making stars. And stars are the only sources of long-term stable energy production known in our marvelous universe. Had the deuterium bottleneck of nuclear instability not existed, the very early universe would have made all hydrogen into heavier atoms, possibly iron, with no possibility of ever forming the long-lived stars needed for the patient development of complex life.

Gravity, mass mutually attracting mass, squeezed the swirling gases of hydrogen and helium into clusters of star-sized nebulae, each so tightly packed that temperatures in the cores reached millions of degrees. Hydrogen began to fuse into helium. But the resulting helium weighed a bit less than the components that went into making it. That lost weight represented the mass that had been converted into heat. The first stars were born. (Our sun converts some 660 million tons of hydrogen into approximately 600 million tons of helium every second, with the lost weight given off as the radiation that warms us and lights our days.) In some cases, not all the matter was drawn into the central star's mass. This excess joined to form planets held in orbits around the star.

Within the cores of these newborn stars, the alchemy of nucleosynthesis could proceed at a pace more leisurely than in the hectic first few moments following the big bang. Heavier elements were now being formed. Concurrently, the star's supply of hydrogen fuel was being consumed. Once the supply of hydrogen within a star was exhausted, the outward flow of radiant heat produced by the nuclear reactions in the core ceased. (Every star, in this sense, is a controlled hydrogen bomb.) It was the pressure of this heat streaming outward from the core that had kept the star in equilibrium with the inward pull of gravity. With the heat pressure gone, gravity crushed the star inward from all sides, forcing its collapse, imploding the star upon itself, imploding and then rebounding in the explosion of a supernova. In that act the heavier elements, such as carbon, nitrogen, oxygen, and an array of others, which were needed for complex life,were seeded into the universe.

In time, these newly formed elements could mix with other clouds of the primordial hydrogen, be pulled together, and form into new stars. That is the necessary history of our star, the sun. Our solar system cannot be a first-generation stellar system. There is too much of us in it. Too much of the heavier elements—carbon, nitrogen, oxygen (the sixth, seventh, and eighth of the elements by proton number), and others on up through uranium, number ninety-two. Our sun, the planets, and we ourselves are the products of bygone stars that blasted their contents into space and reformed to make new generations of stars. We are stardust come alive, and somehow conscious of being alive.

That's the scientific scenario. It's all in the laws of physics. Does a metaphysical theology have anything to say about all this? The first few minutes of history just described cannot be found in the Bible, at least not in this detail. So why turn to theology at all?

There are several reasons. First, consider the laws of physics that made it all happen. Did they precede the universe? That would mean laws of physics existed without the physical material upon which to act. Now that sounds a bit bizarre, physics without the physical. But it is the only solution that we scientists can offer.

Some scientists admit having a philosophical problem with the big bang since such a beginning might imply a Beginner. Sir John Maddox, for example, the avidly secular former editor of the journal *Nature*, one of the two most respected peer-reviewed scientific journals, opined that "the big bang is philosophically unacceptable. . . . It is a theory that will be gone in ten years." He wrote that in *Nature*, in August 1989. Maddox's ten years and more have passed. The big bang theory is now more firmly established than ever. Details are uncertain, but the concept that our universe started hot and dense and is expanding out finds ever more supporting data.

The further philosophical problem of there having been a beginning arises with the idea that the beginning of our universe marks the beginning of time, space, and matter. Before our universe came into being, there is every scientific indication that time did not exist. Whatever brought the universe into existence must of course predate the universe, which in turn means that whatever brought the universe into existence must predate time. That which predates time is not bound by time. Not inside of time. In other words, it is eternal. If the laws of physics, or at least some aspect of the laws of physics, did the job of creation, those laws by necessity are eternal.

There are various scientific conjectures currently popular to help explain how a "beginning" happened. One calls for an infinite, eternal medium that can produce universes, one upon the other, somewhat like bubbles rising in an infinite ocean. Each bubble has its own laws of nature. The bubbles with life-supporting laws have life. We are one such bubble. Of course, we can never see outside our bubble to confirm or disprove this theory. Another conjecture is that a field of "virtual energy," a potential, exists. Under specific extreme circumstances that potential can bring virtual particles into reality.

What all of these conjectures have in common is that something, or more accurately stated some non-thing, an eternal whatever, predates our universe. This whatever-it-is has no bodily parts, is totally nonmaterial, is eternal, and though being absolutely nothing physically has the infinite potential to produce vast universes. Sounds familiar. Kind of like the biblical description of God. In fact it *is* the biblical description of God with one significant difference. A "potential field" doesn't give a hoot about the universes it spins off. The Bible, however, claims that the Creator is intimately interested and involved in its creations. Is there any hint of a metaphysical interest in our universe?

Consider the "coincidences" of the first several minutes, and later as planets formed and cooled, coincidences that led to life and consciousness. From a ball of energy that turned into rocks and water, we get the consciousness of a thought. And all by random, unthinking reactions. Even to an atheist, this line of reasoning must seem a bit forced.

This is not a statement of the anthropic principle: "Gee, our universe is so well tuned for life that there must be a Tuner." No, what we see here is far more significant than fine-tuning. We see the consistent emergence of wisdom, of ordered complex information that is nowhere hinted at either in the governing laws of nature or in the particles of matter that form the brain that lies below the mind's thought.

There is a premise commonly applied in physics: Occam's razor, the idea that, all things being equal, the simplest, most elegant explanation tends to be true. A recent book about string theory even used the title *The Elegant Universe*. Why should the universe be elegant? Why should Occam's razor be true? Why are the laws of nature elegant, and from where did they acquire the wisdom to produce intelligent life? Where indeed? Could it be the metaphysical shining through?

# THE ORDERLY CELLS

# OF LIFE

The opening of the twentieth century marked the onset of the era of physics. Theory and discovery revealed a reality at both subatomic and cosmic dimensions undreamed of just a few decades earlier. Einstein and relativity, Planck and quantum physics, Heisenberg and uncertainty exposed the wisdom within which all existence is embedded.

By the 1950s, especially with the demonstration by Crick, Watson, and Wilkins of the double helix structure of DNA, biological sciences moved to the fore. Almost simultaneously, cybernetics—computer-based information processing—was born. Together with molecular biology, cybernetics brought the world into the era of *information*. For indeed, information lies at the base of both molecular biology and computer science: the former reveals a perplexing depth and breadth of information; the former and the latter both manifest phenomenal ability to manipulate such information.

In the coming chapters we'll take a journey through the wisdom that lies within the cells of life. If scientists had not first discovered such complex processes by which life functions, but merely proposed them as a theory, the scenario of life would be rejected as fantasy. To call the phenomenon of life complex is to trivialize it.

Our study will not be easy. As I discuss the details, some parts will even be tedious. But the wonder of life lies in these details. We'll discern a unifying wisdom embedded in even the simplest forms of life which outshines the wonder of the physics from which it arises. We'll discover that the essence of life, of all life, is the storage, organization, and processing of information. One can only wonder how and from where the complex order of life arose. Life's order is in no way evident either in the atoms and molecules from which that life is composed or in the laws of nature that govern the biochemical interactions among those atoms.

I have no hidden agenda in this tour of the intricacies of life. I make no attempt to deny that life developed from the simple to the complex. Paleontology, biology, and for that matter the Bible each presents its own account of life's flowering. The Bible devotes a mere six sentences to the process. Paleontology records the past in thousands and perhaps millions of fossils. Biology texts fill libraries. Yet all three describe a chain of increasing complexity.

The objective of what follows is twofold: (1) to ponder what processes might have been responsible for life's development in the light of its overwhelming complexity; (2) to discern that the complexity found in life is qualitatively different from that found in the substructures from which it arose. The latter assertion is cause for wonder. Systems can give rise to secondary systems that are more complex than their "parents," but their complexity can be only a fractional extension of that of their parents, an increase in amount but not in type. With life, the increase in complexity from physics to biology seems to be of type as well as amount.

So puzzling is the intricacy of the reactions that power life that at times it seems as if wisdom must be an inherent characteristic of the universe. Our world contains hidden knowledge that is waiting to be expressed. It seems as if a metaphysical substrate is impressed upon the physical.

The Bible, for one, suggests this to be true.

Before dismissing such a suggestion as rubbish, or accepting it as the absolute and obvious truth, let's look at the text. And let's look at it freshly: when a statement is repeated over and over it tends to be accepted whether true or not, and the opening sentence of the Bible has fallen under that spell.

Genesis 1:1 is usually translated as "In the beginning God created the heavens and the earth." Unfortunately, that rendition, which the entire English-speaking world has heard repeatedly, misses the meaning of the Hebrew. The mistake stems from the King James Bible, first published in 1611, based on the Latin Vulgate attributed to St. Jerome in the fourth century and the Greek Septuagint that dates from some 2200 years ago. "In the beginning" is thus three translations downstream from the original.

The opening word, usually translated as "in the beginning," is *Be'reasheet*. *Be'reasheet* can mean "in the beginning of," but not "in the beginning." The difficulty with the preposition "of" is that its object is absent from the sentence; thus the King James translation merely drops it. But the 2100-year-old Jerusalem translation of Genesis into Aramaic takes a different approach, realizing that *Be'reasheet* is a compound word: the prefix *Be'*, "with," and *reasheet*, a "first wisdom." The Aramaic translation is thus "With wisdom God created the heavens and the earth." The idea is paralleled repeatedly in Psalms: "With the word of God the heavens were formed" (Ps. 33:6). "How manifold are Your works, Eternal, You made them all with wisdom" (Ps. 104:24). Wisdom is the fundamental building block of the universe, and it is inherent in all parts. In the processes of life it finds its most complex revelation.

Wisdom, information, an idea, is the link between the metaphysical Creator and the physical creation. It is the hidden face of God.

The human body acts as a finely tuned machine, a magnificent metropolis in which, as its inhabitants, each of the 75 trillion cells, composed of $10^{27}$ atoms, moves in symbiotic precision. Seldom are two cells simultaneously performing the same act, yet their individual contributions combine smoothly to form life. Gridlock is rarely a problem in the human body.

Ten to the twenty-seventh power—a one followed by twenty-seven zeroes, a thousand million million million million atoms—are organized by a single act when a protozoan-like sperm cell adds its message of genetic material into a receptive egg cell. Combined, these two minuscule cells contain all the information needed to produce the entire body at each stage of its growth, from fetus to adult. We are so embedded in the biosphere that the marvel of its organization has become lost within its commonness.

Until the mid-1970s the accepted wisdom was that the origin of this organization that we refer to as life was the result of chance random reactions among atoms, gradually combining, one chance occurrence building upon another over eons of time until self-replication and then mutation produced the first biological cell. Three billion years were thought to have passed between the formation of liquid water on the formerly molten earth and the appearance of the first forms of life.* That was the message of the fossil record of the time, and that was the logic. After all, how else might one account for the origin of biology? Certainly not by spontaneous generation. Louis Pasteur had laid that primitive idea to rest long ago.

Two to three billion years were available for randomness to do its work. "Given so much time the [seemingly] impossible becomes the possible, the possible probable and the probable virtually certain. One had only to wait. Time itself [and the random reactions able to occur within those eons of time] performs the miracles." So wrote George Wald, professor of biology at Harvard University and Nobel laureate. The article appeared in the August 1954 issue of *Scientific American*, the most widely read science journal worldwide, the Broadway of scientific literature.

---

*All references to ages, such as the age of Earth, age of the universe, and so on are calculated from the time-space coordinates of Earth. From other time-space coordinates, these ages could be vastly different. For a detailed discussion of the age of the universe please see the relevant chapters in my book *The Science of God*.

This speculation over life's origins has within it an important lesson: not to confuse accepted wisdom with revealed fact.

In the mid-1970s came the seminal discovery of Elso Barghoorn. He, like Wald, was at Harvard. Barghoorn assumed correctly that the first forms of life would be small, microbial in size. Using a scanning electron microscope, a tool able to identify minute shapes imperceptible to microscopes that probe images with visible light, Barghoorn searched the surfaces of polished slabs of stone taken from the oldest of rocks able to bear fossils. To the amazement of the scientific community, fossils of fully developed bacteria were found in rocks 3.6 billion years old. Further evidence based on fractionation between the light and heavy isotopes of carbon, a fractionation found in living organisms, indicated the origins of cellular life at close to 3.8 billion years before the present, the same period in which liquid water first formed on Earth.

Overnight, the fantasy of billions of years of random reactions in warm little ponds brimming with fecund chemicals leading to life, evaporated. Elso Barghoorn had discovered a most perplexing fact: life, the most complexly organized system of atoms known in the universe, popped into being in the blink of a geological eye.

If you equate the probability of the birth of a bacteria cell to chance assembly of its atoms, eternity will not suffice to produce one. . . . The speed at which evolution started moving once it discovered the right track, so to speak, and the apparently autocatalytic manner by which it accelerated are truly astonishing. . . . [Yet] chance and chance alone did it all. But it is not, as some would have it, the whole answer, for chance did not operate in a vacuum. It operated in a universe governed by orderly laws and made of matter endowed with special properties. These laws and properties are the constraints that shape evolutionary roulette and restrict the numbers that can turn up. . . . *Faced with the enormous sum of lucky draws behind the success of the evolutionary game, one may legitimately wonder to what extent this success is actually written into the fabric of the universe.* (Emphasis added.)

So wrote Nobel laureate, organic chemist, and a leader in origin of life studies, Christian de Duve, in his excellent book, *Tour of a Living Cell.*

One could say that the ancient and immediate flourishing of life on earth is a miracle—for so it would seem. As a reconciliation between the theological and scientific views of the origins of life, one could also assume de Duve's conclusion that chance events are involved but the system is rigged for life. Chance, luck, and randomness pose no threat to theology when the "chances" and "randomness" are "governed by orderly laws and made of matter endowed with special properties," properties instilled in the universe as the laws of nature at its metaphysical creation. Yet even this totally secular view of our universe provides no simple answer that can explain why the system should be so rigged.

The immediate appearance of life on earth, an event undreamed of prior to its discovery, presses the view of nature as being driven by random reactions into a very tight corner. It is reminiscent of an equally tight corner at the other end of the theological scale, in which persons, by invoking preposterously complex equations, demand that the earth be at the center of the universe despite all data to the contrary (the Bible makes no such claim and even mentions the heavens before the earth).

Interestingly, among the most abundant elements within the material fabric of the universe are those that make up the main components of life. Atoms of hydrogen, carbon, oxygen, and nitrogen account for over 96 percent of the human body. These four, plus helium, are also the most abundant elements in the universe. They are also the only elements that can combine to form the long chains and ring-like molecular structures required for life's processes. The big bang via the laws of nature wove a particularly special fabric.

Not withstanding his first name, de Duve writes in what appears to be a totally secular mode. Even Francis Crick, the avowed "agnostic with a tendency toward atheism," to quote his self-description, described the origin of life as nearly miraculous. De Duve's chemist's view of life—that it is written into the fabric of the universe—resonates with physicist Wheeler's concept that the universe is the expression of an idea. For both Wheeler and de Duve, the evidence stems from discoveries in their respective fields of research, not from verses in the Bible, though these also imply that life is an inherent part of the universal plan. On day three of Genesis, when life is first mentioned, we are told the earth brings forth life. Biblically, the word creation does not appear in relation to the origin of life. No creation means nothing new was required. The Bible claims the necessary components were already present.

But can life be the product only of the laws of nature and the characteristics of the matter upon which those laws operate? Or is there an additional need, the imposition of order from an outside source? Order is known to appear spontaneously in chaotic systems via random reactions. Shake a bag full of letters and occasionally clusters will form that spell words, but never ones that spell long sentences. And further shaking always destroys the initial orderly arrangement. When classical systems are far from equilibrium (shaking letters in a bag is a simple example of such a nonequilibrium system), exotic combinations and reactions among the components multiply. Some of those reactions can, by pure chance, produce orderly arrangements. However, unless this order is somehow locked into place, the system reverts to chaotic disorder. This is the demand of the second law of thermodynamics. In any situation where order is not imposed, momentary order always degrades to chaos. The symphony of life decays to street noise–like chaos once the forces of life cease. Dead bodies decay. Order in life is maintained but only at the expense of large inputs of energy-rich foods and oxygen by which the foods are biologically combusted and the energy within extracted. Life in this sense is a constant drive to maintain order in a universe that favors chaos.

We can find order arising spontaneously in nature, but never on the scale of complexity associated with life. And even here, the order is always the result of a force that imposes it. The beautiful crystals of sodium chloride and bromide along the coast of the Dead Sea (the lowest point on the surface of the earth) are a study in order arising spontaneously. But with crystals, their order is not randomly produced. The electrostatic forces active among the ions of sodium, chloride, and bromide force the atoms to form the regular crystal matrix. The laws of nature predictably produce the crystals.

In 1811, the French physicist Baron Jean Baptiste Fourier derived a mathematical expression describing the propagation of heat in solids by which the flow of heat along a rod is directly proportional to the temperature gradient along that rod. An object's temperature is a measure of the intensity at which its component molecules are vibrating. Touching a piece of hot iron hurts because the iron molecules actually slap your fingers. If the molecular slap is hard enough, it will break the bonds among the molecules of your skin and produce a wound. Fourier's discovery provided a simple law that described how this wave of increased molecular motion progressed through an object. As with crystals, one physical statement describes the orderly behavior of billions of individual molecules.

The earth's rotation from west to east and the permanent low-pressure system in the North Atlantic combine to produce the clockwise flow of the Gulf Stream, which carries warm equatorial water north along the east coast of the United States, then west toward Great Britain and south along Europe's west coast. The interaction of this warm water with the cool air over Britain produces the fog in London. The Gulf Stream represents trillions upon trillions of organized water molecules. They follow a single law, first derived in 1835 by Gaspard de Coriolis, that accounts for the behavior of fluids moving on a rotating earth.

The laws and forces of nature, we discover, are able to impose predictable order on a microscopic as well as on a massive global scale, and are describable in logical mathematical terms. But is the biological order within the living cells of the fish that swim in the Gulf Stream merely a more complex expression of nature's ability to impose physical order, or is there a qualitative difference, a difference in type, between the phenomena? We can predict the formation of crystals of sodium chloride, and even the formation and direction of a Gulf Stream. Could we have forecast the advent of life in an initially lifeless universe? Using a reductionist approach, extrapolating patterns up from basic principles, how far can we go in our divination toward the inception of life?

A truly reductionist argument should start at the beginning. Or even before. But there is no way we could predict, from first principles, a universe with the life-nurturing laws of nature by which we function. So take the existence of our special universe as a given. Accepting the big bang that brought with it these life-friendly laws and the space and energy upon which they act, there'd be no simple logic that would predict that some of this energy would change into stable, lasting matter. We'd know that energy can change into matter. Einstein discovered this aspect of nature and made it famous in his equation $E = mc^2$. But that transition of $E$ into $m$ always produces a pair of particles, matter and antimatter, which then mutually self-annihilate, reverting back to their constituent energy. So in theory the universe should be an expanding ball of ever more dilute (cooler) radiation and no particles of matter. Since we are here, we can surmise that some of the matter, about one part in ten billion, somehow survived the annihilation to form the basic particles, protons, neutrons, electrons, and several others.

So let's move the goalposts yet again, and take our universe, its life-friendly laws of nature, *and* it's basic nuclear particles as givens and start our reductionist approach from there.

Let's say we study the individual subatomic particles, the physical properties of protons, neutrons, electrons. These nuclear particles are the building blocks of atoms, and hence form the basis of all matter. We learn completely their physical characteristics, their rest masses, their relative electrical charges, their natural affinity or repulsion, their wave functions. From this and our knowledge of the laws of nature, which in effect direct interactions among the individual particles, we would predict which combinations would be stable, which not. In doing so we could construct a chart of all possible atoms and the periodic table of elements, in essence the atomic composition of the universe in potential. From the physical conditions just following the big bang, we could determine that of these possible atomic nuclei, the energy of the big bang would produce mostly hydrogen, a single proton, the lightest of elements, and helium, the second lightest of the elements, consisting of two protons and two neutrons. By the end of the period of nucleosynthesis, some three minutes after the creation, we would know that the material of the universe would consist of 93 percent hydrogen atoms, 7 percent helium atoms, plus trace amounts of lithium and beryllium, the third and fourth lightest of the elements. All of this is predictable.

Then, discovering that there were slight variations in the distribution of energy in the early universe, we'd be able to predict that as the universe expanded, gravity would pull these ripples into galaxies and then star-sized clusters. Stars would form, and through the temperature and pressures within the cores of the stars and their subsequent supernovae as the stars exploded, nuclides of the remaining eighty-eight stable elements would form. Gold, atomic number 79, would actually be changed into lead, atomic number 82 (not exactly the dream of an alchemist). A study of the properties of these nuclides would reveal that given the threshold energy of the reaction, sodium and chlorine would combine, sharing an electron to form a stable molecule that might be called salt.

We'd look at two gases, oxygen and hydrogen, and realize that if raised to a threshold temperature, they'd combine in a violent reaction, with two hydrogen atoms sharing electrons with one oxygen atom to produce $H_2O$, which at room temperature would be liquid. We could decide to call that liquid water. From the properties of these elements we'd predict that the water molecule would have some extraordinary and unique properties. For example, as the water cools to just above freezing and the thermal motion of its V-shaped molecules decreases, they would separate slightly, expanding the water's volume. In doing so, its density would decrease. The result: the solid form, called ice, would be slightly less dense than the liquid from which it froze. Hence ice would float. We could discover that no other common substance would have this property.

This rare characteristic has far-reaching consequences. In large water bodies, such as lakes and oceans, the decrease in density allows the floating ice to serve as a thermal insulator for the warmer water below the ice, in effect limiting the amount of water that can freeze and thus preventing the oceans from becoming solid blocks of ice. The $H_2O$ molecule, we'd see, would have low viscosity in its liquid form, making it an ideal carrier for circulatory systems. An exceptionally large amount of heat would be involved in any change from liquid to solid or vapor, making it an ideal medium for moderating changes in temperature through the heat absorbed upon evaporation as temperature rises and the heat released upon freezing as temperature falls. In short, we could predict that water could be an unusual liquid base useful for a range of complex reactions.

Through this reductionist study of the universe via its atomic structure, we'd even foresee the production of amino acids, combinations of four or five of the lighter elements, with molecules containing up to twenty-three atoms, though from first principles we'd be hard pressed to find any specific benefit in them.

All told, once we take some givens, we can predict much of the chemical world. But that is where our predictions would cease. We can predict all the elements used in life, but there is no indication that we can predict amino acids joining together in chains of hundreds and thousands of units to form proteins and then proteins combining into the symbiotic relationships we refer to as life. When, in 1953, Stanley Miller, then a graduate student at The University of Chicago, produced a few amino acids through purely random reactions among chemicals found naturally throughout the universe, the scientific community felt the problem of life's origin had been solved. Far from it. Subsequent experiments have failed to extend his results. Thermodynamics favors disorder over order. Attempting to get those amino acids to join into any sort of complex molecules has been one long study in failure. The emergence of the specialized complexity of life, even in its most simple forms, remains a bewildering mystery.

That too may be what the Bible tells us. A superficial reading of Genesis chapter one, the creation chapter, marks the end of each of the first six days with "and there was evening and there was morning. . . ." Because the sun is not mentioned until day four, the logic of kabala takes a deeper view of the words. The root meaning of *erev*, the Hebrew word for evening, we learn is mixture, disorder, chaos—just as vision becomes blurry and chaotic at dusk. The root meaning of *boker*, the Hebrew word for morning, is orderly, able to be discerned, as with vision upon sunrise.

The flow of order out of chaos ("And the earth was unformed

and void," Gen. 1:2) is so surprising, so unusual in our universe, that day by day, six times over, the biblical text tells us of this passage to ever more complex arrangements of the existing matter by the seemingly simple statement of "and there was evening and there was morning." Or in the deeper sense, "And there was disorder and there was order." The saga ends with the appearance of human life: "And God saw everything that was made and behold it was very good. And there was evening and there was morning the sixth day" (Gen. 1:31). The millennia-old kabala reads "And God saw everything that was made and behold it was a *unified order.*" Two thousand years ago, the commentators on the Hebrew text saw within the words a wondrous transition from chaos to cosmos, from a jumble of energy and atoms to the dazzlingly complex and unified order of life.

Life beats the odds of chaos over cosmos, but not by defeating the second law of thermodynamics. Nothing does that. Life wins by *outwitting* the second law. The chemistry of a biological cell is the same as the chemistry that forms sodium chloride. One set of rules for all. But unlike sodium chloride, which follows the rules by rote, life has somehow gotten hold of wisdom, of information, that taught it to take energy from its environment, to concentrate that energy, and with it to build and maintain the meaningful complexity of the biological cell.

Cleverly, life scaled the mountain of complexity. What enabled these complex arrangements of carbon plus a few other elements to become so clever remains an enigma. When, as reductionists, we study the individual atoms, we find a sense of choice but no hint of cleverness. Yet somehow, the dust spewed into space by the nuclear furnace of a bygone supernova has become a human brain that learned to make nuclear reactors here on earth. Philosophizing about the origin of our origins, cosmic or biological, may be fascinating, but rarely does it accomplish more than raising other questions. If we are to search the consciousness of the mind in a rational manner, we had best first study the biology of the brain. That biology starts with its most basic unit, the cell.

GOING inside the body and then inside the cell is a journey to wonderland. Enclosed by its outer membrane, a cell's functions are walled off from the outside. When we look at any structure from outside, we get a highly simplified version of its essence. We decide to pick up a pencil and then do so. Not much to it. But in the path leading from the thought to the act, millions of cells and billions of atoms acting on command were required to accomplish that mundane feat. From the outside it seems so straightforward. Like starting a car: just turn the key. Or a computer: just press the power button. A myriad of hours were required to design the circuits and invent the components so that one simple act will activate the billions upon billions of atoms in just the right sequence needed to ignite the motor or light the screen.

If we could see within as easily as we see without, every aspect of existence would be an unfolding encounter with awe; almost a religious experience even for a secular spectator. A biology text presents a diagram, a cutaway view, of a cell. Within the cell, a dozen or so components are shown and labeled—the nucleus, chromatins, cell membrane, ribosomes, and so on. (See Figure 2.) If this were the reality, our metabolism would be one thousandth of what it actually is. A sketch showing all the cellular organelles would be one smear of ink, such is the density of the parts pumping life within.

In the early 1980s, near Hu Bin in the People's Republic of China, I watched the making of a five-hectare fish pond. Thousands of workers lined the banks and swarmed over the ever deepening space. Some broke new earth, some shoveled. A chain of workers passed earth-filled baskets up the bank, where another human chain dumped and returned the emptied bas-

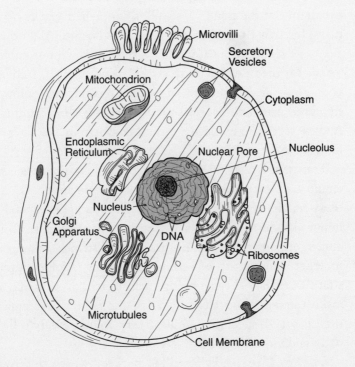

*Figure 2*

*An Idealized Cell with Key Components*

*Note that only representative organelles are shown. In an actual cell, most of these components are present in the thousands, filling the interior space with activity. The cellular components are not drawn to the same scale. Microvilli, the fingerlike projections, increase the cell's surface area.*

kets. The precision was a study in organization. My first encounter with the workings of a living cell was a déjà vu of that experience at the pond. The cell is simply jam-packed, a beehive of activity. Components are stuffed tighter than the circuits of a computer board, and everything is in motion. To get a scale for the rate of activity, consider: on average, each cell in your body, at this second and every second, is forming two thousand proteins. Every second! In every cell. Continuously. And they do it so modestly. For all that activity, we can't feel a bit of it.

A protein is a string of several hundred amino acids, and an amino acid is a molecule having twenty or so atoms. Each cell, every cell in your body, is selecting right now approximately five hundred thousand amino acids, consisting of some ten million atoms, organizing them into preselected strings, joining them together, checking to be certain each string is folded into specific shapes, and then shipping each protein off to a site, some inside the cell, some outside, sites that somehow have signaled a need for these specific proteins. Every second. Every cell. Your body, and mine too, is a living wonder.

The entrance to a living cell is marked by passage through a membrane functioning to keep the bad stuff out, while letting the good stuff in, and expelling what needs to be expelled, waste products and manufactured goods. But who or what decides what comes in and what goes out?

A myriad of portals provide entry, but only if signaled to open and allow entrance. Some of these ports are gated or opened by subtle changes in voltage differences across the membrane. Others open when a molecular key comes and unlocks them, allowing another molecule to pass. The cues come from within the cell if it's a call for the building blocks needed in protein replication, and from outside if, for example, it's a nerve cell coaxing a neighboring cell into action. A vast number of assumptions are woven into the simple act of signaling a membrane port to open. We are so immersed in consciousness, juggling thoughts even when daydreaming, that we project consciousness onto these chemical messengers that control the portals. But where did they get their smarts? Since when do carbon, nitrogen, oxygen, hydrogen, sulfur, phosphorus—the primary building blocks of biology—have ideas of their own, or any ideas at all? They're just atoms strung together to make molecules. Where'd they get the chutzpah to become keepers of the gate?

The laws of nature that govern interactions among atoms are simple and fixed. They produce repetitive formations, such as the crystals at the Dead Sea. Being highly repetitive, they contain little novelty. They do not produce the complex, information-rich molecules we find in life. There is no clue in nature as to how these simple laws could induce the nonrepetitive, multi-faceted information we associate with the genetic code or the proteins made from the information stored in those genetic codes.

Basic cell structure is the same throughout the entire biosphere, from the relatively simple structure of a sponge or algal cell to the complexity of a human. A cell's view of life is awash in a sea of water inside and out. The common denominator of all known life is that it is water based. If a cell is to survive, like a wristwatch in a swimming pool, it must be water resistant, but the cell's membrane must be water resistant both inside and out. Nature accomplished this feat by inventing a molecule with a head end that loves water (hydrophilic) and at the other end, two tails that are rejected by water (hydrophobic).

These phospholipids, as they are called, join tail to tail, forming a double-layered membrane, hydrophilic heads facing the water-based cytoplasm inside the cell and the water-based intercellular fluid outside. Sandwiched between inner and outer water-loving surfaces of the membrane are the pairs of hydrophobic tails. The result of this 20 nanometer (billionth of a meter) thick Rube Goldberg contraption is quite remarkable. Though highly flexible, the tenacity of the bonds between the phosopholipid molecules maintains structure. Pinch some skin. It doesn't break or crack. Release it and it returns to the original shape. Puncture a cell membrane with an ultra-sharp needle and then withdraw the needle. Gloop. The membrane reseals the hole and goes on with its work. Because the membrane has both water-loving (polar) and water-rejecting (nonpolar) layers, very few molecules can pass into or out of the cell without explicit permission. Polar molecules face an unfriendly barrier at the nonpolar tails in the middle, while nonpolar molecules can't get past the polar surface.

Such a membrane might make the enclosed cell a dead end. But nature is clever, somehow filled with wisdom. Thousands of receptor and transporter molecules, special proteins, penetrate the wall, determining what can and cannot pass. Muscle cells and especially muscle cells of the heart have large numbers of receptors designed to pass adrenaline, a stimulant hormone that greatly increases a muscle's energy production. At the sensation of danger (sensation did I say? I wonder just which carbon atom is experiencing this emotional trauma?), our reptilian response of fight or flight stimulates the release of large doses of adrenaline into the blood. Taken up by the heart muscles, the beat increases dramatically, pumping oxygen-rich blood to power-hungry muscles in arms and legs. Cells along the small intestine are constructed to absorb glucose, amino acids, and fatty acids, the products of food digestion, and transport these products to the adjacent bloodstream, where they'll be carried to the membranes of cells.

Membrane design is absolutely brilliant. No wonder all biological cells from bacteria at about one micrometer per cell to human cells at 30 micrometers (approximately one thousandth of an inch) function this way. I've been taught that nature did it all.

But there is a catch to this logic of a laissez-faire nature. It is true that lipids and phospholipids can form naturally. And in the presence of water, they do align to form sheets and even spheres. But a leap in information separates a sphere from a cell. That information is the plan of proteins and other molecules required to produce the portals that allow controlled transport across the membrane. Proteins have never been observed to occur naturally, even in a laboratory test tube. Proteins are the products of a cell's metabolism. There's a chicken-and-egg paradox here. Proteins make cells function, but cells are needed to make proteins. The step-by-step formation of an orderly, organized structure such as a protein is far too complex to defeat the second law of thermodynamics—nature's favoring of mayhem over organization—without help. Nurturing environs isolated from the hostile outside world are essential prerequisites. That is just what a cell provides. We're back to the start of the circle.

Yet somehow nature created both these chickens and these eggs in a geological flash after the appearance of the first liquid water on earth.

If we could walk into a cell, our first task would be to keep from getting bowled over. We'd be faced with a myriad of microsized vessels moving in all directions. Something like crossing Broadway against the light. But this is no hodgepodge of motion. Like Broadway there are well-marked traffic lanes. Picture a three-dimensional intersection of several major, multilane highways—crossovers, on and off ramps, an interlacing of cloverleafs one above the other, traffic moving in all directions. Now take it down, with no loss of the complex weave, to a millionth of a meter and repeat it ten thousand times in a sphere 30 millionths of a meter in diameter. You've got an inkling of a biological cell. As a colleague exclaimed when he first viewed a cell: "My God, it's got more connections than the Houston freeway."

The fiber-like roadways, the cytoskeleton, form a sort of jungle gym. They determine the flexible cell shape, hold the internal organelles in place throughout the cell volume, and provide the tracks on which newly manufactured molecules move from point of production to place of need. The item to be moved is "packed" into a pouch-like vesicle covered with thousands of molecular motor proteins. Each motor protein reaches out (extends) a molecular hand, grabs the selected fiber, pulls itself forward by contracting, releases, extends, grabs, and pulls again. Millions of microscopic hands hauling thousands of microsized parcels in every cell, every second of every day.

To meet the energy demanded for every molecular move, every cell has within it the machinery to take glucose from the foods we eat and to combust it, storing the released energy in a power-packed molecule known as ATP—adenosine triphosphate—a sort of biochemical battery that then makes itself available to whatever power-hungry molecule it might encounter. All life uses ATP as its power package, from dandelions to dangerous lions. ATP is nature's global battery.

The subtlety in ATP production is one more lesson in the wisdom held so modestly within life. If you have ever had the misfortune of experiencing a forest fire, flames shooting ten stories into the air, you have seen the power of combustion. To keep this extraordinary potential under control (in order to avoid burning up the cell in capturing energy), nature has found a way to release energy gradually, step by step, in a series of complex biochemical reactions, rather than leaping directly to the end products of carbon dioxide and water as happens in a forest fire. The cleverness of this elaborate bio-series is simply awe inspiring.

Let's follow the path of an ingested bit of carbohydrate. It's like being at a track and field meet, there are so many activities going on simultaneously. Digestion in the mouth and stomach degrades the large carbohydrate molecules into more manageable–sized glucose, which is able to pass through the intestine walls, diffuse into the adjacent bloodstream, and be swept by the blood flow to some glucose-hungry cell. But glucose is a highly polar molecule and so it cannot cross the nonpolar barrier of the cell membrane. Along comes the glucose carrier protein we call insulin, which attaches itself to the glucose. With insulin on site, the membrane gates fly open and the glucose enters the cell. Here, in a complex multistep process in which a dozen intermediate molecules are formed, free-floating enzyme proteins change the glucose into a substance called pyruvate and then, within the cell's many sausage-shaped mitochondria organelles, oxidize the pyruvate into the end products, carbon dioxide and water plus copious amounts of energy-rich ATP. The first two stages of this process are energy intensive and so require an energy source to power them. This is supplied by none other than—you guessed it—ATP (see Figure 3).

Wait a minute! Something's out of order here. This whole process is designed to make energy-rich ATP, and yet just to get the process started we need ATP. It's the chicken-and-egg all over again.

The "waste" products, carbon dioxide and water, must be dealt with. Water is already the medium of life, both inside and outside the cell, so that waste is not a toxic problem. And carbon dioxide is, as we all know, a gas, and therefore far easier to excrete than if it had been a solid as are most metallic oxides at body temperature. This fact, that $CO_2$ turns out to be gaseous rather than the usual solid oxide, is just another piece of good fortune that we have the choice of taking for granted or, instead, feeling it as a joyful part of the wonderful weave of life; as part of the extraordinary unified order that manifests itself throughout our uni-verse.

One basic cell structure, one basic energy source, one set of organelles common to all life. And one system for regulating this unity, the DNA-RNA team that takes individual lifeless raw materials and organizes them into living, thinking, choosing beings. The complexity in the commonness stretches the imagination. In his book *Life Itself,* Francis Crick was *almost* on target when he observed that the origin of life seems *almost* miraculous.

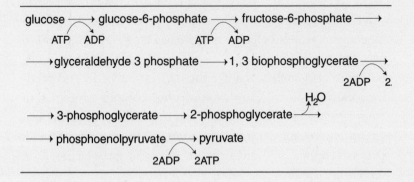

*Figure 3*

*Glucose Metabolism for the Production of Energy-Rich ATP*

*The stages of glucose to pyruvate require no oxygen, use two energy-rich molecules of ATP, and produce four ATP molecules, a net gain of two ATP molecules. This releases approximately 10 percent of the energy within the glucose. The next stage, the oxygen-rich Krebs cycle, omitted here because of its complexity (as if the pyruvate process shown here were not itself a mental wrestling match), releases the remaining energy by reducing the pyruvate to $CO_2$ and $H_2O$. The Krebs cycle takes place within the mitochondria of the cell. One of the many*

molecules essential to the chemical reactions in glucose metabolism, cy-
tochrome C, looks in diagram like this:

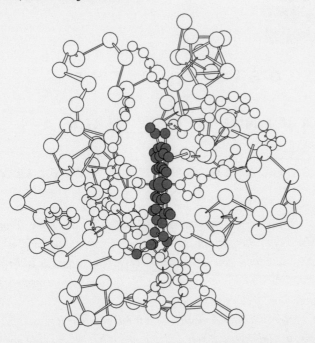

The shape of the molecule is essential for its functioning. With such structural
complexity, one wonders how the cell can tell one molecule from another. And
this is just one example of the dozens of enzymes, all equally complex, all essential, and all performing just the single act of getting energy from food. That energy must then be used to facilitate the functions of life. The implications of this
complexity are worth pondering.

# MEIOSIS AND THE

# MAKING OF A HUMAN

## A STUDY IN SHARED FIDELITY

*This is the story of metamorphosis. In fact it is the story of your and every human's metamorphosis. If it were a screenplay, we'd say it was fantasy, incredible. But it is reality—it is your history. It is the story of cells dividing and producing daughter cells that take paths very different from that of their parents.*

*If a caterpillar could speak, could it tell you that it was about to become a butterfly? And if you had not seen it happen already, would you believe it? Who would believe that from eight identical cells clustered together in the shape of a mulberry could come a human?*

We've studied the making of the universe and the structure of a biological cell. Let's see what it takes to make a human out of those cells.

After life itself, sex may be the greatest invention ever. It has done wonders for engendering the variety found in the biosphere. And yet for all the variety, the most common type of cellular activity in all life, mitosis, is a process that works against variety. With mitosis, a single cell, upon signal, duplicates all of its organelles, including its package of genetic material, the DNA. It then separates the paired products and divides to yield two identical daughter cells, each a replica of the parent cell.

With sexual reproduction, rather than a single cell acting as the originator of the progeny, a pair of cells unites to produce a daughter cell that is similar but not identical to either of the parent cells. To accomplish this, the new cell has a mix of the genes taken from each parent. And in that mix lies the potential for vastly diverse combinations within each species. Instead of a world full of identical clones, sex produces the wonderful variety that is so much of the spice of life.

But mixing genes from two cells poses a technical problem. Each parent cell has a full complement of genetic material, the DNA. For sex to work, and for the progeny to share contributions from both sides of the bond, each of the parents must be willing to relinquish half of her or his DNA, in technical terms changing from diploid to haploid. The process by which the DNA load is halved is termed meiosis.

Meiosis uses the biotechnology developed for mitosis, but at a few key stages it applies that technology differently. In the following description of the process, it may seem as if a human mind is functioning here. Don't lose sight of the fact that it is all molecules at work. There is no brain on site. The wisdom is built-in.

At the chromosomal level, the events of meiosis are similar for male and female, though the timing is vastly different. Each of the cells, the egg and the sperm, that will be active in reproduction, having duplicated its DNA and therefore holding a double complement of chromosomal material, undergoes mitotic division. The duplicated DNA chromosomes align along a central plane of the cell (see Figure 4). In standard mitotic fashion, each parent cell produces spindle fibers that, with their wonder of synchronous motion, reach out from opposing walls of the cell, attach to each of the pairs of duplicated DNA chromosomes, and pull the pairs apart, separating the chromosome sister pairs into two sets of twenty-three pairs. Just what taught the cells to invent and train these fantastically clever fibers remains to be discovered. The sets are now sequestered on opposite sides of the cell. A motor protein then forms, encircles the outer membrane of the cell with a molecular loop, and proceeds to draw in the loop, pinching the cell into two daughter cells. It's hard to keep in mind that it's all just molecules—no brain power directs these events. The two daughter cells, we discover, are not genetically identical. That is because genes devoted to the same function, for example, eye color or ear size, may carry out that function differently in each of the parents. My wife's eyes are green. Mine are brown. Though within each of our five children both eyes are the same color, among our five kids there's a range of eye colors. Why? Because in the separation and division of the chromosomes, genes for eye color are distributed randomly between the two new cells.

This random placing makes possible a multiple of variations not only in eye color. In excess of five million combinations are possible in the complete set of traits housed within the human genome. In addition to this multiplicity, while the pairs are aligned at the equator of the cell, prior to their being pulled to opposite sides, they swap gene pieces in a process termed crossing over. Each of the two new chromosome sets now contains pieces from the mother and from the father, even though there is only one full set of DNA data in each set. By this stage, the number of possible combinations runs into the multiples of trillions. With this vast potential for variety, it is not surprising that no two humans are identical. With all this genetic promiscuity, it's amazing that the vast majority of births produce normal looking kids.

One of the puzzles of meiosis is the altruistic nature of the cell. Why should a cell willingly give up half of its chromosomal information, and thereby essentially guarantee that its progeny will not be an identical copy of itself? I would have thought that altruism stops at self-destruction. A parent's mixing its chromosomes with those of another is, in a sense, self-destruction, since

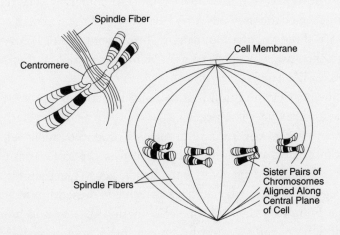

*Figure 4*

*Metaphase in the Cycle of Meiosis*

*Spindle fibers, composed of microtubules, have attached to the sister pairs of chromosomes at their point of attachment (the kinetochore, or centromere). The spindle fibers will now pull the pairs apart and move each chromosome to an opposite pole of the cell. The process in nothing less than an exquisite ballet played out at the molecular level.*

the parent will not be reproduced in the child. Not a very selfish way for a potentially selfish gene to act. The answer lies in the greater wisdom of nature. There is a payoff for the species. Variation aids in species survival by inducing a variety of traits within each species. With variety comes the potential for a wider range of responses to unpredictable environmental changes. But mutations and therefore evolution take place at the level of the individual, not at the level of the species. The individual gains nothing by meiosis. The origin of sex thus remains an unsolved puzzle.

Now comes a further nuance of meiosis. The next cell division in an egg occurs only years later, upon its being fertilized, while in the sperm it occurs immediately. In this division, chromosome replication does not occur. The four new cells (two from each of the previous pair) have only half the complement, twenty-three singular (not paired) chromosomes. These consist of twenty-two plus an X or a Y. Meiosis is complete. Let the cell's sex life begin!

The act of conception is as complex and intriguing as is the meiosis that allows it to occur.

Sperm cells, containing the meiosis-produced single set of chromosomes, are continually produced in the male from puberty through old age. Not so the female eggs. The basis for all the eggs a woman will ever have are already present in the embryo. Following their first meiotic division they wait in limbo until puberty and then fertilization. At birth, some two million pre-eggs are waiting in the ovarian wings, each with a double set of chromosomes—forty-six homologous pairs. These, surprisingly, are not all viable and by puberty only one in five, or some four hundred thousand, remain. At each ovulation, many eggs rupture, but usually only one egg matures. Because of this extravagantly profligate usage of the egg supply, by the time a woman reaches her fifties, the number of remaining eggs is insufficient to maintain the monthly cycle of ovulation.

At ovulation, a follicle in the wall of one of the two ovaries bursts open and a mature egg is released. The previously dormant egg, housing a double set of chromosomes, divides, yielding two cells each with twenty-three pairs (the normal amount). The nuance here is that rather than dividing the cytoplasm equally between the two cells, almost all the cytoplasm and organelles are concentrated in one of the daughter cells, that which is to become the mature egg. The "starved" cell disintegrates. If the egg is fertilized, this excess supply of intracellular material will serve good purpose, providing nutrients and cellular organelles during the early cellular divisions of the embryo.

Along with the released egg, the follicle burst also discharges a cloud of hormones, protective cells that surround and accompany the egg, and a chemical attractant, a sort of perfumed agent that appears to act as a lure to coax sperm to egg. These all are swept into the adjacent fallopian tube, the 10-centimeter channel that winds a path from ovary to uterus. Cilia lining the fallopian tube speedily brush the perfumed come-on ahead of the egg, through and out the fallopian tube, into the uterus. The massive egg lags behind, moving at a much more leisurely pace.

Now for the activity at the other end. As soft fingerlike projections and muscular contractions of the fallopian tube urge the egg along toward the uterus, sperm ejaculated by the male into the woman's vagina start their passage toward the egg. The egg's journey from ovary to uterus takes about four days. The life of an unfertilized mature egg is less than a day, about ten to fifteen hours. Clearly we cannot wait for the egg to reach the uterus if fertilization is to be successful. The sperm fortunately are up to the task. The life of a sperm is about two days, and they are built for traveling (see Figure 5).

Recall that, unlike the mature egg, sperm at the time they are released into the vagina contain only a half set of genetic material. This is concentrated in its head, along with chemicals needed later for penetration of the egg. While the egg is a thousand times bigger than most human cells, comparable in size with the period at the end of this sentence, the head of the sperm is among the smallest of cells, 2 to 3 millionths of a meter (microns) in diameter, 5 microns long. The rest of the sperm is a pure powerhouse designed for motion, a motorized tail, mechanistically a near replica of the cilia that coat our lungs and the flagella of a protozoan, though at some 50 microns in length it is far longer than a protozoan flagellum. Its midsection is wrapped with mitochondria, producers of the energy-rich ATP molecule. These will supply the copious amounts of biological fuel needed to power the flagellum muscle's furious whiplike lashing that propels the sperm during its long journey. Interest-

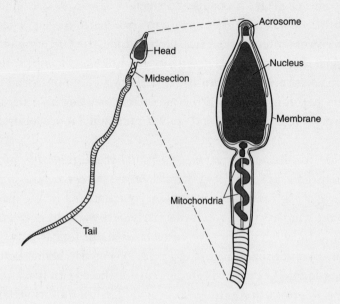

*Figure 5*

*Diagram of a Sperm Cell (Spermatozoon)*

*The term "spermatozoon" is well chosen since the sperm, with its long flagellum—or tail—is very much akin to a protozoan. The head, some 2 to 3 micrometers in diameter and about 5 micrometers long, houses the DNA-containing nucleus and the acrosome, which contains the enzymes needed to penetrate the egg's wall upon contact. The entire head is covered by a plasma membrane. Mitochondria in the midsection supply the ATP needed to power the whiplike actions of the flagellum as it propels the sperm forward at 3 to 5 millimeters per minute. The tail is greater than ten times the length of the head. With these dimensions, the sperm is truly built for traveling.*

ingly, while almost all cells use glucose for the mitochondria energy source, sperm uses fructose.

Somehow sensing the perfumed lure that was released with the egg, the sperm swim from the vagina, across the uterus, into and up the correct fallopian tube. Moving at three millimeters per minute, sperm can reach the egg within a few hours, well under the fifteen-hour life span of the egg.

Approximately three hundred million sperm enter the vagina. They die in droves as they traverse the chemically hostile environment of the fallopian tube. But though hostile, these same harsh chemicals strip the sperm head of its protective coating, preparing it to penetrate the egg. Of the few hundred sperm that finally reach the egg, only one can enter. All batter the egg's protective coating, releasing protein enzymes that work at stripping the egg of its protective coating till one can touch the actual egg wall. A chemical key, a specially shaped protein on the sperm's head, must match the chemical lock, another specially shaped protein on the egg. No foreigners are allowed; a wrong species' sperm would waste an egg. Only if the match is proper can the sperm force its way in, and fertilization occur. A new potential life is under way.

Entrance of the sperm stimulates calcium ion channels in the egg's wall to open. This immediately closes the egg's membrane to all other invaders. One is company. Two's a crowd, genetically speaking. The closure is first induced by a change in the electrical charge across the membrane, and then by a release of a chemical hardener that cross-links the gelatinous outer membrane coating.

The sperm is now within the egg, but it is kept at bay. The egg still has a full set of chromosomes, twenty-three pairs. Half must be discarded. At this stage the egg undergoes meiotic division, division without chromosome replication. Though the egg's chromosomes are divided equally between the two new cells, as before, almost all the cytoplasm is concentrated in one daughter cell, that which contains the sperm's nucleus. This done, the sperm and egg each duplicate their respective twenty-three chromatids, motor proteins move the two nuclei together, the nuclear membranes open, and the chromatids mingle, each finding and pairing with its corresponding partner. Spindle fibers and motor proteins then pull one member of each pair to opposite poles, two nuclear membranes form, and mitotic cellular division occurs. Another bio-ballet, a coordinated dance of molecules within the body, has passed. A few days and several further mitotic divisions will go by before this new bit of life reaches the uterus.

At this point, one has to bend over backward to accept that all of these necessary and interwoven steps have evolved randomly. There are only two forms of reproduction, mitosis and meiosis. There are no intermediate forms visible in nature. Yet somehow, according to evolutionists, a batch of lucky mutations allowed meiosis to blossom on the tree of life.

In human reproduction, meiosis gives way to mitosis, but with an amazing twist to the mitotic principle of daughter cells being replicas of the parent. As the cells divide, some of the daughters discover or interpret how and where to become the cells of a heart, and some a nose or toe. How this miracle of structuring occurs remains a speculative mystery. The knowledgeable differentiation among the newly forming cells appears in principle to be orchestrated by concentration gradients of specific molecules within the cluster of cells, but the wonder of it remains. Who or what supplied the scheme? Wisdom is encoded in the very stuff on which it must act, the blueprint and the builder all in one.

Keeping in mind that the identical DNA is now held in each of the cells, the paradox of differentiation is compounded when we realize that the process appears to be accomplished with no prompting from the mother's body. Fetal development identical to that observed in utero is obtained in vitro through four days post-fertilization with human egg and sperm, and with other vertebrates for much larger fractions of total gestation. Somehow the egg knows what it must become.

By the second day after fertilization, three mitotic divisions have yielded a total of eight cells, though with no significant increase in the overall size of the cluster. Until now the cell divisions have been accomplished through enzymes and organelles already present in the egg prior to fertilization. No new materials have been manufactured. The sequestering of most of the cytoplasm and organelles in only one of the two cells of mitosis now plays a key role. This hoard of material provides the base for all initial activity. Until now the genes of the male have played no part.

Prior to the next mitotic division, eight into sixteen cells (on day three), the organelles within each cell are pulled toward the exterior surface of the eight-celled ball. Then in near perfect simultaneity the eight cells divide. The outer daughter cells receive most of the cytoplasm and associated intracellular structure. But unlike the earlier asymmetric divisions where the "starved" daughter cells were destined for death, the inner hungry cells are to be the building blocks of the fetus, and then months later, the baby.

It's day four. The journey down the fallopian tube is about to be completed. The minuscule embryo, now consisting of between thirty-two and sixty-four cells, enters the uterus, releases an enzyme that dissolves the protective shell within which it has traveled since fertilization, and produces and releases a hormone, human chorionic gonadotropin (hCG), that tells the mother's body she is mothering an embryo. With this message comes a command to terminate her menstrual cycle—an absolutely essential step lest in menstruation the nurturing environs of the uterus be expelled. The appearance of hCG in a woman's urine or blood is the first indicator of pregnancy.

This entire cellular juggling act is preprogrammed within the genes that stay with the fetus and then with the adult throughout life. Somehow, they are cued to turn on only at this stage even though at this stage we scarcely detect any indication that a child is in the making.

The beauty and awe of life's genesis lies in the details. Skipping over the intricacies can lead a person to believe that milk just naturally comes in containers, and is not the end result of a process starting with sunlight shining on grass. What follows is just a sampling of the relevant steps in gestation, each revealing only a hint of the complexly ordered wisdom built into the process.

The morula, as the cluster of cells is called at this stage, floats freely in the uterus until seven or eight days after ovulation, absorbing its needed nutrients from uterine secretions. Internal timing now somehow tells the morula to change shape from a compact ball of sixty or so cells to a fluid-filled sphere, the "blastocyst," housing those previously starved cells destined to become the embryo. The outer cells of the blastocyst are to evolve into the fetus's part of the placenta, the semipermeable interface between mother and "child."

The difference in the genetic composition between the mother and the fetus makes the fetus appear as a foreign invader, and therefore subject to attack by the mother's immune system. The placenta, a product of both the fetus and the mother, has much of the task of isolating the fetus from the mother. In clever fashion, it allows transfer of maternal nutrients and oxygen to the fetus while expelling fetal wastes out to the mother's circulatory system. It's a lesson plan in chemical engineering.

Approximately seven days after fertilization the blastocyst decides to contact the wall of the uterus, release enzymes that literally digest (!) that part of the uterus, and in doing so, mesh with the uterine wall. The entire cell mass consists of a few hundred cells, and is barely larger than the original egg, still about the size of this sentence's period.

Now something quite surprising happens (as if the entire process till now had not also been surprising—but the next step is unique in life). Some of the outer cells divide and mesh together as if many individual cells decided to join forces and do away with the separating membranes. The resulting multicell structure projects capillaries further into the uterine wall, where they exude enzymes that dissolve the mother's capillary-sized arteries. (Don't lose sight of the fact that these enzymes had to be manufactured according to information, wisdom stored in the egg's DNA. Wisdom within wisdom.) The multicell structure has become the placenta. With the blood vessel walls ruptured, the mother's blood floods into the spaces between the fetal capillaries, which absorb the blood's nutrients and oxygen and transport those life-sustaining factors back to the blastocyst. Baby growth can now commence. A bit over one week has passed.

The seemingly dormant cells within the blastocyst awaken and by the end of the second week they align to form a longitudinal axis. Differentiation has begun. Some still unknown mechanism will turn on a set of genes in one cell and then turn it off, while orchestrating a totally separate set of genes in another cell or group of cells (all of which were identical a few days ago). The result is that the originally identical cells become very different parts of the body.

Until now, the cells held only the potential to become the various organs of a fully developed body. Now that potential comes into play, causing differentiation within what was originally a unity. And all according to an intricate ballet.

By two to three weeks, structures that will become the heart and nervous system can already be identified. A bit over a week later the heartbeat begins, though it is not yet palpable to the mother. By week five the key organs are in place: liver, pancreas, thyroid. Beginnings of eyes dot the sides of the head. The embryo is still a mere quarter inch long (a bit over half a centimeter), about the length of the "the" in this sentence, but as full of variety as an encyclopedia.

At the sixth week, precursor cells mark the beginning of the central nervous system. Eyes are now clearly visible. The brain begins to enlarge. It will continue to develop even long after birth. Arm and leg buds each sport five digits fully joined by webbing, somewhat like a duck's webbed feet. During the next two weeks, the cells of the web die, apparently so programmed in their genetic makings, and the fingers separate. Most of the crucial action is complete by the end of the first eight weeks—a transition from embryo to fetus having the beginnings of all the primary organs and limbs of the baby-to-be.

By the ninth week the arms and legs, hands and feet, and most of the body look clearly human. Eyes have begun to migrate from the sides of the head toward their human forward position.

I mention the forming of the central nervous system, eyes, an arm, a finger. The few words trivialize a world of complexity. Each organ is the result of a metropolis of activity.

Consider the eye. Look out the window. Your field of view catches a vista perhaps a mile wide. It all appears projected onto half of a sphere at the very back of your eye, the retina, less than three centimeters in diameter. Yet your brain sees within those three centimeters of information a world a mile wide and knows it is no Disney cartoon the size of a postage stamp. Light from the outside world has reached your retina with only slight distortion. That's because somehow those clever genes in your body produced crystal clear, transparent cells for the eyes' outer casing, the cornea and the lens just behind the cornea, and the thick fluid that fills the globe of the eye between the lens and the retina. Amazing. All those totally clear cells and fluid even though most of our body is opaque or translucent. Some cells of your eyes are yours for life. As you age, more are added, but the ones you were born with are still with you as well.

The iris, which is controlled by an array of muscles, regulates the amount of light entering based on feedback from the retina. Behind the retina is a heavily pigmented layer that absorbs light not captured by the retina. A second array of muscles changes the shape of the lens, bending the light more or less as per the extent of the lens's curvature, focusing the incoming images sharply on the retina. (All land vertebrates use this system to sharpen the image. A fish lens acts in a manner similar to a camera, focusing by moving the lens backward or forward.) Of course, the concept of focusing assumes the brain makes some decision as to what a "sharp" image means. Might the world really be blurry and we just see it as sharp?

All those muscles working in unison with no conscious thought on your part, and all in the blink of an eye, and all originally stored in one fertilized cell. But let us take one final tour-within-the-tour, to show the full miraculousness of this tiny datum within what was once a single cell. Because when the light reaches the retina, a three-ring circus of activity begins.

The light passes through the cornea, is bent and focused by the lens, passes through the vitreous humor, and ultimately arrives at the surface of the retina. But this surface is not sensitive to light! Still the eye can record even a single photon of light. First the light ray or rays must further pass by a stacked array of photomultiplier cells that vastly multiply the signal they receive from the retina. Finally, at the very back of the retina, which makes it the very back of the eye, are the light receptors, some hundred million rods for colorless sensitivity in dim light, and another approximately seven million cones for color vision in bright light. (Notice that in very dim light we do not see colors. The world becomes various shades of gray. That is because in reduced light, only the rods of our retinas function.) (Figure 6.)

We mentally build the range of colors our brain "sees" by mixing the inputs from three types of photoreceptors: sensitivity to low-energy red colors peaks at 630 nanometers (billionths of a meter) wave length; intermediate greenish peaks at 560 nm; and the high-energy end of the visible spectrum, blue, peaks at 420 nm. The visible range of light, 380 nm (deep purple) to 750 nm (dark red), is only a tiny fraction of the total range of electromagnetic radiation from radio waves at 100 meters in length and more to high-energy gamma rays measured in the thousandths of a nanometer.

Housed within the deepest section of the photoreceptors is a pancake-like stack of discs containing light-absorbing pigment specific for one of the three color/photon energy ranges. Ab-

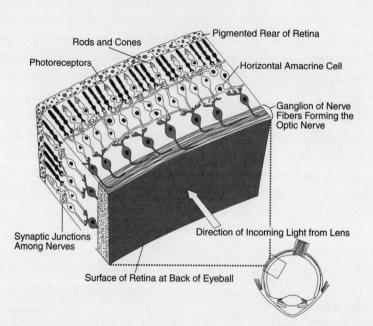

*Figure 6*

*Detail of a Retina*

*Note that the photoreceptors are located at the very rear of the retinal layer and therefore at the very back of the eye. Light reaches these receptors only after it has maneuvered past the multitude of nerve fibers. If we had not learned in an earlier chapter that, according to the Bible, intelligent design, even at the level of the Divine, is not necessarily perfect design, we might erroneously find here an argument against the concept of a metaphysical presence within the physical world.*

sorption of an incoming photon by a pigment molecule starts a chain reaction that heads back up the stack of nerves toward the surface of the retina. In essence, the human retina is designed inside out.

The immediate receptor of the photon is a derivative of vitamin A, termed retinol. A two-stage reaction then occurs that shuts down the receptor after its having fired. In very bright light, this reduces its sensitivity, thus adjusting the eye's signal to the brain. The photoreceptor, now energized by the absorbed photon of light, induces a cascade of as many as five hundred molecules of signal protein, each of which binds to and in doing so activates an enzyme that reacts with up to four thousand secondary molecules.

Sodium ion channels in the membrane, normally open in the absence of light, now close in response to the enzyme activity. This changes the electrical potential (the voltage difference) across the cell membrane and starts the nerve signal on its way to the brain, traveling up through the stacked chain of retina nerve cells that link rod or cone to the long ganglion nerve, the avenue to the brain.

Along the way, while traveling at 100 meters a second, the signal passes by a long, thin, laterally arranged nerve, the amacrine cell. This beauty of engineering compares inputs from adjacent receptor nerves to check for motion in the viewed image. Ever notice how a falling leaf or a moving hand barely visible out of the corner of your eye catches your attention? It's amacrine doing its job, protecting you from an end-run surprise attack. Frogs have the same array but to a far greater extreme than we humans. They respond only to moving objects. A stationary fly arouses zero interest in the frog's brain. If the fly moves, it soon becomes lunch.

Ganglion cells by the millions, leading from all the receptors, join together to form the optic nerve, and being at the forward, that is the lens side, of the retina, must now pass back through the retina, and head for the brain. This puncture through the retina by the optic nerve produces the famous blind spot we all have in our vision. We don't see the blank because our brain produces a conscious image that smears over the lacuna.

The multitude of ganglion cells cannot be treated like strands of cooked spaghetti tossed into a tube. The relative positioning of each receptor must be retained in the positioning of its ganglion so that the signals map onto the brain in accord with the image. That is quite a feat of bio-bookkeeping. But we manage to do it without even giving it a thought. The information these nerves bring to the brain represents points of light. Our brain blends those points into smooth, continuous images. If it didn't, our every view would appear as a Seurat painting, filled with dots of color and shade.

The menagerie of molecules that prepared the body for sight was entirely hidden by my simple words: eyes formed. Sometimes we see the trees and sometimes the forest. With life, there are forests even within the trees. Wisdom within wisdom.

BIRTH is no less complex.

The fetal placenta, "realizing" that the time for birthing has arrived, stops producing a hormone that for nine months kept the uterine muscles relaxed. Without this chemical inhibition, they begin the contractions of labor. As the actual birth begins, the umbilical arteries, for nine months the fetal life line, clamp shut while the umbilical vein amazingly waits to shut just long enough for the fetal blood that happened to be in the placenta at this moment to be returned to the fetus. Is this amazing, or not?! Blood flow is rerouted to pass through vessels of the lungs and intestines as the lungs prepare to breathe in air and the stomach prepares to digest food, the well-balanced nutrients of mother's milk. A hole between the two sides of the heart that had allowed the fetal flow pattern closes as the direction of the flow of blood through the heart reverses.

It's our metamorphosis. We've changed from a tail-bearing organism immersed in liquid with eyes on the sides of our heads, a yolk sac, bronchial arches, a one-chambered heart, a snout-like nose, with webbed hands and feet, and become human. If you could relate this to a butterfly, would it believe such a transition is possible?

It was all programmed into one cell just nine months earlier. Amid the phenomenal complexity of biology, there exists an underlying simplicity so eloquent that it is expressed in a single molecule. In life, oneness is what we represent. A single cell at fertilization contained within it all the potential that you were ever physically to become. And every cell within your body retains that wisdom.

The unifying cohesiveness we found in the laws of physics has come shining through in the workings of biology as well. But in both fields, we see it only when we seek it. A superficial reading of the world teaches of a reality built on entities as diverse and unrelated in their composition as the steel of a gun barrel and the fragrance of a rose. Can they actually have anything in common? Indeed they can and do. Surpassing the harmony of the laws of physics and the workings of biology lies the subtle truth that every bit of existence is composed of a single substrate—energy—created at the beginning. Einstein theorized and a multitude of experiments have since tested the truth of this bizarre and totally illogical reality.

The Bible relates, in the usual English translation, that "the earth was unformed and void" (Genesis 1:2). But the Hebrew of "unformed and void" (*tohu* and *bohu*) has a much deeper meaning. The kabala teaches that *tohu* is the solitary primordial substance, created at the beginning from absolute nothing, the substanceless substrate from which all that is material was to be formed; and that *bohu* is a composite word, as are many Hebrew words of the Bible, meaning *bo*—in it; *hu*—there is. "In it [in the *tohu*] there is [potential]." The single substance filled with full potential from which the world would be constructed was created at the beginning. The hidden potential within the single fertilized cell from which the entire structure of a human arises is a mirror in miniature of the creation.

Yet as remarkable as the underlying unity of the physical world may be, science is on the brink of discovering an even more sensational reality, one predicted almost three thousand years ago, that wisdom is the basis of all existence. "With the word of God the heavens were made" (Ps. 33:6). "With wisdom God founded the earth" (Prov. 3:19). "With wisdom God created the heavens and the earth" (Gen. 1:1). Wisdom is the building block, the substrate, from which all the time and space and matter of the universe were created. Wisdom is the interface between the physics of the world and the metaphysics of creation.

To search out the link between the two will take more than results from the laboratories of physics and molecular biology. Wisdom is not weighable, nor is it easily stained and seen under a microscope. It is going to take a bundle of abstract thought, before which it would be productive to understand how we think. We've reached the workings of the brain.

**6**

# NERVES: NATURE'S

# INFORMATION NETWORK

*The essence of life is found in the processing of information. The wonder of life is the complexity to which that information gives rise. The paradox of life is the absence of any hint in nature, the physical world, as to the source of that information. As reluctant as I am as a scientist to admit it, the metaphysical may well provide the answer to this paradox.*

Wrapped in smiles of love, I hug a giggling baby. A weight lifter exhales as he presses a three-hundred-pound barbell. A composer thinks through the notes of a symphony in the making, her brain on fire with melodies. The brilliant red of yesterday's sunrise emerges from my memory and once again I am filled with awe. All these experiences and a myriad of others that constitute our mental and physical lives rely on the near lightning-fast transmission of impulses within and among the labyrinth of nerves that lace our body—a million miles of bioelectric cables. If we are going to understand how our brains function and, more importantly, how the conscious mind emerges from the physical brain, we must first understand the functioning of nerves, for in essence the brain is a highly structured mass of nerves.

In simple terms, the nerve is the basic structure used in life for sending information from one body part to another. What is surprising is that the body uses the same means of transmission regardless of the type of information. Be it touch, taste or smell, sound or sight, the data related to these sensations or thoughts are encoded as pulses of electricity, voltage peaks, that travel as waves within cleverly insulated extensions of the nerves, known as axons. Upon reaching their targets, be they another nerve or a muscle fiber, the pulses of voltage release a chemical transmitter that passes to the target. The resulting chemical change in the target cell might cause a muscle to contract or a thought to arise. Within the neurological maze of the brain, these identical electrical signals are then categorized among the range of our varied senses.

In some cases the timing of the signals is exquisitely fine. In speech, for you to differentiate between the enunciation of a "b" and a "p," your lips must open some thirty thousandths of a second before you cause your vocal cords to vibrate for the "p" sound to emerge rather than a "b" sound, which occurs when you open your lips and vibrate your vocal cords simultaneously. Thirty thousandths of a second. Consider what this reveals concerning the precision inherent in mental and neurological processing. It's a sliver of time that makes the difference between bat and ball and Pat and Paul. Your brain determines this phenomenally tight timing and you don't even have to "think" about it. It's probably controlled by the brain part located close to the brain stem known as the cerebellum. The entire sequence is encrypted when the signal to vocalize a "p" or "b" arises in your thoughts.

Could this complex yet ordered precision have evolved without guidance? The problem in answering that question is not really knowing what Darwinian theory has to say about the process. While there is a clear description, in the theory and in the fossil record (and in the Bible too) of a development of life from the simple to the complex, there is no statement in any of these sources as to how the development of complexity came about.

A theory sets forth a concept of how specific events occur. It is not merely a description of those events. A careful look at the current theory of evolution reveals not a theory, but merely a description of the "punctuated" jumps in the fossil record.

The evolution of life, if evolution is the proper word, is indeed punctuated. The world awaits a theory for what processes might have yielded those punctuations. It is for this reason that when the London Museum of Natural History, a bastion of Darwinian dogma, mounted a massive exhibit on evolution, occupying an entire wing of the second floor, the only examples it could show were pink daisies evolving into blue daisies, little dogs evolving into big dogs, a few dozen species of cichlid fish evolving into hundreds of species of—you guessed it—cichlid fish. They could not come up with a single major morphological change clearly recorded in the fossil record. I am not anti-evolution. And I am not pro-creation. What I am is pro-look-at-the-data-and-see-what-they-teach.

Darwin realized the problem when he advised us to use our imagination to fill in the punctuations. He insisted in the Introduction to his *Origin of Species* that despite the "steady misrepresentation [of his theory]," natural selection is "not the exclusive means of modification. . . ." In the closing lines of that famous book, Darwin elaborated on this idea. "There is a grandeur in this view of life, with its several powers having been breathed by the Creator into a few forms or into one. . . ." Darwin saw in the wonder of life the need for a ghost in the system, the powers breathed by the Creator. The laws of nature instilled in our universe and the physical conditions on the earth are ideal for sustaining life. But nowhere in these laws or conditions is there a clue for how the organized information of life originated or developed. The wisdom intrinsic to the simplest forms of life is nowhere presaged in the substrate from which life is built.

Don't let the grandeur of your body's origins and organization escape you. No, on further consideration, better not to think about it. You may become tongue-tied. Studying these brain/nerve/muscle interactions has all but tongue-tied my game of squash. I can't swing the racket without marveling at the train of bioelectronic signals that must be pulsing through my body's neurons.

I tense as my wife's shot slams off the front wall. Visual stimulation of eyes produces a chemical reaction in the retinas' photoreceptors, retinas' signals to optic nerves, optic nerves to brain, and brain to motor nerves that will pulse signals through their axons at over 100 meters per second to a phalanx of muscles controlling eyes, toes, ankles, legs, hips, shoulders, arms, wrist, fingers. They're all in on the act as I try to return Barbara's slam from just above the red line. Voltage-gated ion channels in cell membranes open and in doing so allow an avalanche of calcium ions to race through those openings, causing extended muscles to contract. An arm swings and with a bit of skill, and luck, the strings of my squash racket slam into the small black rubber ball.

The process of biological information transfer is a tale of awe. Consider just one aspect of this bodily train of events. How does the brain decide that the two-dimensional image projected onto the eyes' retinas represents a three-dimensional world? After all, the visual image is converted into an array of electrical stimuli, each of which is a one-dimensional pulse of voltage, in essence a single point. These one-dimensional signals, carried by the mass of fibers that make up the optic nerves, are what the brain receives, upside down and reversed just to make the progression a bit more complex. How does the brain know that the identical type of neural signal, having originated in my ears as a reaction to the sound waves produced by the black rubber ball pounding off the back wall, should be decoded as sound? From where does it get its smarts?

For a small number of unfortunate persons, the wiring has gotten crossed, resulting in sight that is heard and sound that is seen. Ever get punched in the eye and you see stars? Why stars, why not just pain? You saw stars because the punch activated nerve endings in the retina, which, as good soldiers, sent an electrical signal to the brain. But to which part of the brain? The part that interprets an electrical pulse from those nerves as sight, not as pressure and pain. Hence the seeing of stars.

Never ever take the simple beauty of sentience for granted. It didn't have to be that way. Complexity underlies even seemingly simple acts.

The technical information on the structure and function of nerves that I am about to present is not intended to encourage you, the reader, to pursue a career in molecular biology or physiology. But it is intended to make you aware that when the alarm clock rings in the morning, and blurry-eyed, you fumble for the button to silence it, there's already in your sleepy body a symphony of cellular reactions in progress, which if you could put them to music, would by comparison make Beethoven's violin concerto seem little more than chaotic street noise. I urge you, skeptic or believer, to acknowledge this astoundingly integrated chain of events that was required to bring this symphony into being. The concept that life may have resulted from inert, dumb, random reactions starting in an undifferentiated ball of energy at the big bang stretches the imagination. Does this prove there's a God active in our world? I personally do not think that the complexity of life proves the existence of the Divine. But it does demonstrate unequivocally that we are missing some basic factors in how the origin and the development of life occurred.

Whether those factors include the metaphysical, or even the Divine, may never be absolutely verifiable. However, the final leap of faith for or against the concept that the metaphysical is active within the physical universe that it created (for both the skeptic and the believer require a final leap of faith) is best made from a position of knowledge. Faith backed by knowledge is much stronger than faith based on an emotionally driven gossamer hope, whether that faith be secular or religious.

From energy to rocks and water, to life, to consciousness. Now that's an impressive and even possibly Divine chain. Let's look at some of the biology that developed from the big bang.

In his book *How the Mind Works*, Steven Pinker decides "not [to] say much about neurons, hormones, and neurotransmitters [those complex molecules produced in nerve cells that allow communication among nerves]." Understand that, according to the title, Pinker's book is supposedly about what goes on inside our heads, that is to say, the output of interactions among neurons, hormones, and neurotransmitters.* Only by avoiding the intricacy of how that output comes about and what limits it is Pinker able to enthuse repeatedly throughout the book about the assumed but untested wisdom that we are purely "the product of natural selection." The complexity of stimulating and transmitting a single neural signal, let alone the functioning of all the other organelles that contribute to the finely orchestrated symphony we call life, belies the assumption. Let's mentally crunch through some of that detail.

In overall functioning, nerve cells are similar to all body cells. A sacklike membrane, consisting of two back-to-back layers of molecules, defines the boundaries of the cell. The water-loving (hydrophilic) heads of these molecules are exposed on the interior and exterior surfaces of the membrane. The central part of the membrane houses the hydrophobic (water-avoiding) legs of the membrane molecules. Channels, which open and close in

*A far better book on the brain/mind interface with much less ego pushing through and far more objectively written is, in my opinion, Rita Carter's *Mapping the Mind*. Both books are 100 percent secular in their approach. Notwithstanding their titles, neither book finds a mind within the brain. Pinker opts that since our minds are limited, we may be unable to solve the enigma of how sentience arises from the brain.

response to chemical and voltage signals from within and without the cell, provide controlled entrance and egress for molecules active in cellular metabolism. A nucleus within the cell concentrates the primary genetic material, the DNA. Mitochondria throughout the cell help convert nutrient glucose into the energy-rich molecule, ATP. Microtubules extend to all regions of the cell, providing structure and, of crucial importance, tracks along which motor proteins can transport needed molecular materials to reaction sites.

Two regions differentiate nerve cells from other cell types: axons and dendrites. The axon is the transmitter. Some 20 microns in diameter and reaching as long as a meter, the axon is the extension of the cell that carries the nerve's message as a bioelectrical signal, known as an action potential, to the target organ. The end of an axon may divide into a bush of terminals, allowing it to stimulate whole groups of targets simultaneously. This plays a central role when signaling a multitude of muscle fibers to come into action.

The dendrite portion of a nerve is the receiver of the axon's message. Fingerlike extensions of the dendrite increase its surface area, allowing it to receive multiple signals simultaneously. In some cases as many as one hundred thousand axon terminals reach a single dendrite. The nerve will sum these individual signals and decide if the total input message is strong enough, that is, exceeds a threshold voltage, to warrant the passing of this information on to another nerve. "The nerve will sum the signals and decide"—it sounds almost mental, almost human, yet it is only a batch of carefully arranged molecules (see Figure 7).

In a sensory nerve, such as those of a fingertip that feel touch, dendritelike extensions respond to stimulation of the skin and send a pulse along that nerve's axon toward the central nervous system, be it brain or spinal cord. Motor nerves then transmit signals received from the central nervous system to the muscles being called into action.

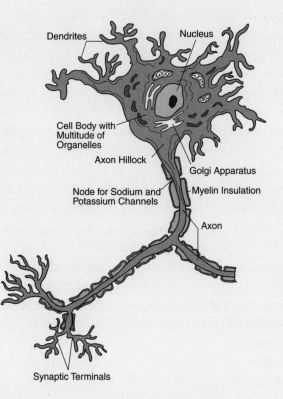

*Figure 7*

*An Idealized Nerve Cell*

*Synaptic terminals by the thousands, coming from other nerves, cover the nerve cell's dendrites, each bringing its own electrochemical output. The resulting signal is converted into the bioelectric action potential at the axon hillock. The action potential then races down the axon to the synaptic terminals, where waiting chemical neurotransmitters will be released, sending the signal on to other nerves or muscle fibers. Axon length can reach a meter when transmitting signals from the tip of a finger or toe to the central nervous system.*

Signal transmission by a neuron is totally dependent upon differences in voltage along the axon and across the neuron cell membrane. Our every thought and deed at some point reduces to a bioelectric signal. The voltage potential is induced by differences in concentrations primarily of sodium, potassium, and chlorine ions. (An ion is an atom with one or more of its electrons missing and so has a net positive electrical charge.) I'll focus here on sodium and potassium.

The concentration of sodium ions inside the cell is approximately ten times less than its concentration in the extracellular fluid just outside the cell membrane. The pattern for potassium is the opposite, being ten times more concentrated within the cell relative to the extracellular fluid. These inequalities are maintained by pumps that operate across the cell's membrane. We house an alarmingly sophisticated technology of molecular dimensions. The differences in ion concentrations produce a negative voltage within each cell of about minus 70 millivolts. Every cell in your body stands ready to give off a biological spark.

Large amounts of sodium can enter the cell only if the cell receives a signal to open voltage-gated channels that span the cell membrane. And that is the key to all neural signal transmission. Upon stimulation, these channels open and sodium ions flood into the cell. Via a positive feedback loop, the initial sudden increase in sodium stimulates more sodium channels to open. In a few thousandths of a second, an ionic avalanche of sodium raises the cell's interior from the original minus 70 millivolts to a potential of plus 50 millivolts. At this voltage, the cell is poised to fire.

Assume a nerve's dendrites have picked up a signal sufficiently strong to warrant its propagation to a further nerve or muscle. The neuron is primed for response. Try to imagine in this discussion how the synchronized, intertwined complexity about to be described could have evolved from inert and unthinking rocks and water, for that is what preceded life on the originally sterile earth.

An adjacent axon has signaled to the neuron that it is to generate a signal, the bioelectric action potential. Perhaps you've touched something too hot for comfort. The heat stimulates the sensitive endings of the nerve, inducing it to send the message to its target receivers, in this case rapidly to the spinal cord and slightly less rapidly to the brain.

The signal, a cascade of ions, travels from the receiving dendrite, past the cell body, and on toward the axonlike extension. At this point the action potential is generated that, as a wave, transmits the signal the length of the axon to the synaptic terminals at its end. Since the axon terminal does not actually attach to the target neuron's dendrite, nature had to invent a method of getting the message across the gap measuring approximately 20 billionths of a meter (20 nanometers) that separates axon terminal from target dendrite. Nature was up to the task. The nerve accomplishes it by having the electrical action potential within the axon stimulate the release of chemical neurotransmitters into the synaptic gap. The electrical signal has become a chemical signal. These neurotransmitters have been "conveniently" stored in organelles called Golgi apparatus near the axon's synaptic terminals. The Golgi are budlike globular beauties of nature that package the neurotransmitters at their point of manufacture in the cell body and then, with the help of motor proteins, transport them and other essential molecules from within the cell body, down the axon, to the location of use near the cell membrane. The Golgi, upon command, release the neurotransmitter into the synapse, where it diffuses across the opening, attaches to the target dendrite, and in doing so triggers a secondary neural signal to start on its way.

Consider the implications of just one aspect of this event. Somehow the Golgi apparatus was positioned at the terminal by nature. Way back at the cell body, near the start of the axon—which in some cases is a meter distant from the terminal—a signal had been given to the DNA to provide the pattern to make messenger RNA, mRNA, which, with the help of transfer RNA, tRNA, and a few other cellular microorganelles, would churn out copies of the one specific type of a wide range of potential neurotransmitters needed for this one type of stimulus, package it in Golgi buds, and via motor proteins that literally walk molecular step by molecular step along microtubule tracks the length of the axon, carry the Golgi loaded with neurotransmitters to the axon terminal area, there to wait patiently in the wings until called upon by the axon's electrical signal—the action potential—to move into action. The trip from the cell body where the Golgi and neurotransmitter are made to axon terminal takes about two days when traveling via motor protein. If nature had relied on diffusion to make the trip, the journey would have taken about two years. When called into action, the Golgi move within a millisecond.

The Golgi bud fuses with the inner surface of the axon synaptic membrane, and then, in a process known as exocytosis, bursts through on the outside, into the 20-nanometer-wide synaptic gap. The electric signal of the axon has been converted into the chemical signal of the neurotransmitter.

Golgi buds might be likened to a vitamin pill. They concentrate the one type of needed molecules and then release them as a batch. For neurotransmitters, Golgi concentrate them at two hundred times greater than would be possible if these molecules were just free floating at the synapse terminal. It's all so clever. None of this wisdom is even hinted at when we view life from the outside. The insights of molecular biology have revealed a complexity at every stage of life's processes such that, if we were forced to rely on random mutations to produce them step by step, in the words of Nobel laureate de Duve, "eternity would not suffice."

The huge concentrated input of neurotransmitters released into the synapse causes them to diffuse rapidly across the synaptic gap. In less than a millisecond they reach the dendrite surface. If the neurotransmitter is the correct one for the job, its shape will complement the shape of a receptor on the dendrite membrane and it will bond. The right key in the right lock— only one fits and nature designed it just so. It is either the result of chance random reactions among rocks and water or the expression of an underlying wisdom poking its head through into the physics of life. Those are the only two choices available.

The amount of released neurotransmitter contains a large excess over the quantity actually needed to bind with the target dendrite. The excess remaining in the synapse blocks the input nerve from transmitting an additional signal. If nature were to rely on diffusion to have the excess molecules drift away, the down time of that nerve would be considerable, measured in seconds and even minutes. But nature foresaw the problem and is neither so lazy nor so patient. In time spans measured in a few thousandths of a second, the excess is absorbed by the axon terminal and the remainder decomposed by enzymes, catalytic proteins fashioned way back in the cell body from DNA-held information that is transcribed via mRNA via ribosomes and tRNA, and so on and so on, and then released into the extracellular plasma to be on hand just in case it is needed to reduce the complex neurotransmitter molecule to its component parts.

Of course the lock-and-key system of the neurotransmitter/dendrite interaction is exactly what a Darwinian-type survival-of-the-fit would favor as a fail-safe system to block out wrongly directed signals. Any survival system would favor it. The question is how did the system arise? In the late 1970s, a symposium was held at the Wistar Institute of Anatomy and Physiology in which mathematicians forced biologists to confront the reality that all calculations of probability say no to the assumption of randomness being the driving force behind life's development. But cognitive dissonance held sway then and still holds today. The conclusion of the biologists was, and remains, that the mathematical assumptions must be incorrect since evolution must have occurred. As I discussed at length in *The Science of God*, both the Bible and the findings of paleontology indicate that life developed from the simple to the complex. That development is not the problem. What is at question are the mechanisms behind the development. The calculations of probability haven't changed since the Wistar symposium, but the understanding of molecular biology has, and that understanding has revealed biosystems far more complex then those that were imagined in the 1970s.

The fastest of mammalian nerve cells can fire a thousand times per second. That requires depolarization of the cell membrane and the restoring of the ionic concentrations of potassium and sodium in less than a thousandth of a second. The ion pumps (molecules that actually pump) that do the job are a wonder of efficiency. Meanwhile the action potential is traveling in excess of 100 meters per second toward its target organ. But we are not dealing here with a copper wire where the electric current is facilitated by readily exchanged electrons within the copper metal. In the biosystem of an axon, the influx of bulky sodium ions through gated channels that pierce the cell membrane must propagate the signal.

The velocity of signal propagation depends upon two factors: axon diameter and how far ahead of the actual signal the sodium channels can be induced to open. The larger the diameter, the more rapid the propagation. To gain this advantage in signal transmission, some animals, such as the squid, have axons a millimeter in diameter, easily visible to the unaided eye. Conduction here reaches 25 meters per second. But axons of this size become prohibitively cumbersome if many nerves traverse the same region, as with the limbs of vertebrates. If this solution had been adopted by human nerves, the diameter of our arms would be measured in meters.

Vertebrates have developed a far more effective, and quite clever, method of increasing the velocity of impulse propagation. Nerves that must stimulate organs requiring rapid response—for example, those that might signal the need to remove a finger from a hot surface, or power a leg muscle to run from a hungry lion—are electrically insulated by a fatty molecular layer, a nonconducting lipid sheath known as myelin. This insulation greatly reduces current leakage (sodium loss) from the axon outward across the axon membrane. It also regulates the positioning of the sodium channels in the membrane.

A typical vertebrate axon is 10 to 20 microns (millionths of a meter) in diameter, fifty times narrower than the squid's nerve, and yet the nerve pulse in the human moves at 100 meters per second, four times faster than in the squid. How is this accomplished? The wisdom of biology doesn't try to beat the rules of nature. It outsmarts them. The secret is in the positioning of the myelin sheathing. Approximately every millimeter there is a break in the sheath, a node, a few microns wide. The sodium and potassium channels are concentrated at these nodes. As the impulse propagates along the axon, it draws sodium always from the nodes just ahead of it and, in essence, leaps forward to that node instantly. The result is signal transmission from the central nervous system spinal cord to a finger or toe in a hundredth of a second.

The action potential (AP), as it moves along an axon, represents a high voltage peak relative to the voltage potentials of the surrounding portions, both before and after, of that axon. In theory that voltage peak should be able to travel in both directions, both down the axon and back up toward the cell body, since in both directions, forward and back, the voltage is lower than the peak. To obviate the possibility of the AP reversing its direction and traveling back toward the cell body, nature has brilliantly invented a double molecular lock on the sodium channels. Once a channel has opened for the sodium cascade into the cell and then closed, that particular channel cannot reopen until the local region has completely depolarized. By that time, the AP has moved out of range, far along the axon toward the nerve's synaptic terminals. The molecular keepers of these gates are very clever. Somehow they have learned their lessons well.

The benefits of myelin sheathing are sadly most apparent in their absence. Multiple sclerosis is a disorder in which the autoimmune system in error destroys the myelin. The result opens the cell membrane to sodium loss across the exposed axon membrane. As the disease progresses and myelin depletion increases, transmission rates slow to a few meters per second. Eventually leakage is so great that the axon can no longer transmit its message. The target muscle becomes paralyzed. The fact that for most of us life's mechanisms work in proper order is a wondrous miracle. When they do not it is a tragedy.

The system described and diagrammed above is an ingenious one for communicating massive amounts of complex information. The parallel processing and perfect timing involved are as elegant as the finest supercomputer. Perhaps some day, in the age of communications technology now upon us, we will imitate and exploit our own design: In the meantime we can only wonder at the workings of our chemistry.

With this understanding of how nerves, the information-bearing cells of the brain, function, we are ready to enter the brain itself. Knowing the complexity of the processes involved, when we see a diagram showing how simple evolution is, how one organ can change into another merely by adding a feature here and there, we must realize that those demonstrations are a farce. As long as the intricate workings of the cell are disregarded, there's no problem for a Steven Pinker, or Stephen Jay Gould, or Richard Dawkins to talk of random reactions producing the goods of life.

It is hard not to be fooled by the foolish arguments when they originate from intelligent foolers. Abraham Lincoln is quoted as having said that while you can fool some of the people all of the time, and all of the people some of the time, you cannot fool all of the people all of the time. The more knowledge one has, the harder it becomes to be fooled.

Those diagrams that in ten steps evolve from a random spread of lines into people-like outlines, and in a few hundred steps simulate a light-sensitive patch on skin evolving into an eye, once had me fooled. They are so impressively convincing. Then I studied molecular biology.

# THE BRAIN

# BEHIND THE MIND

*If the universe is indeed the expression of an idea, the brain may be the sole antenna with circuitry tuned to pick up the signal of that idea.*

When I think a thought, what is understanding what? Just to ask the question I must conjure up internal images. When I think of myself, I recall an image which, if I saw it in a photo, I'd recognize as myself. Actually the immediate image that surfaces is me in my roaring twenties. That's my psyche's image of self. I have to work logic into the picture if I am going to see today's wrinkles and receding hairline.

Surgeons studying activity in various parts of the brain have discovered that as they stimulate regions of the external body, from head to toe, neurological reactions in the brain produce coherent maps of the body, albeit upside down and left–right reversed. Some portions occupy a greater cerebral area than they actually do on the body. The mental maps are, in a sense, reminiscent of a cover that appeared on *The New Yorker* magazine several years ago. It depicted the artist's concept of a New Yorker's view of the world looking west. Manhattan and New Jersey fill the entire foreground, followed by a very thin sweep of the Midwest. Then Los Angeles and San Francisco loom, bordered by a strip of water, the Pacific reduced to a trickle. Hawaii and China fill the horizon. The body, as the brain sees it, also follows the correct sequence of limbs from feet to head, but the emphasis reads like a caricature of the human body— which of course it is. Our psyche is a caricature. Our very, very big feet all but abut our oversized genitals. Legs, having fewer neural receptors, get short shift in the cranial map. Like the Midwest to the New Yorker, the legs are there merely to connect feet to groin. Arms get a bit more space. Hands are huge. The face, especially lips and tongue, are similarly enlarged.

When I think of specific body parts without even physically touching them, the corresponding neurons on the mental map fire. Science fiction and science fact use this thought-to-neuron connection to envision mental control of the world around us. At the neural level, we are actually witnessing a continual display of mind over matter. Or possibly more accurately stated, the consciousness of the mind over the consciousness of matter.

In earlier chapters we've seen that the universe exhibits the essence of a mind, a wisdom behind and even within matter. This universal wisdom, sometimes defined as God's presence, is the essence of the metaphysical as it projects into our finite, physical world. If, in fact, humans are created in God's image, as the Bible claims, the task of this and the coming chapters is to determine if mind—consciousness—precedes matter in man. This will be no small task in a culture steeped in what George Gilder refers to as "the materialist superstition," a worldview in which emotions, mind, and all feelings of spirituality are the products of the physical body. The religion of materialism quintessentially believes that if you can't measure it, weigh it, stain it, and see it under a microscope, it does not exist. No one has yet managed to weigh a mind.

BY birth, many neural connections are already hardwired, set in place by the genetic instructions of the cell. But many are not. We have something in our brains like the fixed read-only memory on a hard disk and then a self-correcting software program that goes along with the ROM. The software of the brain is malleable, reacting and developing in response to ongoing environmental stimulation. Cover the eye of a kitten or allow a childhood cataract to go uncorrected and that part of the brain's body map withers as neurologically adjacent organs take over the mental space originally dedicated to those impaired body parts. As long as there is life in the body, the brain is ever changing in response to that which life brings to the body. Stimulation and love are the recipes the young brain calls for, notwithstanding recent claims that infants can get along without them. Neglected toddlers do survive, but the neural physiology of the deprivation is engraved within their brains.

From all we know concerning the physiology of the brain, the sensory inputs, motor outputs, and interacting neurons, can we tease out the seat of consciousness, that which the mind experiences as self? Whether or not the mind has an aspect that transcends the physical brain is currently a moot point. At the minimum, awareness of what we feel as the mind appears to derive from activity in the brain. This amazing organ of perception, once by necessity treated as a black box—sensory spur in, motor and emotional response out, with no clue as to the intervening processes that linked input to output—is now open to investigation. Revelation of the previously unimagined intricate workings of the brain has challenged the simplistic theory of life's random evolution in a manner similar to the challenge presented by the discoveries of the complexity of cellular molecular biology. The difficulty in displacing the belief in Darwinian evolution, even though the theory fails to describe reality, is that no other materialist mechanism can explain the development of life as displayed by the fossil record and as described in the Bible. In a world so steeped in the physicality of materialism, calling upon metaphysical solutions is out of bounds.

With the help of a range of recently developed instruments,

we can at last begin to define boundaries to the processes of the brain, isolating and identifying those aspects of consciousness that are certainly the products of neural activity, homing in on the field of what might be a mind distinct from the flesh and blood of the brain. All the topics discussed in the foregoing chapters, the detailed structures and functions of cells, the biomechanics of information transmission in nerves, come together in the brain.

The adult human brain has approximately one hundred billion neurons (nerves). So does an infant's at birth, though many of the connections among the nerves have not yet formed. One hundred billion is also the approximate number of stars in our galaxy, the Milky Way, and the estimated number of galaxies in the entire visible universe. In an adult brain, the axon of each neuron connects with as many as a hundred thousand dendrites of other neurons. The branching is stupendous, a million billion connections. That's 1,000,000,000,000,000 points within our heads at which neurotransmitters are racing, sending information from nerve to target nerve. A massive web of activity contained within just under one and a half liters of volume, with most of our conscious thought emanating from a layer 2 to 4 millimeters (about one-eighth inch) thick, the cerebral cortex, which covers the very top of the brain.

Here's a brain test: Wiggle your toes on both feet, and now at the same time wiggle your fingers, extend your arms and move them in horizontal circles with none of these motions being in the same direction, and while doing all this, gently shake your head right to left, left to right and quietly whistle a tune or recite a verse. And now consider all the neurological and molecular complexity involved in every one of those muscular contractions, all of it being induced by a flood of signals processed simultaneously in the central nervous system that started with signals from your optic nerves as you saw and read these words. For each individual motion and thought, the signals (action potentials) travel the length of axons of motor nerves extending from brain to muscle; sodium channels in axon membranes open and close; sodium and potassium pumps restore the electrolyte balance across each cell membrane so the same nerve can fire tens of times each second; neurotransmitters transported by the Golgi apparatus are released into synapses at the axon terminals and bond to adjoining dendrites. Calcium channels in the muscle fibers open to admit calcium ions that bond with proteins that bend and pull back other proteins, contracting a muscle fiber just enough so that the combined action of a million links each cycling five times each second produces the desired smooth force as you twirl your arms, move your head, whistle your tune, move your fingers and toes.

Of course you could never possibly do this on a regular basis and certainly not without concerted and conscious effort. But the wonder that your body represents, driven by a wisdom hidden within every particle of the universe, does the equivalent of it at almost every waking moment of the day. You start to cross a road and turn your head to check for traffic, step out with just the right gait from your leg muscles, judge the distance and timing of an oncoming car. You then turn to look for traffic in the second direction, hear a familiar voice (familiar, meaning recalled from memory), link that voice to a battery of other locations in your brain, the face, personality, and name that go with the voice, feel positive emotions toward the person, call out her or his name adjusting your vocal cord tension and lip shape to the task (without a thought, conscious at least, that it is only a difference of 30 milliseconds in lip/vocal cord timing between "P" for Paul and "B" for Bill), wave hello and yet somehow make it safely across the street, shaking hands with just the proper firmness of grip even though a few moments before you were pressing two-hundred-pound free weights at the gym.

It's called parallel processing, multiple tasking. The brain does it by the millions every waking second of every day, without even thinking—or better said, with barely a conscious thought. I wonder which is more of me: the conscious or the subconscious me? All our actions are preceded by thought, whether conscious or not. We enjoy free will. I cross the street only if I choose to. But is it my mind or is it my brain that chooses? Are they different? An overview of the brain's anatomy and function provides part of the answer.

Let's look at the development of those brainy organs that just orchestrated our walk across the street.

The first recognizable body part in a human embryo is the central nervous system (CNS). Eventually it will become the brain and spinal cord. Its early arrival is not surprising considering the fundamental role the CNS will play throughout life. By just two and a half weeks after fertilization a hollow trough has formed along the embryonic axis as cells migrate in from its periphery. Less than half a week earlier the first longitudinal organization of the embryo's cells had begun. By three weeks, the trough closes to form the neural tube. The entire embryo is still less than two millimeters (about a sixteenth of an inch) long.

During the third or fourth week the heart begins to beat, but not via stimulation from the brain or CNS, though surprisingly it is located near the brain in what will become the head. The heart is still one chambered, like that of a worm. In time the embryo and the heart will fold, moving the heart to the chest and forming the four-chambered heart of an adult human. For now, held within an embryo not quite three millimeters long, it has another two to three billion beats ahead of it, pumping life-giving nutrients to all parts of the body.

Within another day or so, two swellings, like small inverted cups, begin to protrude from the brain, the beginnings of the eyes. Eyes are, in a sense, externally visible extensions of the brain.

By thirty-five days old, the cerebral cortex, the part we associate with conscious thought and intelligence in an adult, has become visible. It's the part of the brain we are referring to when we say someone has a lot of gray matter. Gray is the color of cell bodies and dendrites, the nerve parts assumed to be allied with information storage. The brain starts to enlarge, just the beginning of a process that continues for years. Because the mother's pelvis must support the entire weight of her upper body, it requires considerable bone mass. This limits the size of the opening through which the fetus must pass at birth. The head at birth literally stretches the envelope. The brain at birth is a quarter of the mass of an adult's. Four-legged animals have relatively larger birthing channels since their pelvis need only support half the body weight.

So why, you might ask, are we humans, with our large brains and distinctive abilities, bipedal? Is this a flaw in the wisdom of design? Could we be smarter and mentally richer beings if we walked on all fours and thus had larger brains at birth? Not likely, as we see from the realm of the four-legged world. The wisdom in our design is that higher intelligence requires hands freed from supporting the body. Without that freedom to form complex tools, control fire, develop technology, there'd be no point to a larger brain. We are as large and as small as we need be.

In a way, the brain has outwitted the size restriction. The cerebral cortex and neocortex are wrinkled in a way remarkably similar to the shell of a walnut. This wrinkling has increased the cortex area without needing to expand the total brain volume, and therefore the size of the head. Once again wisdom has bypassed nature's constraints.

The brain produces neurons at an astounding rate in the womb. By birth there are one hundred billion. During the nine months of gestation, that averages out to between four and five thousand new nerves each second. Four to five thousand phenomenally complex axon-elongated cells constructed each second, each one racing out to find its target organ. These are guided on their journey by a trillion (a thousand billion!) structural glial cells. Don't just gloss over these fantastically huge numbers. Each cell houses all the complexity we discovered earlier: nucleus, DNA, mRNA, tRNA, ribosomes, motor proteins, ion channels, and on and on. And all are being manufactured at the rate of five thousand per second. The miraculous nature of life is found in its details.

By eight weeks, the embryo, with all its major body parts now visible, has become a fetus. At this time a very politically incorrect phenomenon occurs in the male, XY genotype fetus. Pulses of testosterone, associated with testes formation, are produced, altering the brain's development relative to the development in the XX female genotype. I dread to write this at the risk of being labeled sexist, but here goes. In general, the difference between XX and XY brain development appears to make girls better in speech and social relationships, boys more advanced in spatial relationships. Not for every human, but in general. Neurophysiologists may have discovered what parents have sensed all along!

At about twenty weeks, 140 days of age, neural connections form between the cerebral cortex and deeper brain parts. Though the difference between the human genome and that of a chimp is estimated to be less than 1 percent, our cerebral cortex has ten times more neurons. Another five weeks and parts of the limbic system, the seat of stored emotions, link with the cortex. This union in adults allows deliberate control of emotions that might otherwise result in automatic explosive response, fight or flight, each time we are stressed.

While I was at M.I.T., two cars got into a tangle on the southeast expressway leading into Boston. The drivers started to argue and one whipped out a handgun, shot and killed the other, and then sped off. His limbic-cortex linkage was in dire need of reinforcement, though I doubt that such a plea in court would have gotten a sympathetic response from the judge.

Myelin sheathing of the axons in the frontal lobes is completed only after reaching full adulthood. Since the frontal lobes are where deliberate throttling of emotion can occur, and since they are not yet fully functional until the sheathing is completed, it is not surprising that teenagers are so often more impulsive than their parents.

We've seen how the brain develops, but that still does not tell us how our thoughts arise. How the organs of a mature human brain interact may provide a clue to the processes of thought. The details you, the reader, are about to endure are admittedly complex. But they are also miraculous wonders in their complexity. To understand what happens when the brain does its work is to understand the beginning of the mind.

Since the only part of the brain visible from the outside is the organ devoted to vision, the eye, that is a reasonable starting point. In a previous chapter we saw how an electromagnetic signal (light) gets from the eye's retina via the optic nerve to the brain. The processing of the signal, breaking it into its component parts, starts immediately at the retina, where some neurons record aspects of color, some motion, some boundary contrast. That is just the beginning of the mental deconstruction of the information.

When thinking about sight, think of the symphony of molecular reactions of each optic nerve as the multitude of nerves communicate impulses, analyzing each impulse, deciding whether or not to pass on the pulse to other regions and other nerves. Think wonder. And ponder how a batch of carbon, nitrogen, oxygen, hydrogen, and a few other elements got together to cooperate so very wisely, thousands, even millions of times every second, throughout the brain and body. Had Darwin known of the wisdom hidden within life, I have confidence that he would have proposed a very different theory.

A scene catches your attention. Perhaps you are watching a

squash game and the black rubber ball has rebounded from the far wall just above the red line. A good shot, your brain tells you. The electromagnetic radiation (light) illuminating the scene bounces off the ball and wall and red line, travels to each of your eyes, which move in unison due to a brilliantly but subconsciously synchronized set of six muscles controlling the motion of each eye. Though they move together, the eyes do not follow identical tracks. The fact that the eyes are separated demands that they move at different angles. This single aspect of brain/muscle linkage is an exquisite demonstration of complexity. If it were absent, we'd have double vision. The difference in the angles of the eyes helps the brain estimate distance between the eye and the viewed object. As the light enters the eyes, the iris adjusts the aperture to regulate the amount of light reaching the lens. The double convex lenses change curvature to focus the image, now inverted, on each retina, where the photons of light are absorbed by photoreceptors. The energy of the incident light rays produces an array of electrochemical pulses, sending information of motion, color, shape, and borders via the optic nerve to the brain. At this stage the entire scene might be reproduced, mapped directly on the cerebral neurons, and then recalled, but life is not as simple as it meets the eye. We don't actually see any of this.

Once the scene has been parceled at the retina into neural signals, the scene is never reconstructed into the light patterns that reached your eye. What you perceive as a fragrant red rose or a black squash ball is, in your brain, a myriad of nearly identical electrical signals devoid of color, scent, or shape, with each signal directed to the specific portion of the brain devoted to the particular aspect of the scene.

Here lies one of the fundamental differences between a mind and a computer. In storing data, a computer, using semiconductors, stores an image as an array of electrical pluses or minuses in voltage, binary data. This is not so very different from the nerve's axon/dendrite storage system. But there is an essential difference in the retrieval of the information. Based on that array of binary data, and emanating directly from it, electric circuits excite pixels on the computer screen that reproduce the original scene.

In the brain, there is no evidence of a "screen" on which, or within which, the original image is displayed for mental viewing. There is no hint in your brain as to how you see the words you are now reading. We know how the inputs of the scene or smell or sound are mapped in the organs of the brain. It's the replay that confounds us.

As the image of the scene passes through the double convex lens common to all vertebrates, the image inverts. What is on the left in the outside world reaches the right side of the retina; the right reaches the left, the top the bottom. Something like heels over head. (Did you ever notice how nonsensical the expression "being head over heels in love" is? Friend, if you are not usually head over heels, you have a very strange way of walking!)

In point-by-point transmission from each retina, a ganglion of a million or so nerves carries the image aft into the brain, where the two optic nerves—one from each eye—meet and divide. All ganglion cells carrying information from the right fields of the retinas of both eyes continue to the right side of the brain. Those from the left retinal fields of both eyes go to the left. This means that approximately half of the retinal input from each eye goes to the corresponding side of the brain. The mental image is still reversed in the brain, since that inversion occurred at the lens, prior to the retina. By comparing the images, now in stereo, the brain estimates distance and texture (see Figure 8).

The first stop on the optic trail is the thalamus relay stations, one on the right half of the brain and one on the left. We've moved a bit over halfway toward the back of the head. At this

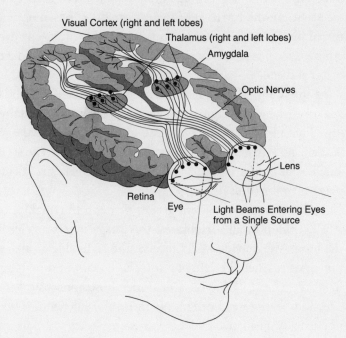

*Figure 8*

*Visual Pathways Within the Brain*

*Note that the signals from the right sides of the retinas of both eyes go to the right hemisphere of the brain. Signals from the left sides of the retinas go to the left brain hemisphere. The amygdala, being closer to the eyes than the visual cortex, receives the signal before the visual cortex and can react to that information even though the amygdala produces no conscious visual awareness.*

point, the signal is routed to the very back of the brain, the location of the visual cortex, and also to a nearby organ known as the amygdala. The pictorial part of vision is processed at the very posterior part of the brain, even though it begins at the most forward portion, the eyes. Emotional aspects of a scene are processed in the amygdala.

All incoming senses, other than smell, are inverted the way sight is and all have the thalamus as the first station. The neural circuitry of the thalamus tackles an intense amount of book-keeping since all the inputs are identical electrochemical pulses. Best not to mix auditory with retinal signals, or you'll be hearing sight and seeing sound. Smell, a most basic of senses, bypasses the thalamus and is routed directly to the amygdala.

It is the amygdala, one of a group of organs at times collectively referred to as the limbic system, that evokes emotional memory even when at times that memory evades our logic. During a recent fall semester, a student confided to me that he wasn't able to concentrate while sitting next to a particular member of the class. He could not understand why. A bit of questioning revealed that the aftershave lotion used by the other man was the one he had favored during a previous summer's escapade. It evoked in him all sorts of bittersweet feelings he had been unable to explain.

While all vertebrates have the emotion-packed limbic system, only mammals have a highly developed cerebral cortex, the site of advanced logic and data processing. The wisdom of the cortex notwithstanding, the amygdala, because of its proximity to the thalamus, induces its response before the cortex can get into the act. The limbic system is the fast route. The cortex is still working on the data when the amygdala is screaming for action. From here arise the responses of fight and flight, and, interestingly, a third emotional choice not usually mentioned, submission. Depending upon one's past experiences, the amygdala quickly chooses a response to incoming sensory information and prompts a response. The first part of us that acts, our first decision maker, is also our least logical self.

Even with the "slow" processing of the cortex, we're talking about milliseconds. Slow is still quite fast. But the difference in timing is unfortunately sufficient to, at times, let us blurt out a response that moments later has us groaning "How could I have said that?"

This fast limbic response ability reveals an important aspect of physiology. The limbic system, and hence all vertebrates, including reptiles, birds, and fish, store emotional information as long-term memory. And against this stored information, incoming data are analyzed. Memory is not the province of mammals alone. All vertebrates house within their brains some aspects of their personal histories.

Having raised trout in my high school years from egg through release into New England streams and lakes, I have no question as to memory in fish brains, notwithstanding the minuteness of their cerebrum relative to the size of their brain stem and limbic system. Just prior to each feeding of the fry, which were housed in outdoor pens measuring some ten to twenty square meters, I would whistle. By the second week, at the sound of my whistle, the water next to where I stood adjacent to the pen would boil as the fry thrashed about in expectation of the feed. Months after being released into lakes, the trout still returned at the sound of my whistle, though all feedings had ceased upon stocking. Similar techniques are used in raising salmon in fjords, and in training homing pigeons

Though long-term memory is not the possession of mammals alone, we still must ask if the response at the limbic level is at all conscious, or are the limbic contributions totally instinctual in animals other than mammals, and possibly in mammals also. Is consciousness solely provided by the cortex, and therefore a domain occupied by mammals alone? Did the fish "realize" that they were expecting food? Did they picture the food? Even without words, did they ponder what might be in store for them at that feeding, ruminating among a range of emotions tied to the flavors of the different feeds they had received in the past? Or did they experience only a direct instinctual response, just as our salivary glands react in a visceral way when we smell pot roast or hot chocolate chip cookies?

As the visual impulses leave the thalamus for the visual cortex, specific aspects such as color, motion, shape have been individually grouped. Now we encounter a lesson in an intricately woven division of labor, and the puzzling reconstitution of that division into a seamless, unified mental image, all within one-thirtieth of a second.

The visual cortex has six distinct layers, each devoted to analysis of an aspect of the image. These were identified over a century ago through the work of two pioneers in neurophysiology. Camillo Golgi, in the late 1800s at the University of Pavia, developed techniques of selectively staining nerve tissue. (Golgi apparatus bear his name.) For the first time it became possible to trace the path of a given neuron. His method is still in use. Santiago Ramón y Cajal, realizing the potential of Golgi's breakthrough, studied the nervous systems, especially the visual pathways, of many animals. In 1906, they shared the Nobel Prize in physiology and medicine.

Neural recycling time is about 30 milliseconds. Changes more rapid than this blur together. That is why a movie or video appears as smooth action. The visual cortex cannot distinguish between individual frames projected at a rate that exceeds the recycle rate. But within each of these slots of a thirty-thousandth of a second, each cortical layer analyzes a specific aspect of the scene as carried to it by the optic ganglion of nerves. In one layer, three-dimensional depth relations are studied by comparing differences from each eye, in another layer colors, in another motion; position; line orientation; and boundaries and contrast between boundaries. The nerve cells in the layer that become active in response to motion show no interest in the color of the object in view. In general an object moving into view elicits a stronger neural response than one moving out. Conversely, those neurons in the layer responsive to color, that is, the specific wavelengths of light, are not "interested" in aspects of motion. The individual layers feed their analyses back to the primary visual region and then via some unknown mechanism, at some elusive site, a unified image emerges.

If the entire visual cortex is destroyed, then no conscious sense of sight remains. The person is blind. However, if the eyes have not been damaged, signals may still reach the thalamus, the way station before proceeding to the cortex. The fast track to the amygdala remains operational, and emotional response to the "view" can arise which the viewer is in no way able to explain since no view has been "seen." It's clear that there's a lot more going on in your head than just what meets the eye. It is not surprising that so many of the texts on evolution eschew any semblance of a mathematical analysis of the theories that random reactions produced this ordered, information-rich complexity. When Lawrence Mettler and Thomas Gregg decided to add a few chapters on the mathematics of evolution in their book *Population Genetics and Evolution*, they brought Henry Schaffer on board. The math Schaffer brings to this totally secular text states clearly that evolution via random mutations has a very weak chance of producing significant changes in morphology. Of course this is exactly why you will search long and hard to find rigorous studies of probability in the works of Dawkins, Gould, or any of the other spokespersons for random evolution. Their approach to evolution is atavistic, a throwback to the time of Darwin, when cellular biology was assumed to be a rather simple affair of slime within a membrane. As we've seen, molecular biology has revealed that it is a mountain more than that.

And as if the complexity were not enough to break through the shell of materialism, there is history as well: the puzzle within the fossil record.

For three billion years, between the oldest fossils of life (bacteria and algae) at some 3.5 to 3.8 billion years ago and the first evidence of animals in the fossil record, 530 million years ago, the fossil record reveals a flow of life that remained one celled or at most groups of cells clustered into structureless communities. No appendages, no evidence of mouth or limbs or eyes.

And then with no hint in the underlying (older) fossils, an explosion of complex animal fossils appears bearing the basic anatomical structures of all phyla extant today. It is what is termed by the scientific community the Cambrian explosion of animal life. Among those structures are eyes. The earliest eyes arrived with stereoscopic positioning, and with lenses that by their fossilized shape appear optically perfect for seeing in water, the habitat of those early animals. We just struggled through the complexity of vision, from the conversion of incoming radiation inducing electrochemical pulses to the analysis of those pulses of information by the host animal. How did all this complexity develop in the blink of an eye?

Especially confounding is the current similarity of the genes that regulate the initiation of eye formation among all five phyla that have visual systems. Were there in the fossil record any hint of a common ancestor of these five phyla that showed a nascent eye, the similarity would be explained as having arisen in that earlier animal. But there is no animal, let alone an animal with a primitive eye, prior to these eye-bearing fossils. Random reactions could never have reproduced this complex physiological gene twice over, let alone five times independently. Somehow it was preprogrammed. This inexplicable complexity arises over and over again.

The deepest part of the brain, the brain stem, is the first stop for impulses rising from the spinal cord. (See Figure 9.) This, in conjunction with the cerebellum, an organ tucked beneath the cerebral mass at the back of the brain, keeps our vital life support systems on track. Here pulses regulate breathing, heartbeat, and in part contraction of the smooth muscles that line the gastrointestinal tract and all blood vessels other than capillaries. With these two organs, we subconsciously tune our muscles. Something somewhere must regulate the force by which we close our jaws, or place our foot on the ground as we walk, or lift an egg. These organs perform those jobs, all with no conscious

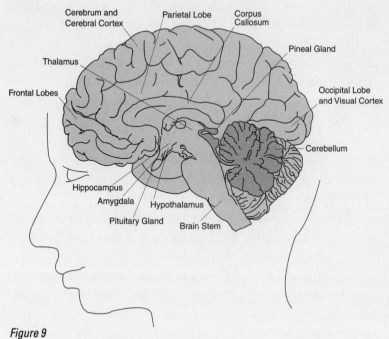

*Figure 9*

*The Human Brain and Its Principal Organs*

sensation of control and yet all precisely controlled. That is why when you take a step and suddenly discover that the ground or floor isn't where you cerebellum expected it to be, you get a jolt and lose your balance. Your foot hits bottom a lot harder than it normally would. The cerebellum had set the motion for what it "thought" was to be a standard step only to find out in the jolt that the situation was far from standard.

Your loss of balance resulted from an upset in the inner ear, another terrific device of ingenious design. Here we find an array of fine hairs set in three fluid-filled semicircular canals, each of which faces in a different direction, set at right angles (90 degrees) to the other two. This arrangement "feels" motion in any of the three directions. Tilting or rotating motions of the head cause the hairs in these canals to bend as the hairs' inertia delays their motion a minute fraction of a second relative to the motion of the head and canal itself. The hair's deflection induces a chemical reaction in a nerve at the base of each hair that initiates the electrochemical signal carrying the information to the limbic system. There the brain subconsciously integrates these data with information related to body position such as angles of leg joints, visual signals, muscle contractions, puts it all together, and learns that its body is falling.

Of course, the simple phrase "the brain integrates these data" hides a junglelike web of elaborately orchestrated neural activity with all the wonder of each nerve's mechanism for signal transmission, a billion and more axon/dendrite terminals in play.

For most of my life I've felt a transcendence within nature, some spiritual rumbling. My family had an apple orchard, and in my youth my father and I would tend it on weekends. The yearly arboreal cycle of winter hiatus giving way to the blossoming growth of spring and summer instilled in me that sense of wonder. But I had not realized the imminence of the marvel, its very presence within my body. Knowledge of molecular biology brought the story home.

Nature is so clever. To intensify the effect of motion, each individual hair in the ear canal is topped with a small node of calcium carbonate that adds weight and hence adds inertia. On a vastly different scale, it is inertia that pushes your head, like a huge node atop your neck, back onto the headrest of your seat as you zoom away from a traffic light.

Interestingly, this arrangement of hairs is insensitive to motions that are uniform, that is, motions without noticeable acceleration. When you wait at a traffic light and then suddenly notice that you are rolling forward, you press on the brakes, only to realize that it was the car next to you rolling backward that gave you the false impression of forward motion. In this case the visual sensation of motion overrode your inner ear's sense that all is stationary. There's more than one voice in each of our heads.

The brain stem is considered to be the oldest part of our brain, oldest in that it is found even in the simplest of animals in our phylum, the chordata. Yet it functions crucially in vision, a most advanced organ, as it activates the twelve muscles, six to each eye, that direct the eyes' movements in precisely synchronized fashion. Of course, recalling that the oldest of animal fossils house fully developed binocular vision, there is no surprise at finding ocular control in this ancient but not so primitive portion of the brain.

The entire brain, including the cerebellum, is composed of two hemispheres, right and left, with all organs but one being duplicated in structure, though not necessarily in exact function. As with nerves from the eyes, by the time we leave the brain stem, crossover has occurred. Stimuli arising from nerves on the left side of the body are recorded and processed in the right hemisphere; stimuli from the right travel to the left. Why this is so is a question yet to be answered, though it may be to match the inversion in vision forced by the shape of our eyes' lenses.

The one organ in the brain not present in a pair is the pea-sized pineal gland, located atop the thalamus. Being unique, it was once thought to be the seat of the soul. Most recent research implies it may be one of the seats of wakefulness. Its product is melatonin, the so-called sleep hormone. Bright light sends a signal to the pineal gland, causing it to decrease its production of melatonin, and in doing so, resets our body's clock to the wake-up mode.

The stimulus to speak, to state what is on your mind, is housed in the right hemisphere. Following a stroke on that side, the person may have much to say but getting it out from the brain into words will now take concentrated and seemingly conscious effort by the injured party. It also takes considerable patience on the part of the listener.

My mother, at ninety-one, could still construct and tell a good joke—slowly—though her left arm and leg were highly impaired by a right hemisphere stroke two years earlier. Formulating successful humor requires relating the logical to the illogical in a surprising way. That skill remained largely unimpaired. It's a lesson in relationships. A person confined to a wheelchair may still be a very full person inside her or his head.

Hearing, the mechanism by which sound waves are converted to electrochemical nerve impulses, is not one bit less complex than the transition of light into neural signals. Sound waves, which are composed of vibrating molecules of air, impact the tympanic membrane, the eardrum, which moves a chain of three tiny bones that move another membrane that sends the vibrations into a fluid-filled canal lined with tiny hairs. These hairs move in response to the changing pressures of the transduced sound wave. Their motion induces an electrochemical signal that races off toward the thalamus. (Notice the similarity between this and the sensing of balance, which takes place in an adjacent portion of the ear.)

Perhaps the most exceptional aspect of the hearing process is the rapidity with which the twenty thousand hairs of the ear can respond. Middle C vibrates at 256 cycles per second. The C above middle C is 512 cps. The C above that is 1,024 cps. The hairs must be able to resolve these fantastically high frequencies in order to differentiate among notes in each of these ranges. This is by far one of the fastest rates of response in our bodies.

In addition to resolving sound wave frequencies, the ear and brain must parse a stream of sounds into words and the words into bits of sentences, notwithstanding that each speaker has her or his own pace of speech, accent, pitch. For the most part we do it seamlessly, with not a conscious thought about the amazing interpreter we have within our heads.

Let's follow the path when, for example, a child's cry is heard at night. The long thalamus route sends the sound signal to the temporal cortex, located on the sides of the brain. The sound is deciphered step by step. Is it a noun or a verb? Is it part of a string of words, or does it stand alone as an exclamation? Then, what sound is it?—A voice. A child's voice. You have a long-term memory of that type of sound. It is familiar. One of your kids. The chain is now complete. The temporal cortex now knows that your child is calling for help and sends this information to the amygdala, which has already received a hint of the emergency via the direct route to the amygdala's subconscious, emotional memory. In response, it has already induced preliminary reactions, such as adrenalin influx to get the body energized for moving. From the amygdala, the call for action heads to another limbic location, the hypothalamus and the cerebellum, which goad your sleepy body into organized motion. You stumble off toward where you think your child's bed is located. Your memory, which is speculated to be a pattern of previously established axon-dendrite synapses somewhere within the cerebrum, having been initially laid down through actions of the hippocampus—also a part of the limbic system—provides this information.

Isn't anything simple in biology? The answer is no. Our every act is comprised of miraculous biochemistry.

And you thought you were just responding to your kid's call. You were, but not in quite the direct fashion your brain told you about it—or more accurately stated, not quite how you consciously perceived it. The brain doesn't bother your conscious mind with all the minute analyses involved in deciphering the signals, analyses that might make cryptology seem simple. All that is performed by the other you, the you you never ever meet. But it is there, housed inside the same head that lets "you" hear your daughter and see a rose and smell its fragrance. All held by what seems to be nothing other than a hundred thou-

sand million axons, each having thousands of terminals, connecting and interconnecting with a million billion
(1,000,000,000,000,000) dendrites. To get an inkling of what
that number, and hence your brain, is all about, I urge you to
count to a billion, a million times. After all, in some sense, the
part of your brain you have not met has done it, so why not have
the you you know do it also? At one number each second, with
no breaks for resting, that task will occupy you for the next
thirty million years. (Hmm...we'd better send out for coffee
and doughnuts and start counting.)

Of course, a nineteenth-century view of the world can justify
the archaic belief that it all evolved by random reactions among
atoms. That belief was conceived before molecular biology
opened the Pandora's box of hidden complexity.

The brain has space for two versions of you: the you you
never meet but that meets with you every moment of your life
as it regulates all the automatic functions of your body; and the
you you know so well, the one that feels as if it is just above the
bridge of your nose within your forehead. The you you know is
also a composite of two: the analog emotions whose source we
often cannot even identify, and the particulate sensory data of
sight, sound, touch, and smell.

Some Eastern religions refer to that spot on the forehead as
the third eye. It might be equally termed one of the three "I's":
the logical I, the emotional I, and the I I never seem to meet.

Is it possible that, in parallel with this bizarre subconsciously
lived multiplicity of our mind, there is also a multiplicity within
the world, a world unrealized at the conscious level, but still
very real in its impact upon the world our conscious physical
senses can access? This would be metaphysical, in the sense of
being outside the physical. Call it, perhaps, an underlying wisdom from which the physical world emerges. A physicist might
call it information. If the parallel is complete, then both the
physical and the metaphysical are together embedded in a

higher singular existence. A theologian might call that singular existence God. A physicist might call it the metaphysical potential field that collapsed and gave rise to our universe. Our brain might be the sole organ by which we are able to sense the metaphysical.

There's anecdotal evidence that the physical is actually embedded within the metaphysical. We all have an unexplainable nebulous desire to reach for some higher purpose, for meaning, in life even after we've satisfied the survival needs of food, clothing, and shelter. The indefinable nature of that sensation is somewhat akin to that which my student encountered when he could not quite pin down why the scent of a particular aftershave aroused in him an avalanche of emotions he thought were long gone.

# THE PICTURE

# IN OUR MIND

*A specific combination of selected atoms allows me to type a letter, say the letter* d, *on my computer keyboard and have the letter* d *appear on my computer screen. With some systems, I can also choose the color of the letter. If I'd majored in computer science or electrical engineering, I'd understand the circuitry that set the path from keyboard to illuminated letter on the screen.*

*A specific combination of selected atoms, mostly different from those of the computer, allows me to think of a letter, say the letter* d, *and the letter* d *appears as a mental image. If I so choose, I can envision the letter in color. But there is a fundamental difference between the* d *on the computer screen and the* d *pictured in my mind. I can have studied neurophysiology for years on end and I still would not have a clue as to how that image gets mentally displayed.*

*The modular construction of the brain has allowed identification of the exact cortical regions from which specific mental images arise. Unfortunately, knowing from where they originate tells us nothing of how they come together to form the pictures we see in our minds. It reminds me of the case with electricity and a host of other phenomena. I learned long ago that protons and electrons have equal and opposite electrical charges. Unfortunately neither I nor anyone else has a clue as to why the phenomenon of electrical charges exists. Yet these forces form the basis for the organization of all material existence. We take them as givens, as inherent characteristics of nature, something like the images in our minds.*

We've reached the last stop in our investigation of the brain: how we physically perceive consciousness.

For most of the miracle we refer to as life, the brain is an on-going story, one of continuous neurological growth and attrition, or "pruning" in the jargon of the profession. Within the first few months following fertilization, genetic coding establishes the basic structure of the brain. That fixes the locations of its organs. The genetic hegemony over brain structure is clear. All humans have their vision processed at the back of the brain, language on the side, logic in the front. But the absolute genetic control stops at the level of the modules' locations. Once those are established, a lose-win arrangement begins in earnest. A sort of mental Malthusianism, if such a word exists, takes over. Each of the modules produces a surfeit of neurons and axons, sending them out to the general target areas, but not hardwiring them into specific locations in those targets. Of course, just to reach the general area is a wonder yet to be understood considering the distances these axons must traverse relative to their micron-sized dimensions.

Once in place, stimuli generated spontaneously by the organs themselves and also received from the environs—visual, audi-tory, tactile—cause these axons to fire action potentials. As this neural activity forms patterns, some connections are reinforced, others are ignored. Those axons that consistently fire in unison, and therefore are likely to be receiving similar stimuli from a common source, wire together. This enhances their common synaptic connections. Those axons that are not being actively stimulated are "pruned." Use it or lose it is an adage that applies to the brain as well as to muscle tone. We may have close to six feet of body below our heads, but it is through the sum product of those connections in the brain that we know our bodies and ourselves.

Let's take a quick overview of the cerebral terrain, and then return to crunch through the details. It's the wisdom implanted within the biology of the brain that lays the basis for the wisdom of the mind.

The brain stem, a three-inch-long organ, joins the spinal cord with the rest of the brain. A slight shortening of the stem has been related to the occurrence of autism. The lowest part of the stem, the medulla, regulates such thoughtless processes as blood pressure and breathing. Just imagine what life would be like if every breath of air were a conscious effort. It's interesting that during speech we subconsciously override the brain stem's autonomous impulse to take a breath. This allows us to control air flow over the vocal cords. Perhaps it is because controlling our breathing touches such a deep part of the brain that conscious deep inhaling and exhaling is such a relaxing exercise. It is a very rare person who can consciously control the other domain of the medulla, blood pressure, but it has been known to happen.

Next up on the brain stem, the pons handles much of the change in body functions between sleep and wakefulness. Next in line comes the midbrain portion of the stem. Here the muscles of the eye are activated. Because this function is so very deeply seated, the eyes of a nearly brain-dead person can follow someone as she or he walks across the room.

Once above the brain stem, we reach the limbic system. We've already dealt in detail with its components and their functions. Briefly, the thalamus receives all incoming stimuli other than smell and routes them to the nearby amygdala for rapid emotional response and also to the cortex for slower logical consideration. As a crucial relay station, the thalamus connects with all major cortical regions: vision, sound, and motor (motion), helping to integrate the emotional with the rational. Adjacent to the thalamus, the hypothalamus, a cluster of pea-sized nuclei, connects with the frontal and temporal lobes as well as with the thalamus and the brain stem. Here hormones are secreted and, among other processes, feelings of hunger and satiation are induced.

The cerebellum, located at the lower rear of the brain, has its own cortex, though it is only three layered, whereas the cerebral cortex, the seat of our logic and conscious intellect, has six layers. The cerebellum helps regulate balance and the inherent sense each of us has of where the various parts of our bodies (limbs, head) are located relative to one another at each moment. Do you just take it for granted that you know where your feet are right now? Don't. It takes a brain to keep track of what is doing what and where it is doing it. If not, we'd be stumbling over our own feet. The brain stem gets most of the cerebellum's output.

Over the limbic system and cerebellum come the two hemispheres of the cerebrum. Outermost, covering the entire cerebrum, is the gray matter, the cerebral cortex, an eighth-inch-thick landscape of rolling hills (known as gyri) and valleys (sulci). Though the simple surface area of the brain is about a twentieth of a square meter, the folding brings its effective surface area to 1.5 square meters, a thirtyfold increase in brain power with no increase in head size. A brilliant solution that human logic might have overlooked. The cortex is the brainy part of the brain, the location that adds up the multitude of neural inputs, and, depending upon their summed magnitude, determines if they warrant further thought or action. Moving into the cerebrum, below the cortex, the gray matter gives way to white, colored by the myelin sheathing of neural axons that fill the entire region. Here, within the individual axons, nerve pulses are shuttled at breakneck speed in all directions, linking the various cortical lobes and joining the emotion of the limbic system with the logic of the cortex.

Deeply buried nuggets of gray matter, the basal nuclei, lie adjacent to the limbic system. Their axons extend to the nearby thalamus as well as out to the cerebral cortex. The basal nuclei help coordinate physical movement, monitoring neural initiation and termination of muscle contraction. When they are unable to fulfill their tasks, simple tasks of life become emotionally painful burdens. Malfunction induces the erratic, uncontrolled tremors known as Parkinson's disease.

From the deeply buried region of the basal nuclei and limbic thalamus, a veritable fountain of white-sheathed axons wells outward, radiating to and from all regions of the cortex. This is the internet of the brain. Much of the interior mass of the cerebrum, which itself makes up most of the brain's volume, is composed of these axons. Information transfer is a major job of the brain.

Neurons in the cerebral cortex make our decisions related to judgment and voluntary movement. They are our intelligence, all housed within those few surface millimeters of brain tissue. They decipher sounds into language, radiation into vision, store particulate, factual information—the cologne was by Armani— and via axons reaching up from the thalamus, they integrate this factual information with the amygdala's emotional memory of the smell. "The scent makes me happy but I can't say why" is your amygdala speaking to you via your cortex. Every thought, though not every emotion, emanates from the cerebral cortex. Each process has its own particular region within the powerhouse of cell bodies and dendrite synapses we refer to as the cerebral cortex.

The brain is a wonder machine of intelligence based on biochemical reactions, each expressing wisdom that in no way is presaged by the components from which it is built. At every level of life, from the isolated cell to the interaction of nerve and muscle, through to the $10^{15}$ neural connections within a brain, a depth of information surfaces that annoyingly has not an iota of justification being there. Nature, left to itself, favors disintegration, homogeneity. But the saga of life is a puzzling story of increasing complexity, of uniqueness, of order being locked in place, defying nature's degrading pull. And the brain is the top-of-the-line example of this successful struggle against oblivion.

The frontal lobes, encompassing the forward half of the cerebral cortex, account for the intellectual you. It's what tries to make you think as a human. Associative reasoning, the ability to form analogies (which is possibly the key ingredient of genius), conscious thought, speech, control of impulses—all find their origins here.

Then at just about the midpoint of the head, at the aft portion of the frontal lobes, the precentral gyrus controls voluntary movement. Integration of what we see with how we reach for what we see takes place here. So much of what we assume as natural, such as picking up a cup of water and moving it to our lips, emanates from this region of the brain. Every act we do, trivial though it may seem, must be controlled, and for the most part the cortices of the cerebrum and the cerebellum orchestrate that control. A deep cortex-lined valley, the central sulcus, marks the end of the frontal lobes and the beginning of the parietal lobes. Here touch sensations received from the body are recorded on one of the several mental body maps, each of which is inverted, heels over head. From here also arises the ability to recognize and manipulate symbols such as numbers for math and words for language. Along the lower sides of the cerebrum, at about the level of the temples and ears, the temporal lobes interpret the words heard in speech. The left temporal lobe deals with word recognition; the right with the ability to say the words—speech. It was the stroke's damage to my mother's right hemisphere that made it difficult for her to tell the jokes that she held within her left hemisphere.

Because of this division of labor between the hemispheres, there is a crucial need for transfer of information from one side to the other. The cross-linking of information between the hemispheres is accomplished by a massive bridge of axons, some eighty million, the corpus callosum plus additional axon bridges among the individual lobes within each hemisphere. Were it not for the intense cross linkage, the two selves within each of us would be battling for control at the conscious and subconscious levels. But transfer alone is not sufficient. To avoid potential right-side/left-side conflicts, the information must be shared in a timely fashion. This requires synchronization to within some sixty thousandths of a second. Considering the complex, multi-faceted signal processing intrinsic to vision and hearing, this precision in timing is nothing less than astounding. For vision, if this were not the case, the reconstruction of what one side of our eye "sees" would be out of phase with that of the other side. The brain would find an overlapping or doubling of the view. For speech, we'd be stumbling over our own words, hearing echoes generated within our heads. Once again we come face to face with the brilliance of nature.

Seeing, reading, hearing, thinking words each has its own dedicated section in our heads. When speaking, which involves motor control (mouth and tongue manipulation, vocal cord tension, breath control), neural regions in the precentral gyrus and parietal lobes fire. Seeing words in a written text engages the occipital (visual) cortex. Thinking and mentally forming words activates the frontal lobes. Clearly in the design of the human brain, language carried a lot of weight (see Figure 10).

Major cortical areas used for motor control (movement) and sensation (especially touch) by chimpanzees and bonobos, the animals genetically most similar to humans, in humans are devoted to language, especially that part of language related to formation and implementation of speech. Language is, in essence, the encoding of information. The intellectual processes of language in humans have replaced the physical robustness of movement found in other primates.

It's by no coincidence that the oldest extant interpretation of the Hebrew Bible makes the quality of communication the distinguishing characteristic between humans and all other animals. This relates to work in the second century, when the Bible was translated from Hebrew into Aramaic by the scholar Onke-

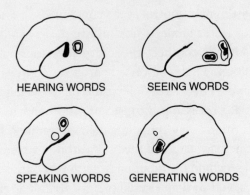

HEARING WORDS      SEEING WORDS

SPEAKING WORDS     GENERATING WORDS

*Figure 10*
*Responses of the Human Brain to Different Tasks Related to Words*
*The most active areas, according to brain scans (PET), are shown as dark areas*
*in this figure. As you read these words, this is what's happening in your brain.*
(Figure after Marcus Raichle, School of Medicine, Washington University)

los. In Genesis 2:7, we are told that the soul of humanity, the *ne-shama* in Hebrew, was instilled in humankind. In modern terminology, we'd refer to this event as the change from homonid to human. The verse concludes in its usual translation, "And the Adam became a living soul." The Hebrew however contains a subtle difference, reading, "And the Adam became to a living soul." Based on the superfluous "to," which in Hebrew signifies a transition, Onkelos wrote, "And the Adam became a speaking spirit." Not merely speech marked the transition from the pre-Adam, humanlike hominids discussed in the Talmud to the spiritually full human Adam. Anyone who has raised animals realizes that animals communicate. The change with humans was the special spirituality that entered the communication, a spirituality that could only be communicated by speech as we know it.

The essential role of the frontal lobes in character determination was demonstrated through a tragic accident to a nineteenth-century railway workers. Phineas Gage and his colleagues were laying tracks, using explosives to help clear the roadway. Explosives in the wrong hands can yield unexpected results. As they set a charge in place, premature ignition sent a steel rod smashing through the front of Gage's face. It entered just below his left cheekbone and continued upward through the top front of his skull. Miraculously Gage survived—physically. Mentally, he was a changed person. His prefrontal cortex, the area just behind his forehead, the part that in essence is the logical self, was mangled. And the self Phineas now confronted was no longer the self he and his colleagues had known before the tragedy. Formerly a responsible, even industrious employee, he became a compulsive drinker, most often incapable of completing tasks that required attention for more than a few moments. Yet his self-awareness remained, indicating either that the rod had missed some specific patch of the frontal cortex devoted to self or—more probable considering the extensive damage—that the projection of self is like that of a hologram. Destroying part of the lobe still leaves enough information to create the whole, though in a highly degraded form. Emotions became the major driving force in his behavior.

Today, with sophisticated brain imaging, we realize that Gage had lost a part of the frontal cortex having intensive linkage with the unconscious impulsive self of the limbic system. Since the latter remained unscathed (physically, though certainly not emotionally), the emotions of the amygdala could now at last run the self with considerably less moderating interference from the cortex's logic. In a word, Phineas Gage became somewhat of an impulsive child housed in the body of an adult male. Childlike because the axons that join the cortex with the limbic system complete their myelin sheathing, and hence their information-carrying ability, only during the late teenage years. Until then the impulses of the limbic have a major say in behavior. Gage had lost the target nerves in the cortex of those connective axons.

Children are usually treated more leniently by the law than are adults who commit similar crimes. There is a neurological basis for their social behavior. But what of Gage? Is he responsible for impulsively breaking the law? His will had been so severely compromised by the physical damage that, while it might be advisable to isolate him from open society, it would not seem proper to punish him. We have learned from Gage and others with brain lesions in the frontal area that much, perhaps all, of one's personality is composed of neurons and the multitude of axon/dendrite synapses they form. The private, individual self we each hold so dearly as our very being is sequestered in quite discrete portions of the physical brain. It is not some mystical metaphysical cloud that surrounds our head.

The arrangement of those neural patterns through which our conscious mind reads the self is built gradually from each life experience. These form our memories. Simply said, in the words of psychologist Hanna Shir, "You are your memories." From studies of brain-damaged persons, a map elucidating the physiology of memory is slowly emerging.

Memories and the emotions they bring to the cortex provide the information upon which we base the choices of our free will. According to the Bible, there is a third input to the choices we make. That is the soul of humanity, the *neshama*. The *neshama* looks at the choices presented by the logic of the brain and analyzes each in relation to whether the result of the choice will move us closer to or further away from the universal consciousness, the unity that pervades all existence. It's the *neshama* that urges each of us to seek meaning and purpose in life even after we've satisfied our natural drives for survival and pleasure. The *neshama* is the creation mentioned on day six of Genesis that changed an animal, amoral Adam into an Adam with moral responsibility (Genesis 1:26, 27). Phineas Gage lost none of his *neshama* in the accident, but he had lost his ability to formulate within his window of free will those choices based on logic. The *neshama* can act only on the choices with which it is presented. Phineas remained a human, though he had to be treated as a human child housed in the body of an adult.

CP routinely drove her young daughter to school. Monday, the 13th of October 1986, was no different. By coincidence the date coincided with the 10th of the Hebrew month of Tishri, the holiday of Yom Kippur. A car swerved, a collision occurred, a seat belt failed, and CP's head slammed into the windscreen, bounced off, and smashed onto the side window. Nine-year-old Leah was unhurt. Following the accident, CP remembered nothing of her life that preceded the crash. To her, Leah was just another child. She had to relearn speech. Interestingly, most of her motor skills remained unimpaired. How to walk, the use of cutlery, were all still there. These types of voluntary movements are coordinated unconsciously, like the skill to ride a bike even after not having been on one for years. They are mediated in the cerebellum. This "little brain," located behind and below the rear of the cerebrum, has its own cortex, and hence its own in-house capability for dealing with routine actions. Memory of love was a different matter.

The actual cause of CP's impairment was not clear from brain scans. What we do learn from experiences such as hers is that aspects of memory are distributed over the brain, and that while access to parts of long-term memory may be lost (though possibly still present in neural connections no longer able to be reached), new memories can still be deposited and then stored as long-term. CP was able to remake a full life, though as a person quite different from that which Leah and her former friends had known. Not everyone is so fortunate. A classic case is that of Barry Tiller (not his actual name).

Barry had suffered for over a decade from serious debilitating seizures caused by erratic neural activity in the cerebral area in and about the temporal lobes. At the age of twenty-seven, Tiller in desperation opted for experimental surgery. (The operation was performed during the early 1950s.) The medial portions of both temporal lobes and the forward (anterior) two thirds of the hippocampus were removed. His seizures ceased. But so did his ability to store all new memories for more than an hour. Unlike CP, who lost all record of her former life but could build new memories and a new personality, Barry Tiller retained access to the old but could add nothing new. He was totally unable to transfer short-term memory to long-term. Each day's experiences were brand new. Doctors who visited him daily were always new acquaintances. A novel was out of the question since it would take longer to read than Barry could mentally store the plot and characters. An hour after lunch there was no memory of what had been eaten.

Every new experience we have, whether it's choosing tonight's menu or looking at a work of art, is embedded in a seamless flow of consciousness that relates the past to the present. In our thoughts and fantasies, awareness of self and the passage of time are always present. For Barry Tiller, the passage of time ended in the 1950s. He could not fathom that forty years later he had become an old man. After all, in his mind he was still twenty-seven.

But Tiller retained the ability to store one type of newly acquired information. That was the ability to learn new motor skills, skills that did not require conscious or cognitive awareness in their execution, like riding a bike. He could even learn to play tunes by rote on the piano. Once newly learned, they were his thoughts to keep. Just as CP had not lost her subconscious voluntary motor skills, Barry Tiller could learn newly acquired skills of this type.

From these and other similar (though less dramatic) cases, two distinct regions for laying down long-term memory have been identified. For those memories requiring strong cognitive, cerebral cortex involvement, the hippocampus is essential. The memory does not reside in the hippocampus, but the hippocampus is the way station that actively takes the short-term memory and transfers it to long-term storage. With the hippocampus gone, as in the case of Tiller, the brain has no way of transferring short into long term. The exact location of the storage bin or bins is still a matter of conjecture, though it is thought to be somewhere in the cortex. The route for implanting cognitive memory must not be the route for its retrieval. Tiller could retrieve with full capacity events that occurred prior to his operation, even though he no longer had most of his hippocampus. For motor skills that are memorized and then performed later by rote, the cerebellum with its own cortex appears to be up to the task of storing and handling their implementation.

Between 90 and 95 percent of all persons are right-handed, right-eyed, and right-footed. The hemisphere skewing toward the left does not appear to be socially inflicted. By the fourth month following fertilization, most fetuses already suck their right thumbs, a left-hemisphere activity.

The right brain is not without its skills. It knows its way around. Lost in a city or a forest? Listen to the right lobes for spatial clues as to where you are and how to get out. Severely damage the right side of your hippocampus and you will get lost inside your own home. Facial recognition is also a subconscious facility of the right hemisphere. It is in fact quite similar to using spatial clues for navigation. But, as with navigation, though the ability is built-in, the implementation for a given region or face requires learning.

For several years beginning in 1981, I traveled frequently to the People's Republic of China as an adviser to their government. In those early years just after China was reopened to the West, my flight from Israel was quite circuitous, the requirement of the diplomatic reality of the time. (It was common knowledge that I had Israeli citizenship. China had not yet recognized Israel politically, though it admired Israel's technology.) I had no trouble envisioning the plane's roundabout flight plan, but disembarking was always for me a moment of extreme embarrassment. I found it a near impossibility to recognize the persons whom I had met time and time again. My mental plan for facial recognition was, and remains, so firmly fixed into the Caucasian mode that I never learned to distinguish clearly among the dozen or so smiling Asians who waited to whisk me through customs. They had no trouble spotting me on the tarmac. In those days I was the only Caucasian in the crowd. To my Chinese colleagues I was the different face in the crowd.

The right brain has more extensive long axon structures than the left brain. It integrates the whole, while the left brain likes details, calculations, abstraction of mathematics. The right brain seems to be the more emotional side, picking up rhythm as opposed to the digital notes and words. As with language, music is also processed in parts: rhythm, pitch, loudness. The reconstructed piece yields an emotional as well as an informative response. It's Beethoven (fact) and it's beautiful (emotion); it's Satchmo and it swings. The complexity of the processes lends credence to the notion that our brains are wired for music.

In the same temporal region in which the right brain hears music, the left brain is devoted to language comprehension. The location, known as Wernicke's area, is named after its discoverer, the nineteenth-century German neurologist and psychiatrist Carl Wernicke. In the same period, the French surgeon and anthropologist Paul Pierre Broca identified the location of deliberate speech generation. The area, now named after Broca, lies in a posterior portion of the left frontal lobe.

Logically, language comprehension is processed in a region adjacent to the left hemisphere auditory cortex. This shortens the physical connection between the cortical region where incoming sounds are recorded and the place where those sounds are deciphered into language.

The right brain seems to tap emotions related to pessimism, depression, and fear, while the left brain actually has a region—in the mid upper frontal lobe—that when electrically stimulated evokes a feeling of amusement to the point of laughter, unhumorous as the current situation might be to the unstimulated brain. The temporal lobe may be the site of spirituality. At the least it is the site of perceiving spirituality. Stimulation there can yield what has been described by patients as a divine experience. Neurologically speaking, the reaction is similar to temporal lobe epilepsy. Vincent van Gogh is thought to have experienced the affliction intermittently.

Is a joke truly funny when we laugh at it, or is it merely some aberration of our frontal lobe? My guess is the surprise juxtaposition of the logical with the illogical is actually that which our brain experiences as funny, notwithstanding the ability to get a laugh electrically out of a totally deadpan situation. We see sparks of light when we bump our heads into the door in a pitch-black room, and also when the visual cortex is electrically stimulated. But we also see light when photons strike our eyes' retinas. Like the joke we find funny, light really exists, notwithstanding our ability to induce the sensation of light by a bump to the head or a shock to the cortex.

The same is likely true for feeling the divine spark. That we are physically wired in our brains to experience spirituality, that a spiritual sensation can be induced by electrical stimulation to a part of the brain, makes an inspirational moment or event no less real spiritually. The biblical message of Eden is that humankind was created for pleasure. That we blew it is another matter. If pleasure is part of the message for life, being wired to appreciate a good joke fits right in. If apprehending the spiritual in the commonplace is also part of the message for life, then it is equally logical that we would be wired to apprehend the divine.

. . .

THE effects of testosterone on the fetal development of the brain become evident in adult behavior. Though both women and men have the same general brain structure, and also left-hemisphere dominance, processing by the parallel lobes differs significantly. Neural connections between the two hemispheres have been reported to be larger in women than in men, both at the corpus callosum and at the connections between the two lobes of the cerebellum. The junctures facilitate the integration of the emotion and rhythm of the right hemisphere with the logic and language of the left hemisphere.

In standardized tests of skills, women in general do better at matching items having common characteristics. Men excel at rotating three-dimensional objects mentally. The language skills of women exceed those of men. Women develop language at a younger age. Men generally prove better at abstract math-like reasoning, while women are better at the details, at arithmetic, and at highly precise manual tasks. Men are better at archery, at hitting a target, catching a ball. Men can learn a travel route faster. Women are better at remembering landmarks along the way.

From twenty years of experience as a teacher and supervisor of prekindergarten play groups in villages and kibbutzim, Yael LaHav has found that from the first day of attendance, the girls head for the dolls and playhouse, the boys for the jungle gym. Of course by the time they are three or four years old, society may have imprinted this behavioral pattern. It stays for life. Just picture a hunting trip. Do you envision men or women? And whom do you mentally see at home, tending the kids, keeping house, men or women? The answers are obvious. In the few hunter-gatherer societies still remaining, the male-female split of tasks is the same. Molecular biology and brain imaging show that much of the mental difference between the sexes is nature, the imprint of hormonal activity on the prenatal brain, and not nurture, the pressures and morays of society. If IQ is a measure of overall abilities, then women and men are similar.

Considering the spread of individual talents between the sexes, if each of us is willing to accept that being different does not mean being less valid, less valuable, less intelligent, then it seems a woman-plus-man team makes a very good combination for survival in a world that moment by moment poses a multiplicity of problems each begging its specific solution. That seems to be what nature had in mind with sex and also what the Bible had planned from the start. "And the Eternal God said, 'It is not good for the Adam to be alone; I will make a helper opposite him. . . . Therefore a man shall leave his father and his mother and cleave to his wife and they shall be as one flesh" (Genesis 2:18, 24).

# THINKING ABOUT THINKING

## TAPPING INTO THE CONSCIOUS MIND OF THE UNIVERSE

*Within the brain we perceive the consciousness of the mind, and via the mind we can touch a consciousness that pervades the universe. At those treasured moments our individual self dissolves into an eternal unity within which our universe is embedded. That is the message both of physics and of metaphysics.*

*Is a single molecule of water, one unit of $H_2O$, wet? Not sure? Well, what about two molecules? Are they as wet as water? What's the point of transition at which a batch of $H_2O$ molecules become wet? What about a single neural axon/dendrite synapse? Does it contain a mind?*

*As for the wetness of $H_2O$, from its chemical properties, the hydrogen bond, and the 104-degree angle at which the two hydrogen atoms orient as each shares an electron with the single oxygen atom, we could predict that if there were enough $H_2O$ molecules present, we would find the properties of what we refer to as wetness. Wetness arises from the interactions, the weak bonding, among billions of billions of $H_2O$ molecules. Each drop of water falling from a faucet contains $10^{21}$— that's a thousand billion billion—molecules of $H_2O$.*

*And what about the axon/dendrite synapse? Is it a bit of a mind?*

*Research has proven that a synapse can be the site at which a specific piece of information is stored. If we knew all the properties of this single synapse, we might be able to predict the vast analytical potential of joining similar synapses by the billions into an interactive, associative network. We could foretell that the primary capabilities of such a union would be the storing and processing of information. We might even call that union a brain. But unlike the prediction of the wetness of water, we'd have not an inkling that this brain, at some level of complexity, say at a concentration of a hundred billion neurons, each exuding a thousand or more axon terminals, would go critical, like a nuclear reaction, and give rise to the mind we see within our brain.*

*The mind is our link to the unity that pervades all existence. Though we need our brain to access our mind, neither a single synapse nor the entire brain contains a hint of the mind. And yet the consciousness of the mind is what makes us aware that we are humans; that I am I and you are you. The most constant aspect of our lives is that we are aware of being ourselves. Even in the illogical jumble of a dream, filled as it may be with fantasy, the constant is that we are ourselves.*

The brain is amazing. The mind even more so.

A few years ago, IBM's supercomputer Deep Blue was taught the ways and bi-ways of chess. The AI folks at IBM had built a very clever machine. In the spring of 1997, Deep Blue challenged chess master Garry Kasparov to a match. It was a battle for both of "them." For a while it looked as if Kasparov might win. But Deep Blue carried the day. In time, and likely in short time, an invincible program will beat all human challengers, always, and probably with relative ease.

That victory neatly laid to rest a claim made by Berkeley philosopher and professor Hubert Dreyfus. In his book *What Computers Still Can't Do*, Dreyfus predicted that AI—computer-generated data manipulations—would never reach a level of skill able to outwit and outplay a chess champion. It took less than five years for Silicon Valley to prove Dreyfus wrong. Kasparov's defeat carried a strong message to Dreyfus and others of his ilk. To paraphrase the late Harry S. Truman, if you don't know how to cook, stay out of the kitchen. If you're not skilled in computer science, don't make predictions about what AI can and cannot do. If you want to philosophize about the capabilities of AI and the wonders of the human brain, then first learn the technology and the biology. It is for exactly that reason, learning the terrain, that I took you through the workings of the brain. I want us to address the brain/mind interface from a stance of knowledge, not from some deep-seated commitment to an emotionally charged heritage, be it materialist or metaphysical.

For all the well-deserved hoopla over Deep Blue's impressive victory, a major accomplishment for AI, for all its phenomenal data-crunching, algorithmic power, I doubt that Deep Blue was aware of even a pinch of the emotion that Mr. Kasparov experienced as he sweated through the tourney. And much of the brain's function is exactly this type of mathematical robotics. We can make a machine that laughs just as we can find within the brain the cortex module that when stimulated forces a guffaw. The mechanical operations of the brain can be matched and outmatched by AI. What makes the accomplishment of Deep Blue brainlike rather than mindlike is not the skill of a chess move. It's the lack of the experiences that Gary felt as a human battling a machine. The brain takes facts and integrates those facts with emotion. The mind takes the product of that integration and experiences it. A computer will probably never reproduce what Heidi felt as she watched the sun say goodnight to the mountain snow. That's the work of a mind.

SCIENTISTS may someday prove that the mind is totally a flesh-and-blood, physical phenomenon, or, quite possibly, they will find it emergent, not defined by the physical. From what we have seen of brain function, whether the mind is purely physical or partly metaphysical—call it divine if you wish—the brain's very existence is quite simply mind-boggling and quite possibly miraculous. There is brilliant design in the brain, and to make it requires the nature of our universe, which means we need a metaphysical force, a potential not composed of time, space, or matter that created the time, space, and matter of our universe. It's worth reemphasizing: the inequality between cosmology and theology is not whether there was a metaphysical creation. That is a given. The debate is whether the metaphysical whatever-it-is, or was, that produced our universe manifests interest in the physical reality it created.

The entire processes of the big bang and the evolution of our universe are more fantastic than any science fiction buff ever imagined. It sounds almost biblical. Not just the elusive physics of the process, but the fact that the effects of this single event are felt till today, fifteen billion years later. The entire universe, we included, formed from, and runs on, the energy brought forth in that ancient flash of creation. That galaxies and people and orange juice and oatmeal could come from a ball of energy should tell each of us something about the metaphysical force that brought it and us into being.

When our kids were younger, they would ask Barbara and me about our lives in the United States, before we moved to Israel. They wanted so much to rummage through the steamer chests of my parents and grandparents, piled in the attic of what is now a two-hundred-year-old farmhouse on rural Long Island. We, as adults, are not so very different. As we now contemplate the universe's birth and growth, we are rummaging through what are the remnants of our cosmic childhood. And at each stage we meet head-on a reality that is so brilliantly designed and inexplicably complex that we simply have to take it as a given.

Why should matter attract matter? Why doesn't matter repel matter? Why should protons and electrons have their equal and opposite charges, even though their masses differ by a factor of over one thousand eight hundred, and a neutron have no charge even though it has a mass quite similar to that of a proton? Why the Pauli exclusion principle and the quantized characteristics of orbital electrons? All these realities are humanly illogical and totally arbitrary, but without them there'd be no molecules, no rocks and water, no brain, no mind.

What does it take to make a brain? It takes a big bang producing a universe guided by laws of nature somehow tuned to lead energy into rocks and water on a user-friendly planet that can take those rocks and water and change them into a marvelously complex, data-crunching, algorithmic, sound-, sight-, touch-, and smell-sensitive wonder capable of processing thousands of inputs in parallel with a cycling time of thirty thousandths of a second. When contemplating the amazing complexity of the human life form, don't just think of the entire body encased in a smooth flexible layer of skin. The skin hides the wonder within the body, just as visible nature masks the metaphysical within which it is embedded.

Think of the inner workings. Think triple-layered cell membranes with voltage and protein-regulated channels for getting nutrients in and products out. Think RNA polymerase receiving a signal from some remote region and then searching for and opening just the correct spot on the three-billion-nucleotide-long helix of DNA, pulling complementary nucleotides from solution to produce mRNA that, when transported out of the nucleus, will find a ribosome that will decode it, pulling molecules of tRNA bearing just the right amino acid out of the twenty variations of amino acids in the cytoplasm, all at fifty shots a second. Each cell producing two thousand proteins every second. Think motor molecules carrying their protein cargoes step over step along microtubules. Golgi apparatus, neurotransmitters, and more and more. Multiply that complexity by a billion written out a hundred times for the brain alone, and then for each of those hundred billion nerves, sketch out a thousand axon/dendrite synapses.

The wind blows and thousands of leaves shimmer in the sun. Your eye sees them all. A million, more probably a billion, ion channels opening and closing along the ganglia of a million optic nerves leading from retina to thalamus and on to the visual cortex, cycling thirty times a second, as bioelectric signals, the information that records the motion of each of those leaves, reach into your brain. A myriad of chemical reactions, all in parallel, simultaneously recording the data.

Trees, eyes, the brain, from inert rocks and water via dumb unthinking random reactions? Logic alone tells you it could not have happened by chance. But the materialist superstition of our culture, the idea that if you can't measure it, it's not there, insists that chance be the explanation. And once a fact is imprinted on a mind, like the song a sparrow learns in its youth, that fact is yours for life. Believe it or not!

And yet here we are. A small part of a vast universe thinking about its origins, rummaging through steamer trunks in the attics of space and brain, trying to find the meaning of that which we call the mind.

FOLLOWING his defeat by Deep Blue, Garry Kasparov is quoted as having felt he was being challenged by a "new kind of intelligence." Indeed he was. The intelligence that confronted Gary was all calculation and no emotion. It was the cerebral cortex and more, but with no amygdala getting in the way. For Deep Blue, the match was truly no sweat. It's the perfect approach for chess, but for love it's a loser.

It's no secret that we can isolate within the brain those areas responsible for given activities. We can even measure the gradual increase in size of specific cerebral lobes we associate with cleverness in living vertebrates as we move from representatives of the oldest of vertebrates in the fossil record, aquatic life (some 530 million years ago), to reptiles (first fossil appearance 320 million years ago), then opossum-sized mammals (250 million years), primates (60 million years ago), and finally Homo sapiens (sixty thousand years before the present).

But the mind is very much greater than a layering of holistic feelings of self and awareness onto the observable facts recorded by the brain. True, the conscious mind arises from the brain. Destroy the cortex and you destroy consciousness. Destroy the brain and the palpable mind goes with it into oblivion. But the physical organs of the brain may be only the circuitry that makes the mind humanly perceptible. In that case, a form of consciousness may remain. Smash a radio and there's no more music to be heard. But the radio waves are still out there. We just don't have the apparatus to change the electromagnetic radiation into mechanical sound waves. The brain does for the mind what the radio does for music.

Brain-mind questions have all the trappings of disputes in which illogical solutions are required, yet resisted, to explain data that contradict established, out-of-date theories. The persistence of theories for a randomly driven evolution of life in the face of data from molecular biology and the fossil record, both replete with evidence against it, is one such example. Another, now largely settled, was the yearning in 1960s and 1970s for an eternal universe, one that would obviate the need for a metaphysical beginning.

Once a paradigm is established, only two avenues exist to depose it. Either a more logical paradigm is formed that fits the data satisfactorily, or overwhelming evidence demonstrates that the existing model is wrong even though no new model is known. The erroneous concept of an eternal universe fell under the weight of data that consistently indicated a big bang start of time, space, and matter. Just what the nature of the metaphysical force that produced the creation might be is still being debated. The tree of Darwinian evolution has bent and bifurcated to become a bush of evolution. That too fails to satisfy the known data, but there is no other *physical* model even remotely possible as an explanation and so randomness persists as the catechism of the school faithful to neo-Darwinian belief.

Modeling the brain/mind interface, forming a paradigm for how the brain gives rise to the mind, suffers a similar challenge. There is no hint of how we physically view and hear and smell the messages of the brain. Yet a metaphysical solution is untenable to a materialist school steeped to believe only that which can be seen or measured, the summing of the individually observable parts. Unfortunately, at the brain/mind interface, this reductionist approach misses the crucially holistic nature of the mind.

Quantum mechanics required a paradigm shift from classical mechanics, a shift even more extreme than accepting a universe with a beginning. Quantum mechanics necessitated replacing logical observable processes with the "illogical" phenomena of the subatomic world. The very existence of our universe calls out for a metaphysical explanation, an explanation that by definition is illogical in physical terms. The undisputed yet enigmatic existence of our self-awareness, our consciousness, does the same. The mind may be our only link to the reality of the metaphysical.

J. A. Wheeler likens the universe, all existence, to the expression of an idea, to the manifestation of information. This has the ring of quantum mechanics. The widely discussed quantum wave function that is attributed to each and every entity in the universe contains the information that totally describes that entity. The implication of Professor Wheeler's statement is not only that the wave function, which in street language refers to information, is a fundamental property of existence. The implication is much more profound: that information is the actual basis from which all energy is formed and all matter constructed. It sounds, at first cut, bizarre. But then, before Einstein's $E = mc^2$ who would have guessed that the basis of all matter, solid, liquid, gas, in every corner of the universe, is something as ethereal as energy, the totally intangible, completely massless, wave/particle duality we label as a photon? The massless, zero-weight photon gives rise to the massive weight of the universe.

That bit of strange, illogical physics is a proven fact. An equally illogical aspect of our amazing universe may be on the horizon. As specious as it may seem at this moment, if Wheeler and others are correct, information may be the fundamental substrate of our universe, a substrate made visible when expressed as the energy and material and space of the universe. In a strong sense, our universe may be the manifestation of information.

That swings back to a thought presented much earlier. Every particle is an expression of information, of wisdom. The self-awareness we experience is the emergent offspring of that wisdom. The more complex the entity, the more complex the information stored within. We tap into it via our brain. Because information is present in all existence, the consciousness I feel as my self-awareness has a cosmic history. It does not arise from my brain de novo. Aspects of it have been present from the start, the very start, the big bang. Consciousness, as wisdom, is as fundamental as existence itself.

The basic question is, just as ethereal energy can give rise to that which appears to be solid matter, can information provide the basis for that which precedes the energy? Such a suggestion is not so very outlandish. Rest assured, no one has a clue as to what makes a photon, the basic "particle" of energy.

The idea that a universal consciousness, emergent from this wisdom, might be present in what we habitually refer to as inert matter finds support in a range of scientific data. Entangled particles act in concert even though separated by distances that obviate the possibility of their communicating in the reaction time required. It is as if each particle is simultaneously aware of the other's action at the instant of the action. A classic example of this linkage is the famous double slit experiment in which waves or particles passing through an open slit, call it A, know whether slit B is open or closed. The particles behave as if they are conscious of the condition at B even though B in no way affects conditions at slit A.

In the Drakensberg in South Africa, my family and I watched from a distance as one baboon preened another, and then after a quarter hour or so at this activity rolled over and had the other baboon preen her (or him?). That certainly smacks of some sense of self-awareness. And it makes all the sense in the world, if consciousness at varying levels is ubiquitous.

A grain of sand contains the slightest hint of the skyscraper of which it is to become a part. Do the very elements of the brain, the carbon, hydrogen, oxygen, nitrogen, have within them the barest trace of consciousness, which will combine and emerge as the complexity of a fully functioning brain? The same elements that build our brains and bodies form stars and galaxies. In that case our minds are but a part of a vast and conscious universe.

The elusiveness of the mind finds a parallel in that of other fundamental givens in nature. The properties of gravitation, electric charge, and magnetism can be measured only via their interactions with or relative to other matter, never in isolation. Unlike a cup which when held remains a cup, we can't grab a photon and observe it as a photon—hold it in our hands and examine it from all sides. Every catch of a photon changes the photon into something else—heat, an energized particle. Whatever emerges from the interactions, the photon is gone, and like the smile of the Cheshire cat, only its interaction remains. Likewise, though we can examine and identify the various lobes of our brain, we can only sense our consciousness. We can't isolate it.

In classical physics, all future events are deducible from the initial conditions of the situation. Quantum mechanics revealed a contradiction to this age-old assumption, a contradiction highly illogical and equally highly relevant to thought. We can accurately predict the general flow of a reaction, but not the exact path. Because the flow of all future paths is what is called probabilistic, known in general but not exactly, the exact future is not contained within the present. This is one of the reasons for assuming time's arrow moves only forward and never reverses. The probabilistic nature of nature, the spread in possible outcomes, means we cannot reconstruct the exact past from the present. And so we cannot go back to the past. This is not chaos theory, a theory based on our inability to measure exactly the present conditions. The quantum reality is that there is no exact present to measure.

In our brains, this probabilistic spread relates to the range of thoughts from which we select our next action. In a sense, the divergence opens a window of options from which we may choose, either via the logic of our frontal lobes or the raw emotion of our amygdala.

Plants and humans are sensitive to sunlight. But plants always move toward the light. Their DNA programs produce this reaction. Humans usually prefer being warm and therefore enjoy the sunshine—usually. That is, though humans often prefer being warm, some choose to move toward and some away from the sun. Identical sensory input, sunlight, yields very different reactions in different persons. Current environment plus memory produce the decision. That's our frontal lobes at work, integrating the fact of heat, plus emotion, the feeling of being too hot or too cold. The decision is selected from among the possibilities stored within the synapses of the brain. The choice is ours.

Is being aware of the feeling that it's too hot or too cold different from the feeling itself? The puzzle is not that we react to the sensory inputs of a situation. The puzzle is how we conceptualize in mental terms what seems to be actual physical reality. When I think ice cream is cold, or chocolate chip cookies smell wonderful, I neither feel bodily cold nor smell the cookies, but I do recall the cold and imagine the smell.

Conscious self-awareness is what makes me know that I am a specific human. I know it is I and not you who am in my mind. At about two years old, children begin to realize self and that others have minds too, and that those minds can be manipulated. (Raising kids is so instructive as to how the mind works.) There are aspects of our selves that we can describe, such as professional skills, joys, fears. Looking at the data from a brain scan, a skilled technician may be able to locate the parts of the brain active during specific mental and motor tasks. A neurologist, knowing the brain terrain and seeing the data output, may be able to tell what the activity was that induced each specific brain lobe to fire. For one who knows the visual cortex very well, the data can even reveal if the scene being observed is in color or black and white. But the experience of the vision is something else. That escapes the machine's reading.

We talk about missing links in evolution. We have a missing link right in our heads at the brain/mind connection. The move from brain to mind is not one of quantity—a few more neurons and we'll tie the sensation to the awareness of it. It's a qualitative transition, a change in type. The mind is neither data crunching nor emotional response. Those are brain functions. Mind functions are self-experience, seeing, hearing, smelling. The replay of what came in. These are phenomena totally different from the acquisition of the information. That is why adding up the synaptic data would predict a brain, but not a mind.

There is an aspect of science known as Fourier analysis. When presented with a very complex mix of data, such as several individual waves superimposed one upon the other, using Fourier analysis we can strip away those waves that represent noise and be left with the pure information that we seek. Can we strip away the sensory and emotional data of the brain, and be left with the experience of those data, the mind?

Sir John Maddox, former editor-in-chief of the renowned journal *Nature*, summed up our knowledge of consciousness in a piece featured in the December 1999 issue of *Scientific American:* "Nobody understands how decisions are made or how imagination is set free. What consciousness consists of, or how it should be defined, is equally puzzling. Despite the marvelous successes of neuroscience in the past century, we seem as far from understanding cognitive processes as we were a century ago." It's not likely that we'll find a mind by looking in a brain, though we require a brain to contemplate the mind. Fourier analysis does not help in the realm of sensation.

Defining an aspect of our world in an arbitrary manner rather than in absolute terms is not new to science. Essentially all of scientific inquiry is based on fundamental "givens," values that just are what they are for no discernible reason, such as the electromagnetic charge. These facets of nature are observed as being intrinsic to the working of nature, but arbitrarily set by the laws of our universe. Gravity is not matter and protons are not charge. Yet gravity emerges from matter and charge from protons. The universe might have been created without gravity, without electric charge. It would be a very different universe, but a universe nonetheless. Some givens are totally illogical in human terms of logic.

The quantized nature of radiation is but one example in which a new and fundamental component, the quantum, had to be introduced (by Max Planck in 1900) to explain an observed aspect of reality that has no a priori explanation. That the flow of time should pass at different rates because of relative velocity or differences in local gravities is patently ridiculous. Time is time. The duration of my minute should take exactly as long as the duration of your minute no matter where you are. Makes all the sense in the world. But as Einstein predicted in his revolutionary theory of relativity, it just turns out not to be true. The rate of flow of time varies from place to place in the universe. Sometimes we just have to fold our reductionist tents and realize the whole may not be totally contained in the sum of the parts. Consciousness has all the trappings of another nonreducible element of our universe. The conscious mind is not mystical, but it may be metaphysical—meaning out of the physical.

The discovery of nonlocality, of action at a distance, illogical though this phenomenon is, has revealed the linkage of disparate parts of the universe. The infinitely extended wave characteristics of all matter give physical basis to the metaphysical claim that the entire universe is entangled. Taken together, these seemingly illogical but validated insights point to something quite logical and marvellously wonderful. The universe is truly a uni-verse. All existence is joined through the expression of information, an idea, wisdom. Our mind is the emergent link that occasionally taps into that unity. You know when it happens as the surge of exhilarating emotion envelopes your entire body. At those moments, as one's local individuality dissolves into the unity that embraces all existence, we realize the full meaning of "the Lord is one."

# ILLUSIONS

## GAMES THE BRAIN PLAYS

## WITH THE MIND

*My computer seems to have a mind of its own. When I press the return key, it insists upon indenting the next line of text even though this time I want lefthand justification. The brain outdoes the computer by a mile in getting its own way. Like a computer, the brain is a victim of habit. View two or three dashes and the brain images a line. The brain always wants to fill in the dots, even when dots are all there are. The difference between a computer and a brain is that you can eventually get your own way with a computer if you persevere. With the brain, there are situations when there is no solution short of surgery!*

*In essence the brain says to reality: Don't bother me with the facts. I have my preconceived notions. Ideas triumph over the physical world—even when they are mistaken.*

We say seeing is believing, but what is red to your eyes may be maroon to mine. The difference in opinion is not merely one of definition. The difference is in how the brain images reality. The brain may actually "see" the colors differently, integrating its nature (DNA instructions for how the brain's neurons were distributed, largely before birth) and nurture (personal history; how those neural connections were altered and reinforced by

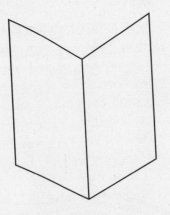

Figure 13

Then there is the question of is it a cup or two faces? A duck a rabbit? Which of the lines is shorter, which line longer? (See Figure 14.)

14 (continued at top of facing page)

experience). Together, nature and nurture conspire to provide consciousness with a finished product, the inside view of the outside world.

The breakthrough of the brain is that nurture actually can alter what nature gave us. What might take genetic mutations a multitude of generations over millennia to accomplish, nurture can bring about in seconds. The images and sounds and smells and touches that you let into your brain today actually determine how you think about and see the world tomorrow. What you think (or see or sense) determines how you think. In other words, don't expect to philosophize in the morning, if you orgy in the evening. As Louis Pasteur observed, chance favors the prepared mind.

But some mental prejudices seem oblivious to the nurture aspect of life. They are hardwired from the start. We are used to seeing illumination from above. That's the sun's usual position relative to us. For reading a book it would be hard to light the opaque pages from below. But what about face recognition? Lighting a face diagonally from below might do just as well as top lighting. But the brain's cortical regions for face recognition are strongly prejudiced toward shadows produced by light arriving from above, as with the sun. Lighting of even the friendliest of faces from below can produce a sinister, even grotesque expression, and sometimes a shape that is not even recognizable as a face.

The phenomenon known as lateral inhibition is equally hardwired in the brain. Look at the grid below of black squares on a white background. Notice the gray clouds that keep popping in and out at the intersections of the white lines? There is absolutely no gray present. None. But try as you may, you cannot stop your brain from "seeing" those gray clouds. To your brain, the pure white paper at those locations is gray. You recognize this as an "artificial" effect in the grid, but realize that at each letter you are reading on this page exactly the same phenomenon is taking place, making the letters appear more sharply in your mind's eye than they are printed on the page (see Figure 11).

Figure 11

The effect arises from the eye/brain system wanting to present a mental picture in which contrasting borders are artificially accentuated. The brain is trying to help the mind see the boundary. As the visual signal travels from retina to thalamus and ultimately is dissected and recorded at the visual cortex, there is a progressive increase in border contrast at each stage.

The nervous system accomplishes this contour enhancement via lateral connections among adjacent neurons. As the action potential races along neuron A, which has "seen" light, it inhibits adjacent neuron B from firing. B not firing tells the brain that at the retinal start of B there was no light—hence the gray. Even if there happened to be some light at B, the brain will never find out about it. If an area is equally lighted, then all the neurons are equally excited and equally inhibited. The result for that situation is a mental picture of flat light. Only when there are inequalities does the inhibition become visible.

The subtlety of the process is amazing. Notice that the gray clouds appear only at the intersections of the white lines, and not along the white lines between the intersections. That's because the neural cells receiving stimuli from the intersection are

surrounded by lighted cells on four sides, [...] inhibition. Neural cells getting informatio[...] lines alone are inhibited from only two si[...] than surgery, which I am not suggesting, wil[...] It's built in from birth. Lobsters have the sam[...]

But it's not really a problem. In most cas[...] helps us avoid bumping into things. The re[...] tern below teases the brain in the same way[...] a sense of virtual motion between adjacent [...]

Figure 12

Ernst Mach in the late 1800s prod[...] ping effect, in his famous Mach fig[...] 164). One moment the vertex poin[...] out. It could almost fly away, but it's[...] moving. Just a two-dimensional d[...] implied perspective and total lack [...] brain think it's three-dimensional. [...]

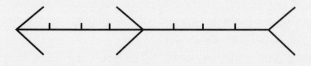

And don't tell me there is no triangle in the Kanizsa drawing below (Figure 15). It may be a phantom, but my brain tells me it's there. But what sees the triangle, the visual cortex or the visual mind? The former is physical, the latter virtual. The visual cortex records the data. The mind does the seeing.

So much for the naive thought that seeing is believing. The brain uses its past experiences, integrating them with current points of reference, to form its opinion. When those points are missing, the brain may fill them in. Data from brain scans indicate that it is the frontal lobes, the logic areas of the brain, that take the data from the visual cortex and decide what to forward to the conscious mind. In some cases the brain cannot make up

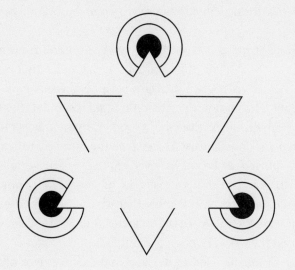

Figure 15

its "mind." Physicist Roger Penrose and artist M. C. Escher exploit these elusive neural effects to produce their famous drawings of figure/ground reversal, or a staircase that ascends and descends simultaneously.

Intellectually we live with figure-ground reversal continuously. The world looks so physical and yet I know with absolute certainty that its creation was the result of a metaphysical force. But what is physical? What is metaphysical? Where's the divide? Just what are protons made of? Don't tell me three quarks, because then I just have to ask what quarks are made of. It all reduces to fields of force, which means that for all the tightly packed hardness of a piece of rock or iron or any matter, there really is nothing there at all. Finally I can give a meaning to one of the more famous statements attributed to Yogi Berra: Nobody goes there anymore. It's just too crowded.

There are logical relationships for which the brain hunts, and when they are not there the brain does the best it can with the facts it has. Be they visual or verbal, the brain strives to have them fit within the context of what it perceives as logical. And logic is based on deductions drawn from the brain's personal history.

In front of me is a tree. As I change my gaze, rotating my head slowly to the left, the information that my retina receives of that tree moves toward the right peripheral visual field. That is the only information that the retinas get to send to the brain. So what should the message be? Good grief; the tree is moving. But the brain does not try to convince the mind that the tree is actually moving to the right, even though its image falls ever further to the right in my eyes' view of things. Why not? The cerebellum lets the cortex know that it's the head, and not the tree, that's on the move. In this case the brain corrects for the illusion that the world is moving about.

But, far from any view of the land, stand on the deck of a boat rocking so gently that the ocean swells are barely perceptible, and on a clear moonless night you'll watch the stars swaying back and forth. It's quite a shock at first, since as every landlubber knows, the stars are the "fixed stars." When you can't feel the motion, it's hard to tell what's actually doing the moving. I feel from every sense I have that it is the sun that circles the earth. Yet the earth on which we are standing is rotating at close to a thousand miles per hour, and we can't feel a bit of it. Inertial motion, motion without changes in velocity, eludes those specially weighted hairs near the inner ear that we studied.

Hold your hands steadily out to the side, fingers spread apart, pointing upward. Now try to count them while looking straight ahead. Tough going. You can't count them. Only a very narrow central retinal field, the fovea, has enough receptors to differentiate among items even as large as fingers. Objects off to the side become hazelike. But peripheral motion, that's a different story. Move your fingers one at a time and you can count them all, which means the information is there, but blurred together when not in motion.

Try reading a book while lying on your side, or watching a video while stretched out on the couch. Not easy unless you align your head with that which you are reading or watching. When a slide at a lecture is projected incorrectly, on its side, the entire audience as if by command tilts their heads to compensate.

Yet we can learn to read upside down. Jewish communities of European extraction read from the Torah with the scroll laid flat on a large table. The reader stands at the base. Kids in the congregation, mine included here in Israel, crowd at the opposite side of the table and read along, but to them the text is wrong side up. They read with equal facility from both views.

Perspective lets the brain construct the impression of 3-D images from the 2-D scenes viewed on a TV screen or gallery painting (see Figure 16). Nothing like a few points of reference to make depth plunge into a flat piece of paper. The brain does it all because it knows that as things get smaller, one behind the other, each is progressively further away. It's still a flat piece of paper no matter what the brain tells the mind.

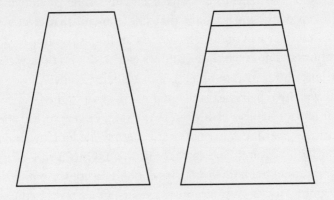

*Figure 16*

Perspective is what makes the sun and moon appear larger at the horizon than when overhead. The brain needs a clue to tell it whether these celestial bodies are nearby or far off. Only as they approach the horizon are such markers available. Being behind the terrain of the horizon means that they are beyond the horizon, and therefore they must be far away and therefore large. Swimming off the Israeli coast near Dor, at full moon I could see, with the turn of my head, both the moon just above the eastern horizon (moonrise) and the sun just before setting in the west. Both being near a horizon, they both looked large and of equal size. The sun's actual diameter is four hundred times greater than the moon's, but the sun is 93 million miles distant from the earth while the moon's distance is some 238,000 miles. This four-hundred-fold difference in distance accounts for the similar perceived size. The horizon presents no clue of these measures and therefore no telltale sign as to whether the sun or the moon is further or larger.

The eye has one obvious flaw. The nerves of the retina must

pass from the front back through the retina, in order to proceed to the brain. This punches a hole in the retina, a place where there are no light-receiving cones or rods, and in doing so produces a blind spot in the signal sent to the brain. But no such lacuna appears in your mental view of the world. That's because the brain fills in the space by assuming that what it sees around the spot continues into the blank. It's quite amazing to find this blank.

# O                                              1

Hold the book about 9 inches from your face. Shut your right eye. While facing straight ahead, look diagonally at the "one" with your left eye and slowly move the book toward your face. The "O" will disappear. It does so when the image of the "O" falls on the blind spot of your left retina. As the book continues in, the "O" reappears as the image moves to an active site on the retina. You'd never know the blind spot was there without checking. The brain keeps this defect a secret.

Our brain gives us the illusion that we feel our toes in our toes. But every study makes it clear that all physical sensations, be they touch or smell or sound or sight, are registered in our heads, projected onto one of several body maps laid out on the cerebral cortex. So why do my toes feel like they are way down on my feet? Because that's the illusion my brain presents to my mind—a necessary and powerful example of the triumph of ideas over physical reality.

Daydreams may be wishful thinking, but they always make sense. Sleep dreams are altogether different. No logical processing occurs here, so any image we can conjure up slips by the brain's logic censor and appears on the screen of the mind. I wonder if when we are consciously aware, the mind may be driving the brain even as the brain presents its data and emotions. When we dream, day or night, the brain may be driving the mind. That we have several levels of consciousness is clear. How many times have you driven home, only to arrive and not remember much of the trip, but be mightily aware of the thoughts you were mulling over in your mind during the trip?

The ultimate illusion is one that nature plays on brain and mind. We look out at the world and see a myriad of solid objects. Scattering of high-energy particles by atoms tells us that the nominal radius of an atom's nucleus is $10^{-15}$ (one million billionth) meters. The nominal radius of the surrounding electron cloud is $10^{-10}$ (one ten-billionth) meters. That means the fraction of the total volume of the atom that is solid, a cubic function of the radius, is one part in a million billion. That number is a billion written out a million times. For every one speck of space in a solid iron bar that is filled with a nuclear particle, there are a million billion specks of space that are wondrously and gloriously empty. Like the blind spot in vision, there's nothing there no matter how solid it feels. Impossible to internalize, this fact is contrary to every experience we have. But it's a verified aspect of reality. Ethereal force fields among the atoms and molecules penetrate the voids, giving the vast open regions the impression of "solidity." If our eyes could perceive to the minuteness of the subatomic scale, we'd see we are walking on a very sparse grating. The illusion of solidness is due to the "weak" resolving power of our eyes.

Even the particles that make up the atom, the protons, neutrons, electrons, may not be solids after all. They may all be extended forms of energy. If indeed matter is the conscious expression of information, then the idea of mind over matter requires a revision. It must read the consciousness of the mind over the consciousness of matter. Anyone who has witnessed the holder of a black belt in karate shatter a brick while barely touching it (referred to as a soft break) will find nothing new in this idea. It's done more by concentrated thought (Chi) than by physical force.

Fighting it all the way, we are being dragged, kicking and screaming, into accepting the truth that our material existence is more fiction than fact. I say it. I teach it. The logic of my frontal cortical lobes analyzes the data and believes it. But in my limbic emotions, I fight it all the way. I want nature to be natural, natural by human, physical standards, and it doesn't seem to be turning out that way after all.

Even without cranial illusions, we find that the world gets curiouser and curiouser with each new discovery. Physics has touched the metaphysical realm within which our physical illusion of reality is embedded. In crossing the threshold from the physical to the metaphysical, science has discovered a reality it had previously relegated strictly to the mystical. It has discovered the presence of the spiritual, for that is really what the metaphysical is, within the land of the living.


AND what about the land of the nonliving? Though there is no explicit mention in the Torah of life after death, there are oblique references to what appears to be an aspect of life that extends beyond our physical existence. When the death and burial of a biblical personage are discussed in detail, the sequence is always the same: death, gathering to the deceased person's people, and then burial. Gathering to the people always occurs immediately after death and always precedes burial. In some cases, such as with Jacob, burial was months after death and gathering (Gen. 49, 50). Clearly, the gathering was related to death and not to the physical act of burial in a family sarcophagus. There's an intriguing aspect to this description of being gathered to one's people. This is exactly the experience that persons who have survived clinical death report. Deceased relatives reach out in vain to greet them as they return to the land of the living. Certainly this similarity may be mere coincidence. At near death, the brain must be encountering drastic chemical stress that might trigger these exotic "dreams." The concordance between the biblical account and the human experiences is however worthy of consideration, even though there is no way of proving which explanation is correct. (As an aside, I have interviewed several persons who had such experiences. Four claimed never to have heard of the effect prior to their encounter with it. When I asked one what she now thought about dying, she said she was not looking forward to it, since "it might not be as pleasant the next time.")

IF you take one thing away from this chapter, let it be that illusions are powerful. Sometimes the brain tricks the mind in its perception of reality, and sometimes the mind realizes the brain is up to tricks. Be that as it may, the brain and the mind working in concert have deciphered the most startling of all illusions: that the physical world we see about us is a structure so ethereal that were it not for forces produced by theoretical, never observed, virtual photons, we would fall right through, evaporate into nothingness, dissipate into a metaphysical cloud.

I think it was in *Duck Soup* that Groucho Marx declared: "Are you going to believe me or your lying eyes?!"

# READING BETWEEN

# THE LINES

## WHAT DOES IT ALL MEAN?

*There's an ancient tradition that when Divine revelation comes into the world, only one part is given as prophetic writings. The words are only a part of the message. The other part is placed within nature, the wisdoms inherent in the Creation. Only when we understand those hidden wisdoms will we be able to read between the prophetic lines and fully understand the message. With the help of science, we are learning to read between the lines.*

We look at the world and see a marvelous creation. A myriad of forms fill the landscape. Diversity seems to be the message. But when we look below the surface, we discover a world made of a mix of identical particles that are actually waves and then realize that the waves are massless expressions of information. Physics has exposed the metaphysical basis of existence.

We study the physiology of a cell and the neurology of the brain, mapping the flow of each sound and each sensation as it makes its way from the world around us to its being recorded as synaptic unions among the multitude of nerves that store our intelligence. And then we envision that which we have seen or heard or smelled or touched. Activity increases in specific parts of the brain. A flow of images passes before our mind's eye. And therein lies the rub. We have not a clue as to how those images form in our heads, or from where they are played back. The mind is as ethereal as the seemingly solid floor upon which we stand.

We pick up a pencil, scratch an ear, and don't even dream of the billion and more complex biochemical/electrical reactions that secretly metamorphose our thought into neural activity and finally muscular contraction. We join with our spouse and nine months later, from information held within a cell and a half, the miracle of life repeats itself. But all the wonder is hidden. Why? Why aren't the subtleties of physics and the phenomenal symphony of which life is composed apparent for all to see, visible, right up front? Why are they sequestered beneath an exterior that looks so simple?

The implication is that we can settle for a superficial reading of nature if that's all we want, but the ultimate reality of existence lies below the surface, between the lines. The very fact that the Bible opens with the creation, followed by a detailed account of the physical development of the universe, carries a pivotal message. A single creation yielded all of nature, the energy, the matter, and the space for all future existence. It's not by chance that the only name for God used in the creation chapter (Gen. 1) is Elokiim, the biblical name for God as made manifest through the forces of nature. The message is that through nature, of which we are a part, we can discover the immanence of the metaphysical Creator within the creation.

That the essence of this metaphysical presence is not always obvious does not mean it is not actively there.

> And the Eternal passed before [Moses] and proclaimed: the Eternal the Eternal, God compassionate and gracious, slow to anger and abundant in kindness and truth. (Exod. 34:5)

That is a reassuring claim by the Bible, but is it true?

Purim is a holiday of costumes and feasting, of sending gifts and reading the Book of Esther. The holiday falls in that changeable time of year when winter and spring argue over whose turn it is to run the weather. Josh, aged five, dressed as a king, felt great as only a five-year-old can when he's sure his costume makes him look like the real thing. During the night before Purim, a rare snap of freezing rain had sheathed the trees of Jerusalem in ice. Frozen droplets hanging from the branches splashed the morning winter sunlight into a rainbow of colors. The air was like crystal, the sky blue.

We're not used to icy sidewalks in Jerusalem. The first I knew of the fall was when Josh's siblings came crying to tell me. Someone had left the door to our bomb shelter open. (Bomb shelters are a reality in parts of the world.) Josh had slipped on the ice and fallen down several of the rough cement steps. The disappointment of his torn costume matched the pain of the bump on his leg. As I hugged him, feeling his hurt and trying to make it all better, in my mind I raised a voice toward heaven and demanded in anguish: "Why did You let it happen?"

I wasn't expecting an answer and only later did I focus on the significance of my question. Not why did You make it happen, but why did You let it happen. Why didn't You step in and save the situation? Get involved, God. We just survived the twentieth century. Two world wars, the Holocaust with twelve million humans burned to death under the guise of a nationalistic ideology. Only to be outdone at the other end of the spectrum by the universalistic ideology of communism, with Stalin sending in excess of thirteen million persons to perish in the frigid wasteland of Siberia. From such horrors, it might seem that the putative Creator merely wound up the universe, gave it the energy to run, and then let whatever emerged play itself out. Such a scenario does not sound very much like the Sunday school version of the biblical God. If You are really there, God, what is Your role?

It is a question worth asking even by a believer. I wonder if

Abel equally pleaded with God as Cain beat him to death. In the famous "Am I my brother's keeper?" sequence, the Book of Genesis records God confronting murderer Cain: "Your brother's bloods cry to me from the ground." Bloods, in the plural. Apparently Cain didn't know how to kill Abel and so had to beat him repeatedly. There were many wounds. As Abel's life ebbed, blow upon blow, he must have called to God for help. But none came, though God had just accepted Abel's offering.

The Bible teaches reality, not some fantasy. Notwithstanding the biblical description of God as "compassionate and gracious, slow to anger and abundant in kindness and truth," the Bible knows that bad things happen to good people and God lets them happen. By the fourth chapter of Genesis, Abel, the good guy, is dead. Our impulse might be to flee from such a God. Jonah tried and failed. We are in this universe and that is the way it operates. If we were gods, our concept of how to run the world might be different. But we are not gods. (My guess is that if we were, considering the political and social problems we've generated, we'd make a much worse mess of it.) The biblical challenge is to understand the apparent randomness in nature, the acts that bring at times joy and at times tragedy, within the context of the claimed compassion and graciousness of a Creator involved in the creation.

Resolution of the apparent inconsistency between the immanence of the Creator and the presence of grief lies in the biblical definition of creation. In my previous book, *The Science of God*, I discussed at length the biblical description of creation. As learned from Isaiah (45:7), creation is the Divine act of *tzimtzum*, God's spiritual contraction. In the resulting spiritual space, the undifferentiated simplicity of God is fractured, yielding in its stead the variety of our universe. With this Divine contraction comes the release that allows for our choices of free will and the leeway for the seeming imperfections and meandering courses found in nature.

In the Book of Genesis, in a passage related to the patriarch of Judaism, Christianity, and Islam, we read, "And the Eternal said 'Shall I hide from Abraham that which I am doing. . . . For I have known him for the purpose that he may command his children. . . .'" (Gen. 18:17,19). Commenting on these verses, the thirteenth-century kabalist Nahmanides asked why is it written "I have known him"? Doesn't God know all persons? Nahmanides answered his own question. God knows all life, but the degree of Divine direction to an individual person depends on that person's individual choice of how close to God he or she wishes to be. A half century earlier, the medieval philosopher Maimonides made the identical observation in his *Guide of the Perplexed* (Part 3, chapter 51). Only the totally righteous have one-on-one Divine direction, and even that guidance may not ensure a life free of pain and suffering. For the rest of us, chance and accidents do occur. It's our choice as to where we, as individuals, fall within that spectrum of behavior that stretches from intimate Divine direction to total random chance.

Our difficulty in perceiving the metaphysical reality within which the universe is embedded is that we view creation from the inside looking out. We try to envision a metaphysical reality totally void of the distinctions of time, space, and matter. But all our thoughts are built of the physical images we have experienced, images of time and space and matter. Creation, however, is the exact opposite. It starts with a transcendental simplicity and gives rise to complex physicality.

Moses asked to see that transcendence but was denied permission. "And [Moses] said: show me please Your glory. . . . And the Eternal said you cannot see My face for no human can see Me and live. . . . You will see My back but My face you shall not see" (Exod. 33: 18, 20, 23). God's back is the imprint of the Divine within the world.

Einstein is widely quoted as having said, "The most incomprehensible thing about the universe is that it is so comprehensible." It is Einstein's discoveries that let us comprehend facets of the universe previously inconceivable. And it is those discoveries followed by the revelations of quantum physics that have clarified why the universe is in fact so very comprehensible to the human mind. It is because we are a part of the universe that has become aware. And in being part of it, we have come to comprehend ourselves. Our newly found self-understanding has revealed that wisdom is contained within even the simplest of particles, and an unfathomable and even perplexing amount of wisdom is present in the complexity of life.

The mystery of life's origins and its ordered complexity is not simply one more difficult scientific roadblock waiting for a physical explanation. Life, and certainly conscious life, is no more apparent in a slurry of rocks and water, or in the primordial ball of energy produced in the creation, than are the words of Shakespeare apparent in a jumble of letters shaken in a bag. The information stored in the genetic code common to all life, DNA, is not implied by the biological building blocks of DNA, neither in the nucleotide letters nor in the phospho-diester bonds along which those letters are strung. Nor is consciousness implied in the structure of the brain. All three imply a wisdom that precedes matter and energy.

Consistently, at every level of complexity, the information that emerges from a structure exceeds the information inherent in the components of that structure. This is true from subatomic electrons, the lightest of known particles, to the human brain, the most complex structure yet encountered in our universe. This ubiquitous emergence of information, of wisdom, cries out for explanation. The range and depth of it are fantastically unlikely to have happened by chance. Its origin is told in the three-thousand-year-old opening word of Genesis, *Beraesheet*. Not the superficial reading, "In the beginning," but the far deeper reality, *Be' raesheet*, "'With wisdom' God created the heavens and the earth." The substrate of all existence is wisdom.

As physicist Freeman Dyson stated when accepting the Templeton Award: "It appears that mind, as manifested by the capacity
to make choices, is to some extent inherent in every atom. . . .
God is what the mind becomes when it has passed beyond the
scale of our comprehension."

And again, physicist R. B. Laughlin in accepting the Nobel
Prize: "I myself have come to suspect that most of the important
outstanding problems in physics are emergent in nature."
Emergent, meaning not implied by the individual parts of the
structure, not merely quantitatively different, but qualitatively;
a difference in type from their components. The whole is not
equal to the sum of its parts. The whole is greater than the parts
could "imagine." And Professor Wheeler: the "bit" (the binary
digit) of information that preceded and gave rise to the "it" of
matter.

This emergence of wisdom in the material world appears as if
it is de novo only if we fail to realize that an aspect of mind is
"inherent in every atom." In a sense, mind has been present
from the beginning. We might had known that all along. After
all, with wisdom God created the heavens and the earth.

The link between the spiritual and the physical finds expression in the details of the biblical Tabernacle said to have accompanied the Israelites during their forty-year trek in the desert.
For all the multiple listings of the items kept in the Tabernacle,
the first item mentioned is always the ark within which lay the
tablets Moses received on Sinai. The ark represented revelation, contact with the Creator. The second item is not the can-
delabra with its pure flame of spirituality, and not the altar with
its message of sacrifice. The second item is always the table on
which twelve loaves of bread were placed each Sabbath. The
table represented the physical world, with bread as the penulti-
mate symbol of the wisdom inherent in nature. To harvest a loaf
of bread from the energy of the big bang requires far more than
ideal laws of nature and an earth with a life-friendly environ-

ment. To get to a loaf of bread out of a big bang requires the emergence of the complex and orderly interactive system of information processing we find first in the simplest of biological cells and ultimately in the human brain and mind. Rather than explaining the appearance of this wisdom by attempting to establish the validity of the spiritual at the expense of the material, or the opposite, the material at the expense of the spiritual, the Bible in its wisdom by establishing a link between the ark and the table speaks of an integration of the two, the imprinting of the metaphysical upon the material. The physical and the metaphysical become one.

Humanity has spent and continues to spend billions of dollars on ventures such as the Hubble Space Telescope. We are seeking our origins with the passion that one lost in a desert searches for an oasis. "As a hart pants after water so my soul yearns for Elokiim. My soul thirsts for Elokiim" (Ps. 42:2, 3). Again, the Bible chooses the name of God as Elokiim, the aspect of the metaphysical Creator made manifest in nature. Our hope is that Hubble data will provide us with a clue as to the origin and place of life in the scheme of the universe.

Christian de Duve articulated the conundrum of life: "Faced with the enormous sum of lucky draws behind the success of the evolutionary game, one may legitimately wonder to what extent this success is actually written into the fabric of the universe." Being written into the fabric of the universe implies something other than merely having the created laws of nature amenable to the flourishing of life. Being written into the fabric of the universe tells us that we are made of the stuff of the big bang. We were present at the creation.

THE mind presents each of us with two inner lives. One deals with the demands and joys of daily life, and one seeks the transcendental, the eternal aspect of our finite existence. In all the thought experiments we might compose, fantastic though those imaginings might be, the one feature that remains is the flow of time. Without time all events cease, and without our memories, our concept of time vanishes. The world becomes a still life in the fullest meaning of the phrase. Time is the continuing reality of life. The relativity of time, discovered by Einstein, was the first of the steps that moved physics into the realm of metaphysics. I find it intriguing that time is also the first item that the Bible makes holy, holy in the biblical sense of being separate from the remainder of existence. Not a place, not a person, but totally abstract, intangible time—the seventh day, the Sabbath.

The Sabbath predates Moses, Abraham, Noah. Only Adam and Eve, the biblical parents of all humanity, predate the Sabbath. Long before the ritual of religion made its way into theology, the Sabbath was established. The Sabbath is the Bible's gift to all humanity; the crown of the six days of creation. It is the undersold superproduct of the Bible. It ritualizes contemplation, fits it into a timely rhythm, superimposing its cycle onto the other cycles that nature has imprinted through light and dark, satiation and hunger, phases of the moon.

The word Sabbath comes from the Hebrew *shabat*, meaning to rest, to cease from work. The essence of the Sabbath is rest. It stands in juxtaposition to the previous six days of work. Erich Fromm, in *The Forgotten Language*, described it perfectly: "Rest is a state of peace between man and nature. . . . Rest is an expression of dignity, peace and freedom."

The Bible understands the human psyche. It realizes that harmony between the two lives we live, the temporal wants of the body and the transcendent needs of the soul, is rarely a spontaneous happening. Without a ritualized, established routine there is always a reason for the tangible immediate demands of life to take precedence over our more abstract spiritual desires. There's no difficulty in being "holy" in a church or mosque or synagogue or temple. But the aspirations of theology far exceed our behavior in places of worship. The inherent aim is to bring the holy, the metaphysical, into the daily life of the marketplace. Bringing the spiritual into the tasks of the work week takes practice. Religion provides that practice. It's the pumping iron that gives us the spiritual strength to make theology a part of our mindset. The Sabbath is the day of practice. It's Eden, the message of which is that humankind was created for pleasure. The Sabbath returns to us a taste of Eden and helps us spread it through the entire week.

The absence of work introduces a formalized time for family intimacy on all levels. Not by chance, the structure of the Sabbath makes it the most physical and the most spiritual of days. A receptive spirit requires a happy body. Spirituality can't be weighed, but our emotions tell us there is an aspect to life that transcends the physical. Even after satisfying our physical needs, we humans feel there is still something missing. It's meaning. We seek purpose in life. The Sabbath opens space to consider what that purpose might be. The reality of quantum physics has proven that the future is not merely an extension of the present. There is a slack, a leeway in the direction taken by the forward flow of events. That margin of variability opens a window for how and what we choose, how we use our free will.

When we exercise our will, we choose from within a window of possibilities built from our cumulative knowledge and experiences. That window is not static. In a feedback loop, the range of our current choices is skewed by how we chose in the past. How we thought in part determines how we think. As we grow, our window of choices shifts in accord with our new experiences. Subconsciously, the brain presents a range of possibilities, some of which rise to the conscious. We choose from among them.

There are tools to help link physical needs to spiritual desires, our "is" with our "what ought to be," without denying either. If, for example, loving one's neighbor is a goal, then there must be some hints as to how to achieve that goal. There are, and they are listed in that same biblical passage of Leviticus (19:18) that insists we can learn to love: Don't take vengeance and don't bear a grudge. In daily terms, don't keep accounts of what your spouse did or did not do. If you want some poison for marital bliss, remember each time your partner was late or forgot your birthday or didn't compliment you on your new garment. That's keeping accounts. Acting on those accounts is vengeance. The Bible's suggestion is to define love as focusing on the virtues while acknowledging the shortcomings. Identify your loved one with those virtues. At the end of a tough day, consider that it may have been tough for your spouse too. Just before you walk in the door, review why you married in the first place. Your spouse is the same person, but now with the additional demands of family life attached, not all of which are filled with fun. Cut through and see beyond that baggage. A happy spouse is a happy house. Do it just for the totally selfish reward of having a happier life. Forget the spirituality altogether. That will come naturally, from the bottom up.

The human soul, the *neshama*, is our link to the transcendent. It knows the meaning of "the Lord is One." It realizes that there is a unity that pervades all existence and from which all existence is composed. The *neshama* looks at each potential act and quietly asks, will this act move me closer to or further from the joy of touching the oneness of creation?

As we learned from the tragedy of Phineas Gage, the brain can build upon and feed the mind only from that which it has stored. If we choose to feed it with trivia and violence, then rest assured trivia and violence is what will appear within our window. The tranquility of the biblical Sabbath prepares the mind for the spiritual, that part of our lives so often masked by the noise of the work week. One day a week, it says, expect pleasure. And the amazing truth of the human mind is that if you expect pleasure, you'll find it.

Fifteen billion years ago all of us and all we see were part of a compact homogeneous ball of energy. That was the entire universe. Once we were all neighbors. Then as the ball expanded outward, differentiation occurred, masking the underlying unity. Some minute fraction of the energy went into making the atoms of the ninety-two elements. I doubt if it could have been predicted which primordial atoms of carbon would end up in my brain and which in a leaf of the eucalyptus tree across our courtyard. There is plenty of room for chance and choice in the world. If I had been born a few hundred miles to the east, I'd be aiming missiles at where I am now.

Yet behind the chance, there's an absolute truth. It's the unity underlying all existence, the singular wisdom from which we are constructed. Your soul knows it. Listen carefully. You can hear it. With training you can learn to enjoy it. Sabbath R&R, rest and re-creation, is a good primer.

PLATO likened our perception of life to persons viewing shadows on a wall while unaware of the far grander reality that produced those shadows. Science has revealed part of that larger reality. In the wonders of nature, we have discovered the imprint of the metaphysical within the physical. As one who sees the wake of a boat that has passed by, so we encounter the hidden face of God. And the hidden face is indeed grand in its collective simplicity. We live in a time when finally we can begin to read the text that lies between the lines. That text is yielding to us the secrets at which the written words had only hinted.

> Then the eyes of the blind shall be opened and the ears of the deaf unstopped. . . . The people that walked in darkness have seen a great light; they that dwelled in the land of images, upon them the light has shone. (Isaiah 35:5; 9:1)

# EPILOGUE: HINTS OF AN

# EXOTIC UNIVERSE

*Teacher of physics par excellence, the late Robley D. Evans, said that if you have key points you want to get across to an audience, first tell them what you are going to say, then say it, and then at the end tell them what you said.*

We are at the end.

From the traditional view of classical physics, the universe seemed so logically constructed. Principles of determinism taught the obvious fact that identical causes produce identical results. There were the distinct and separate natures of particles and waves, and of energy and mass. All this fit well within our human psyche's opinion of how a world should work. And then came Einstein and Planck and de Broglie and Heisenberg, and the assumed logic underlying existence tumbled into an exotic, unpredictable fantasy that turned out to be true. Their discoveries revealed relativity, quantum physics, uncertainty, particles with fuzzy extended edges, particles that aren't ping-pong-ball-like entities but are actually waves. And then most bizarre of all, that these waves might actually be representations of something as intangible as information, as wisdom.

Enlightenment is far easier when we have no preconceived notions of what must be true. The social and professional pressure to conform to accepted ideas can be monumental even whenever

mounting data contradict their validity. It's called cognitive dissonance. As great a mind as Einstein's bowed to convention when
his cosmological equation of the universe showed that the universe might be in a state of violent expansion, that the galaxies
might be flying out from some moment of creation. Rather than
publish this astonishing prediction, he changed his data to match
the popular, though erroneous idea that our universe was static. In
doing so he missed the opportunity to predict the most important
discovery science can ever make relative to the essence of our existence: Our universe had a beginning; there was a time before
which there was neither time nor space nor matter.

Today we have another seemingly logical, but quite likely erroneous, piece of accepted wisdom forcing itself upon our paradigm of existence: that the physical world is a closed system;
that every physical event has a correspondingly physical cause
preceding it. It's not a question as to whether or not we can predict the exact effect of a given cause. Quantum physics says we
cannot. But our logic insists that each physical effect must be
initiated by a physical cause. How could it be otherwise?

The very knowledge of the big bang provides proof otherwise. The physical system we refer to as our universe is not
closed to the nonphysical. It cannot be closed. Its total beginning required a nonphysical act. Call it the big bang. Call it creation. Let the creating force be a potential field if the idea of
God is bothersome to you, but realize the fact that the nonphysical gave rise to the physical. Unless the vast amounts of scientific data and conclusions drawn by atheistic as well as devout
scientists are in extreme error, our universe had a metaphysical
beginning. The existence—if the word existence applies to that
which precedes our universe—of the eternal metaphysical is a
scientific reality. That single exotic fact changes the rules of the
game. In fact, it establishes the rules of the game. The metaphysical has as least once interacted with the physical. Our universe is not a closed system.

But does that which created the universe still interact with the creation? That's a question each person must confront. Without solid scientific knowledge, philosophizing about this puzzle is very much a nonstarter. We've surveyed the science and discovered a complexly ordered wisdom expressed in the molecular functioning of life nowhere evident in the structures from which life is built or in the laws of nature that govern the interactions of those structures. That wisdom in life is the imprint of the metaphysical.

This may have the ring of a radically new idea. It's not. Thirty-four hundred years ago, the opening words of the Bible set forth the identical concept. *Be'raesheet*—with wisdom—God created the heavens and the earth. With a bit of intellectual endeavor, we can avoid the type of error that Einstein made as he bowed to a popular prejudice of his day.

Almost a millennium ago, the medieval philosopher Moses Maimonides wrote that we can know *that* God is, even though we cannot know *what* God is. "The Eternal knew [Moses] face to face" (Deut. 34:10). It is not written that Moses knew the Eternal face to face. "The Eternal said [to Moses and to us today]: I will make all My goodness [the wonders of creation] pass before you. . . . You shall see My back [as one finds, in the wake left by a boat, evidence of the boat's passage], but My face shall not be seen" (Exod. 33:20, 23). Even in the closest of encounters, the face of God remains hidden.

# DNA/RNA: THE MAKING

# OF A PROTEIN

*Other than sex and blood cells, every cell in your body is making ap-*
*proximately two thousand proteins every second. A protein is a combi-*
*nation of three hundred to over a thousand amino acids. An adult*
*human body is made of approximately seventy-five trillion cells. Every*
*second of every minute of every day, your body and every body is or-*
*ganizing on the order of 150 thousand thousand thousand thousand*
*thousand thousand amino acids into carefully constructed chains of*
*proteins. Every second; every minute; every day. The fabric from*
*which we and all life are built is being continually rewoven at a most*
*astoundingly rapid rate.*

The flow of life, like some cosmic ballet, is staggeringly dy-
namic. But the marvel life presents is held not merely in its vi-
brancy. Its wonder couches most deeply in the extent of the
information that directs this vitality and in the fact that this information is qualitatively different from all other laws of nature.

To understand why your child resembles you, and why a stalk
of wheat grown from a seed resembles the plant from which the
seed was taken, is to understand how information is transferred
biologically. That transfer rests in the exquisite functioning of
two molecules, deoxyribonucleic acid—DNA—and its messen-
ger/partner, ribonucleic acid—RNA—the superstars of all life.
For three billion years, this molecular dream team has worked
to produce every known life form, plant and animal alike.

Encoded on these two molecules as strings of specifically arranged nucleotides is the information dictating that humans have ears on opposing sides of the head, a face with a nose in the middle, a pancreas that produces insulin in response to the presence of glucose in the blood, and every other physical characteristic of our bodies. Humans have approximately eighty thousand such information packets, each packet known as a gene. Considering the intricacy of a human body, the number of genes is surprisingly small. And equally impressive is that all this is orchestrated by a set of instructions initially stored in a single fertilized egg.

The genetic revolution is still in its infancy and so we can only speculate on how the complexity of the human body is actually encoded. It would seem that for such a small genetic package to produce such meaningful structures, each gene must somehow provide a general plan as well as specific instructions. There may be codes within codes. The system by which genes are programmed in the DNA suggests this plausibility.

The DNA of each human cell contains approximately three and a half billion nucleotide bases. Clusters of these bases form the eighty thousand genes each of which in turn produces a specific molecule, most commonly a specific protein. Since some three thousand bases are used to code for a typical protein molecule, with over three billion bases available for the coding, this could produce more than a million proteins. Yet we make our living using approximately eighty thousand proteins. This means that the space occupied by the genes accounts for less than 10 percent of the total number of nucleotide bases. So why the other 90 percent? Is it just excess cellular baggage? Possibly—but equally likely is the possibility of genetic codes not yet discovered tucked within the remaining 90 percent that might instruct a cell how to build a human from the eighty thousand proteins. (That number is an approximation. The number is thought to be between 60,000 and 100,000, with a few estimates based on RNA sequences reaching 140,000.)

In this appendix I discuss the nuts and bolts of nature's greatest encrypted source of information, DNA, and how that information is held and read by the cell within which it lies. The process is nothing short of amazing. Along the way we'll discover some of the riddles that genetics has uncovered. Interesting though these puzzles may be, a much larger mystery looms below the surface: the source of the encoded information and the source of the awesomely complex, yet nearly error free, system that deciphers and translates this wisdom into the structures of life. In general, simple laws, such as the laws of nature, cannot give rise to complex information that exceeds their own unless that complexity is a fractal extension, a duplication in number and type of the base law. This is simply not the case with the genetic code. The information therein is apparent neither in the atoms and molecules from which DNA is formed nor in the laws of physics and chemistry that govern the interactions among these molecules. And yet if the fossil record is correct, the endowed wisdom of DNA seems to have been present from the very earliest stages of life on earth. How the coding that drives all life sprang into existence remains a mystery. The scale of the mystery is best realized by the complexity of its product.

Each human cell contains about two meters of DNA. Considering that a typical cell is nominally 30 millionths of a meter in diameter, squeezing in the two meters of DNA requires a feat of compression on a scale of one hundred thousand to one. Nature has the answer. Two complementary strands of the molecule twist and fold into a supercoiled helix, a helix within a helix, yielding a ball less than 5 millionths of a meter (approximately one one hundred thousandth of an inch) in diameter. To read the information stored, the helix is opened. But opening such a tight helix can cause damage. Therefore the bonding of the helix must be strong enough to maintain stability but sufficiently motile to allow rapid opening when the signal is given for information to be retrieved. A combination of strong covalent bonds maintaining the integrity of the basic molecular structure with weaker, more easily broken bonds, holding the strands in the helix, solves the problem.

As with the fine-tuning of the four forces of physics, the relation between the so-called strong and weak chemical bonds is equally precise. For basic stability, DNA is a winner. It has been recovered from hundred-thousand-year-old Neanderthal bones. The biological system of information storage is phenomenally dense. If all the information in all the libraries in all languages were transcribed into the language of DNA, it could be recorded within a volume equivalent to 1 percent of the head of a pin.

The entire encoding alphabet for the eighty thousand genes consists of four molecular letters, the nucleotides, laced along a repetitive molecular backbone. This greatly simplifies the production and selection of the materials needed to construct the DNA helix and its RNA messenger/partner. How these molecules work together to produce life is intimately related to their almost-too-good-to-be-true chemical structures.

Of this pair, DNA is the primary repository of the bio-information. To get a glimpse of this remarkable molecule of life, we must pass from the outer cell into the nucleus, a structure that appears as a cell floating within a larger outer cell. At some 10 millionths of a meter in diameter, the nucleus occupies slightly over 10 percent of the total cell volume. Its walls are perforated by thousands of porelike openings, each approximately 100 billionths of a meter in diameter. The micro-scale of it all is awesome considering that from this micro-world emerge fully formed humans and elephants and even great blue whales.

The nuclear pores are designed to keep DNA inside the nucleus while allowing ready access and egress for a host of other molecules. Interestingly, though the nucleus is built for production, an energy-intensive activity, it has no autonomous source of energy production. It relies totally on energy supply imported from the outer cell in the form of the familiar ATP.

And here's that enigma again. It's worth contemplating throughout the discussion because it shows its head in a dozen

e steps and the two spines the outside handrails. The DNA is
w in position to provide the information needed in protein
thesis. (See Figure 17.)

roteins are strings of amino acids, but genes are strands of
eotides. Combinations of twenty different amino acids are
in the proteins of life; there are four bases used in the nu-
des. Potential for a language problem exists here. Nature
ates the translation from bases, similar to the letters of the
o amino acids, the "words" of the protein (with the pro-
eing the sentences) by using groups of three bases as a
r each specific amino acid. GCG, for example, codes for
no acid alanine, CGG for arginine.

immediate question that arises is: Why did nature
ith the coding to store the secrets of cellular life? Since
ment a cell is organizing amino acids at the incredible
ousands per second, why not store the information in
s directly as a string of amino acids, or have twenty
aselike molecules, each one indicating a specific
, rather than the three bases that nature actually
nch is that eighty thousand genes are simply not ad-
uctions to produce a human from an egg. The com-
at leads from DNA to RNA to protein allows for
f nucleotides to contain information other than
nes. If so, then the amount of information able to
on these chromosomes can be vastly expanded.
er limited to the genes per se. The multiple lev-
ch aspect of the physical world can be read, such
ticle and energy/matter dualities, may find par-
ing of the genome.

c protein is needed by a cell, a chemical messen-
e outer cell, through a pore in the nuclear mem-
cleus. How the messenger knows to go to the
remains a mystery. This messenger finds the
ne (one of the twenty-three pairs), locks onto
and moves along, nucleotide by nucleotide,

different ways, the problem of how the entire process originally got started. The first stage in making ATP requires a dozen or so intricately dovetailed protein enzymes, each one picking up the action just as the previous one leaves off. These enzymes are manufactured using information stored in the DNA. Retrieving and deciphering the wisdom held within the DNA in order to make the enzymes requires a good deal of energy. Get the problem? For the energy we need ATP. For ATP we need enzymes. But to make the enzymes we need the information held in the DNA and to get that information we need the energy supplied by the ATP. I guess if you buy a car at the top of a hill, you can drive away without a battery. It runs by itself. But something had to get the car to the top of the hill in the first place.

Of course the naysayers will call for another source of energy that started it off and evolved into ATP. But those pleas begin to ring thin. There is no hint of this evolution within the cell and usually nature keeps its ancestral record of development, such as the fetal heart "evolving" in the womb from an initial single tube to the mature four-chambered organ, or the human fetus at an early stage bearing a yolk sac (such as do fish larvae) that is then absorbed by the fetus.

And then there is the problem that, according to the fossil record, it all developed so very rapidly, almost simultaneously with the appearance of liquid water on earth.

I wish I knew what I meant when I agree with my colleague Dennis Turner that there's a ghost in the system. I'm a scientist. Studying nature is what has put bread on my family's table for a good number of decades. I want nature to work like nature. But at several key stages in the development of our universe, nature seems to have behaved most unnaturally. It's what Nobel Prize–winning physicist, and avowed atheist, Steven Weinberg referred to in his excellent book *The First Three Minutes* as the "embarrassing vagueness . . . the unwelcome necessity of fixing initial conditions," of having to accept a batch of initial conditions simply as "givens." "Givens" in scientific jargon is the sophisticated way of saying that that's the way it is and so let's start the discussion from those givens without understanding how they got there. First we start with the need to accept a universe housing forms of matter and the laws of nature as givens, for no a priori reason. And now at the other end of the dimensional scale we have DNA simply showing up on the scene with immediate complexity and working marvelously well.

The relationship between DNA and RNA is even more of a marvel. Through the pores of the nucleus, a continual supply of raw materials arrive bringing copious amounts of ATP and the building blocks needed to manufacture the information messenger, mRNA. In the opposite direction, passing from the nucleus into the outer cell, flows a steady stream of power-spent ATP, now referred to as ADP, and newly manufactured mRNA molecules, each carrying with it the design for, and the command to make, a specific protein.

Though DNA stores the information that makes the cell run, it functions in a purely passive mode, providing the patterns for mRNA molecules. The command to read those patterns comes from instructions brought to it by inducer proteins (again, proteins needed to make proteins, like a story without a beginning) arriving from the outer cell, and sometimes even from outside the cell. To get a feel for the almost unreal complexity of the cell's mechanics, let's first look at the structure of a DNA molecule, and then survey how it works. The basics are the same for all plant and animal life.

DNA comes in sections called chromosomes. Humans have forty-six: twenty-three from mom and twenty-three from dad. Fruit flies have eight. Potatoes have forty-eight. Don't be fooled by the name. Chromosomes are white. They got their name—chromo, the Greek word for color—when, in early genetic research, dyes of different colors were used as means of identification. Each human cell, other than blood and gamete (sex) cells, has four sets of its eighty thousand genes held in the

two pairs of double helix. In theory, could reproduce your entire self. All t in any given cell only five to ten thou tively being expressed. This partial type of cell from another, which in cell happen to know where it is s that form an ear where ears are to

The DNA helix is some 25 reaches a tenth of a meter (ab wound. If we were to unwin up, one cell's worth would Put all the DNA in your b that reaches to the sun an is not some abstract fact.

DNA codes its infor nucleotide molecules. (A); guanine (G); cyto replaced by uracil (U confusing RNA, th source of the info keeps the number

DNA consists along two spine groups. Here mentary base ing with C. ( two strands the discov Wilkins t

Wher the dou the tw off at beer

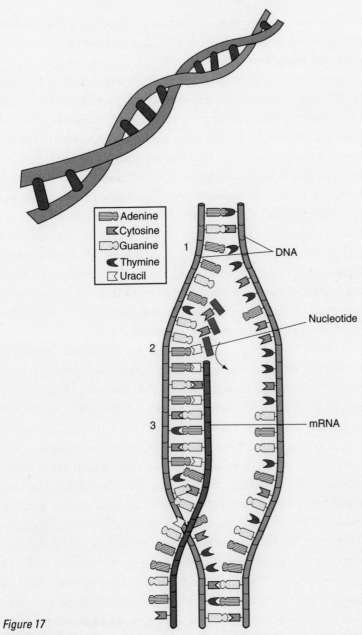

Adenine
Cytosine
Guanine
Thymine
Uracil

1

DNA

Nucleotide

2

3

mRNA

Figure 17

Synthesis of mRNA

Matching and joining of the nucleotides to form the mRNA takes place at the rate of 50 substitutions per second, a feat of molecular gymnastics.

until it comes to the specific sequence of bases that marks the beginning of the gene that codes for the desired protein.

At this stage, the signaling molecule changes shape, and in doing so allows—or causes—an enzyme called DNA-dependent RNA polymerase (I'll call it RNA-P) to join the action. In fact it really starts the action. This amazing molecule is a phenomenon in itself. It binds to the promoter region of the chromosome situated at the start of the gene and at that point breaks the hydrogen bonds holding the parallel strands of DNA (recall that these bonds are strong enough to pull the DNA into its helix spiral, but not so strong that they obviate the rapid opening of the rungs on the DNA ladder). Once the helix is opened, the nucleotide bases are ready to be read. The RNA-P now moves along the exposed nucleotides, identifies each base, and assembles a chain of their complementary bases. For base A, a base U is selected from the surrounding cytoplasm; for G a C is selected and joined to the previously selected U.

To get a feel for the complexity and speed of transcription, please take a deep breath and read the following sequence of events in one burst. It's what your RNA-P is doing inside every cell of your body, a thousand times over, right now. The RNA-P opens the helix, reads each nucleotide base, selects the correct complementary base from among the four types floating in the intracellular slurry, concurrently selects from the same slurry the molecules that make up the spine of the lengthening strand of mRNA being manufactured, trailing behind the RNA-P, joins the just-selected base to the spine, takes the portion of DNA that has been read and reseals it to the parallel DNA strand from which it was separated, opens the portion of DNA to be read next, reads it, and continues this juggling act till it reaches a coded stop order. Exhausting just to describe it. And RNA-P does this manufacturing at fifty bases a second, a lot faster than you read the words I used to express just one step of the fifty. ATP is there to provide the needed energy at each junction. Keep in mind, this entire sequence is performed by molecules reading molecules, molecules selecting molecules, molecules walking along with other molecules. Don't project too much brain power or body power onto the system. It's not little people in there. It's simple molecules that somehow seem to act like little knowledgeable people, as if they had a wisdom of their own. Which they do.

Because the demand for proteins is so great, as many as thirty RNA-P molecules may be at work simultaneously along a single gene, churning out multiple copies of the single-stranded mRNA. Electron micrographs of an active gene show strings of partially completed mRNA extending at right angles from the DNA spine, with increasing lengths of the mRNA as we move along the spine toward completion of the gene. The impression is something like the branches of a balsam fir seen in silhouette, short near the start of the gene and broadening as we approach the end.

With the mRNA now copied from the DNA, a further complication arises. If DNA is 90 percent "junk," then so is the complementary form of mRNA. This "junk," known as introns, must be removed, snipped out by some very wise snipper and discarded. Then the remaining sections, known as exons, the sections of bases that code for the wanted amino acids, must be spliced together by a very clever splicer. With that completed, the mRNA is ready for its task, making a protein.

Through a process not yet understood, each completed, snipped, and respliced mRNA is transported through a nuclear pore and into the outer cell cytoplasm, where two parts of another of the cell's wunderkind, the ribosome, join together, clasping the mRNA within. The business of a ribosome is to read the information sequestered on the mRNA in groups of three bases, call for another type of messenger, transfer RNA or tRNA, which brings the amino acid that corresponds to the just-read three bases (one of twenty types of tRNA in the surrounding fluid, one type for each of the twenty different amino acids used in proteins) to the ribosome and then to join the amino acids one to another, as the bases are read, thus forming the needed protein. Easier said than done. And yet with each of your seventy-five trillion cells incredibly producing two thousand proteins each second, including this second within your body, nature makes it seem easier done than said.

The detail continues. The cellular mechanism literally swims into action. It makes our heads spin and it should. The wonder of life is in the dovetailing intricacy. If the requested protein is for export, manufacture occurs at an organelle known as the endoplasmic reticulum (ER). Here thousands of mRNA are at work making proteins that upon completion are swallowed by the ER, a portion of which then pinches off, forming a pouch called the Golgi apparatus. Golgi are carried to the cell membrane on the backs of motor proteins (molecules that actually walk along the microtubule tracks that lace the cells and know where to and when to walk). When signaled, the Golgi fuse with the membrane, pop open on the outer side, and spew their contents of newly made proteins into the extracellular fluids and blood. I wonder just how each Golgi knows which direction to send the motor protein that is carrying it. Marvelous.

And there are the huge chaperone molecules which engulf the entire newly made protein, check it for errors by examining the protein's charge characteristics, and then, if an error is found, send it back for retrofit. There are a hundred and more functions that in the interest of brevity I have not described, such as shopkeeping, diverse manufacturing, transporting functions, and others. All are going on simultaneously in each cell, each function equally crafted like a fine Swiss watch. We take it all for granted. We see the system from the outside. But from the inside, like turtles, it's a wonderland all the way down.

Five minutes ago I couldn't even spell "endoplasmic reticulum," and now I know I'm full of it.

# MITOSIS AND THE

# MAKING OF A CELL

*"Acquire wisdom and with all your resources acquire understanding." (Proverbs 4:7)*

In the text we skipped one major step in the making of a person. We covered meiosis, the sexual reproduction that starts the process. But immediately thereafter, a fetus needs a plenitude of proteins sequestered in a universe of cells. Lots and lots of cells. In fact, about seventy-five trillion for an adult. And they all start from one single cell that somehow knows to divide, and to differentiate into a variety of cell types, and to divide again.

Aside from the brain, biological cells are probably the most complicated entities on the face of the earth, and possibly, hubris notwithstanding, the most complicated entities in the universe, far more complicated than we would imagine from an external view of the organisms they combine to make. The complexity of life's biology is held within its internal processes, but dressed in the simplicity of our outer bodies. Anyone who thinks nature is simple should take a short user-friendly course in how just one cell "evolves" into two cells, and then marvel at the vast amount of information required to bring about this single step and then wonder at its source. Let's have that course now. The hidden face of God is found in the details.

Packed into a membranous ball a few tens of a millionth of a meter in diameter are close to ten thousand proteins and another thirty thousand molecules, each busily occupied with its own intricate functions, processing the information encoded for life. Information is the key ingredient that shows up in every aspect of life.

To reach the 75,000,000,000,000 (i.e., 75 trillion) cells in an adult body, only forty doublings of the cells would be required if all the cells divided at an equal rate. That is, 75 trillion is about $2^{40}$, two to the power of forty, two multiplied by itself forty times. The entire process of cell division, referred to as mitosis, takes only about an hour. Which might imply that if we can move from fertilized egg to full adult in forty generations of cell divisions, the entire process should be finished in a mere forty hours. But it isn't so easy. Between each session of mitosis, a cell goes through an interphase lasting some ten to twenty hours, depending upon cell type, during which (among other activities) the cell is occupied with protein synthesis—that amazingly complex flow in which individual amino acids are brought into the cell, strung onto specifically ordered chains, snipped, respliced, and then shaped into structures of life.

Still, even at twenty hours per division, forty generations would occupy a mere eight hundred hours—about a month. But of course, twenty years is closer to the reality in the making of an adult member of our species. So what's taking so long?

First, most cells don't live for the entire life span of the body. They age, die, and must be replaced. The vast majority of newly manufactured cells are of this replacement variety. Hence the total tally of cells present at any one time pales in comparison to the numbers that have actually been, or are being, made.

Right now, within your body, new cells are being produced at a clip of four to five million this very second, and every other second, too. A typical cell, 20 to 30 microns (millionths of a meter) in diameter, has an approximate weight of a billionth of a gram. At four million new cells per second, that equals four milligrams of cell weight each second, 400 grams a day, or about 140 kilograms (300 pounds) of new cells produced each year! No wonder it is hard to keep our weight in check. But don't blame it on the cells. It is safe to say the you that was you a year ago, the atoms and the molecules of your body, is not the you you are today. Your body sloughs off, discards, close to the entire 140 kilograms of body tissue each year. Epidermal skin (the outer layer) is continuously being replaced. These cells last about a week before being traded in. In your mattress there's an entire ecosystem of micro-mites eating what was you yesterday.

Red blood cells account for 40 to 50 percent of total blood plasma. They are packed with oxygen-carrying, iron-rich hemoglobin—hence the red color of the oxidized iron—and live for about one hundred days. Almost a million die and are replaced each second, the product of our red bone marrow. Cells in the less user-friendly environs of the stomach and intestinal lining, where the acidity (pH = 1) is ten times that of lemon juice and a hundred times that of vinegar, coke, or beer, survive only a day or two. Intestinal juices will put a pucker on your lips if you taste them. (Blood is about neutral as far as acidity is concerned; pH = 7.)

Some cells, such as nerves and skeletal muscle, in theory can survive for the entire life of the individual if not damaged, which sounds good but has concurrent disadvantages. A severed nerve, unable to be replaced, may result in permanent loss of function in the related muscle. While muscle cells do not divide, damage triggers a process that can reconstruct the injured tissue.

In addition to the demise of cells as a limitation on mitotic accumulation, cellular metabolism plays a key role. In the life cycle of essentially every biological cell, there is a phase or stage known as GO. Contrary to what the name might suggest, the GO phase signals the cell's exit from the reproductive flow. While continuing to be metabolically active—that is, producing enzymes, hormones, and other proteins upon demand—a cell in the GO phase becomes nonproliferative, extending the entire mitotic cycle well beyond the ten-hour minimum.

Part of the tragedy that leads to cancer is an inopportune mutation in a chromosome that signals a cell to bypass the GO phase. Since cell division involves replication of the chromosomes, the mutation is passed on to the cell's progeny. Generation by generation, the number of these wayward cells increases, and each one replicates without pause. Without GO to extend the duration of the cycle, mitosis occurs each ten hours. Within four days the errant cell has laid the basis for a thousand cells, and in eight days the tally is a million. If the cancerous cells can spread from their initial site, they soon tap the resources of the entire body in their unchecked lust for growth and reproduction.

Errors in the genetic code can arise from a variety of causes, including a simple mistake in copying the DNA during its replication. Hence cell types that divide frequently, such as skin and stomach, have high incidences of carcinoma. Cancer among long-lived muscle and nerve tissues is far more rare. Some diets correlate positively with cancer. Fat intake, for example, correlates with breast cancer. It tends to occur once for every ten women who consume on average 150 grams of fat per day, but has almost no occurrence in those rare populations consuming less than 40 grams of fat per day. Diets rich in smoked and pickled foods correlate positively with stomach cancers.

As we progress in our journey through the complexity of a cell's reproductive cycle, reflect upon the fact that within each healthy adult body, cell division is happening at a rate of four or five million times per second. Don't lose sight of the fact that you represent an awesome wonder.

The most crucial stage in cell development, the actual replication of the genetic material, occurs long before any externally visible change in the cell wall is noted. This came as a surprise to biochemists, since logic would have it that the entire mitotic process would be one integrated affair. Not so. Though the duration of total cell cycles varies widely among tissue types, the time required for the actual mitosis, the division of the parent cell into two daughter cells, is similar for all—about one hour.

If we take sixteen hours as a typical cycle for the entire metabolic procedure in cell growth and function, DNA replication starts a full ten hours prior to the mitotic division of the cell. The process is quite similar to that which we encountered in RNA production.

The primary factor facilitating the accurate replication of the long molecules of DNA is that each of the four nucleotide bases will bind properly only with its complementary base: A with T, G with C. The chemistry behind this exclusive pairing serves as a spell checker: T and A join together with two hydrogen bonds; G and C bind with three hydrogen bonds. If by error a T binds to a G, the G is left with one hydrogen unpaired, a signal that an error has been made.

Proteins, named DNA polymerase (I'll call it DNA-P), enter the DNA double helix, unravel a portion of the helix, separate the strands, and then move nucleotide by nucleotide along each of the two now-separated chromatin (DNA) fibers. On each fiber, the DNA-P draws from the surrounding slurry of the cell's nucleus the nucleotide base complementary to the nucleotide on the parent chromatin strand and then inserts it onto the newly molded chromosome skeleton, the backbone of the DNA strand, the string of alternating sugars and phosphate molecules that is also being simultaneously formed. Since each of the two newly formed DNA strands is a copy of the older strand of DNA that previously occupied that same position, as the DNA-P moves along the parent DNA strand, the new daughter strand mates and helically winds with the original parent. It sounds almost oedipal.

Unlike RNA transcription, where only one strand of the

DNA is copied, in cell division both strands of the double helix are being copied simultaneously. Therefore, upon completion, the cell has two complete sets of chromosomes, or ninety-two in all.

The cleverness of the system is subtle. Not just how it learned to get a protein to open the helix, or how DNA-P is made and then knows to come on the scene, or how it finds and joins the correct base. Those acts in themselves are near-wizardry and plead for explanation. The cleverness here is that each new strand winds helically with the parent from which it was, as a complement, copied. Brilliant! Now quality control proteins can check the new work directly against the original template, the parent strand. Had the two new strands wound about each other, new to new, and the two parent strands re-formed their original helix, the quality control would have been a far more difficult and far less efficient task. With the new DNA bound to the old, if the quality control protein finds for example a base A in the new strand at a site where in the parent there is a G base, it "knows" an error has been made, and it knows the error lies in the new strand. So it clips out the base A, draws a base C from the slurry, and splices it into the strand. The cleverness of these proteins sounds almost human, but they are only molecules that together can make a human body.

In DNA replication, the DNA-P reads about fifty nu-cleotides per second, the same rate as in mRNA production. Even at that amazing clip, considering that there are seven billion nucleotides that must be copied, we're talking a very long time to complete the project. Seven billion copies divided by fifty copies per second equals 140 million seconds. With about 30 million seconds per year, we need over four years to copy one cell! Nature gets the job done in about ten hours by having a few thousand DNA-P enzymes working simultaneously, each copying a portion of the parent, and then splicing the parts to make the whole chromosome. In human terms, the feat of completing the copying job in ten hours is as if a team of readers would plow through ten books, each four hundred pages long, every minute, nonstop for the ten-hour period and, while reading, would simultaneously organize and coordinate the information into a single coherent text.

We've made the double load of DNA bursting with potential, all sequestered in a single cell's nucleus. The cell waited patiently ten hours or so as the bevy of DNA-P feverishly turned out the second copy. Notice that though there are thousands of DNA-P enzymes working in unison, we don't get, let's say, forty-eight pairs of chromosomes or forty-two pairs. The tally always comes out just right, forty-six pairs. They seem to know not to re-recopy that which has been already copied.

Mitosis can now begin. And once it starts the process is nonstop activity until the end. In the hour that it takes, it is fantasy brought to life, pure and simple. Disney could not have done it better. Forget religion, forget theology, universal consciousness, all the esoteric stuff. Just immerse your mind in the awesome biological reality of you.

Outside the cell's nucleus (which now houses a double set of the chromosomes), two organelles referred to as centrioles migrate to opposite sides of the cell. The membrane surrounding the nucleus starts to disintegrate and at the same time the centrioles begin organizing microtubules, each some 25 billionths of a meter in diameter, into spindle fibers that extend between what is shaping up to be two opposing poles of the cell. Keep in mind, something has to cue this synchronous dance of the molecules. Somewhere inside the cell there's a molecular mind in tune with a molecular clock.

While this is happening, what appeared at first to be a spaghettilike jumble of the ninety-two chromatin strands condenses into chromosomes that upon close examination are actually twenty-three pairs of pairs, that is forty-six pairs or ninety-two chromosomes in all. Each pair is joined at a single point. Some pairs appear as Xs and some as Vs, depending upon the location of the constricting bond. Each pair consists of two identical copies of a given DNA/chromosome. Now that might represent a problem. Cells don't want double pairs. They want single pairs. And that is what mitosis is all about. (See Figure 18.)

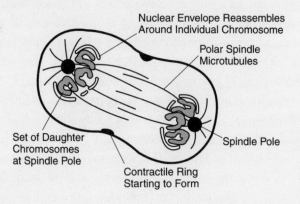

Figure 18

*Telophase in Cell Division: The Fifth Stage of Mitosis* (Figure after B. Alberts et al., *Essential Cell Biology,* Garland Publishing, New York, 1998)

Some thirty-plus minutes have passed since the start of mitosis.

The spindle fibers, extending from the poles of the cell in a manner not totally understood, now draw the chromosomes to the central region of the cell, setting them in a line perpendicular to the fibers that pass between the poles—something like a map with a central equator made of the chromosomes, laced with fiber lines of longitude running to the north and south poles. Spindle fibers have bound themselves to each chromosome at the point of constriction. And then, wonder of wonders, miracle of miracles, a motor protein associated with each chromosome grasps with its little molecular hands the associated spindle fiber, contracts pulling its particular chromosome pole-

fully—never partway. Yet the wisdom-packed design of muscle fibers clustered into woven ropelike groupings enables us to crook a finger partway, to open our mouth a bit to whistle and a lot to bite. Does design of this level imply a designer? Some very smart persons say yes; some equally smart persons say no. After decades of research in the physics and biology of the world, I opt for yes as the answer to that question.

All three classes of muscle, indeed all muscles in all eukaryotic life, function in similar fashion. The main players are cables of actin and filaments of myosin arranged in overlapping parallel fashion that pull against one another—something akin to the myosin shimmying up a rope of actin. Myosin is equipped with hands with a powerful grip that swing forward and grab the actin. By bending its "wrist," the myosin pulls itself forward along the actin. That's it in brief. The details hold the wonder, and to me at least, another indication of "the Force" active in the system.

The molecular chain known as actin forms in pairs that twist into a double helix (as does DNA). Each twist of the helix is a mere 70 billionths of a meter in length and approximately 10 in diameter. Binding sites for the myosin are closely spaced along the actin molecule, but due to the helical twist of the two woven molecules, some of the potential sites are masked. The result is an exposed actin site that myosin can grab only ("only") every 3 billionths of a meter, more than enough for the needs of a body.

Myosin wants to grab. So the impulse must be regulated, controlled by our free will and choices that in turn control the neurons that signal for muscle contraction. The regulation is accomplished by a pair of proteins, one of which blocks myosin access to the actin-binding sites. The second binds to the first and has a site for calcium ions. The action potential of a nerve signaling a muscle to move into action reaches its axon terminal and causes the Golgi apparatus to release its neurotransmitters into the nerve/dendrite synapse. These, in milliseconds, diffuse across the synapse and bind to receptor molecules on the mus-

ward, releases, reextends its hand, grasps and contracts again in a manner similar to the functioning of muscle proteins, each stage pulling its burden a bit closer to the pole. Tugging, straining, reaching out again, and pulling some more until finally the single link at the point of constriction breaks and the sister chromosome pairs move apart, each toward its respective pole. Five to ten minutes pass as the motor proteins reel in their cargo through the ten or so microns of cytoplasm to reach the pole. As always, energy-rich ATP is there to power each molecular effort.

About forty minutes have passed since the start of mitosis.

With the chromosomes separated into two identical sets of twenty-three pairs each, one set at each of the poles, the final phase begins. New nuclear membranes form, individually packaging the two sets of genetic material safely away from the ensuing activity. The helical chromosomes can now unwind, exposing their genes for reading. At the equatorial line of the cell, a ring of muscle protein forms. This, as with all muscles, consists of fibers of actin and the motor protein, myosin, with its hands that grasp, contract, release, extend, and grasp again. As the myosin pulls the actin ring ever tighter around the cell's middle, the cell membrane begins to contract at what was formerly the equator. The cramping continues and ultimately the cell divides in two, with the original cytoplasm and the organelles therein being shared between the two halves.

One hour has passed since the start of mitosis.

Over the next five or so hours, the identical genes in each of the two new cells will produce more organelles (ribosomes, mitochondria, ER, etc.), making each of the cells full fledged and full sized. The cells now have the choice of continuing toward another round of division or entering the GO phase of nonproliferative metabolic activity.

But there is a bit more to the story. At two stages, one before and one after the actual mitotic phase, the cell checks itself for overall size and DNA functioning. Then, during the one-hour mitotic phase, the spindle fiber network is monitored, determining if the fibers have joined properly to the chromatids. Any malfunctions are repaired. If not repairable, the cell self-programs for cell death and disintegration. Its usable parts are recycled in other cells.

That's how we work, but how did we evolve? Not through random reactions over billions of years in energy-rich ponds, thermal or otherwise. The fossil record has laid waste to that false assumption, so popular prior to the discovery of the earliest fossils. As so many secular scientists active in the search for the origin of life have expounded: The data indicating the rapid emergence of life on our initially lifeless planet teach that life is inevitable in our universe. True enough. Life seems inevitable. The question is, what made it inevitable?

If life is inevitable, I have to ask why? But try as I might, I cannot envision this complexity evolving without powerfully guiding restraints and catalysts propelling it toward completion from its beginning. Call those restraints and catalysts nature if you like. But it is an amazing nature operating in a manner never dreamed of as being naturally possible just a few short decades ago, a time when life seemed to function with an almost childlike simplicity. Then came the discoveries of molecular biology, the cellular mechanics of life, that allowed us to peer beneath the veneer of simplicity and discover the depth of information sequestered therein.

# MUSCLES: MOLECULE

# IN MOTIO

*The text covers the complex orchestration of cellular processe sends a nerve's signal careening down its axon. This wonder chemistry is matched by the intricate functioning of the muscl those signals set into action. Muscles provide life with the option tion. They account for almost half of an adult's body weight close to a quarter of the oxygen that we breathe, even when u rest. During exercise, oxygen consumption in some muscles crease greater than tenfold. Studying muscle dynamics c breathtaking as using them.*

Humans have three basic categories of muscle: sm involuntary, such as those that line the walls of blood v intestines (no conscious thought is associated with t ing); cardiac, which are also involuntary for all of t three billion beats of a lifetime, but are structured in similar to the third muscle type; and skeletal, which p voluntary motion we usually associate with muscle bu

It seems paradoxical that although we flex our li out, up and down, muscles produce force only by by pulling in. They pull, but never push. By the bri of joints and levers in our limbs we can do "push"-u haps equally surprising, when a muscle fiber gets from a motor neuron to contract, the fiber can

cle fiber's outer membrane. The signaling nerve's neurotransmitters, upon binding to a muscle fiber, stimulate calcium ion
channels in the muscle cell membrane to open. Calcium ions
flood into the cell and bind to the receptor sites on the secondary blocking protein, causing it to flex. In the flex, it pulls back
the primary blocking protein and, in doing so, exposes the actin
sites. Myosin makes its grab.

Is this a Rube Goldberg machine? Couldn't nature just as
simply have had a nerve send a signal to the myosin telling it
when not to grab and when to grab the actin and contract the
muscle fiber? Perhaps in this complexity we are seeing the
structures that led to the system we see today, something like
the scaffolding required to support an arch while the arch is being constructed. Could the complexity be evidence of evolution,
rather than design?

One could imagine the "evolution" of an arch. First there'd
be the need to span a space between two pillars or two ridges
over a narrow channel. A straight, horizontal beam between the
two, each end resting on a side, would do as a starter. Then, to
gain more height, two beams set in a V-shape, but pointing upward, might accomplish the goal. In time one could envision
adding further pieces to smooth over the sharp "V" vertex and
add strength. This could eventually allow the wood pieces to be
replaced by an arch. In this scenario, we avoided the severe snag
of first having to "evolve" the complex biomolecular mechanisms of photosynthesis that produce the cellulose and the simpler carbohydrates and then from them "evolve" the tree from
which the necessary beam was taken. We took the tree as a
given. In its development or evolution, the tree trunk grew in
order to span space, to allow its leaves to rise above the forest
canopy. We merely adapted that vertical spanning feature for
our horizontal needs.

With an arch, once completed, the scaffolding is removed;
no trace of it remains. Nature tends to be conservative, keeping

the "scaffolding" in one form or another, such as the yolk sac or the one-chambered heart or the primitive tail we all had while in the womb. So if muscles did evolve, their mechanism should show signs of the scaffolding—the equivalent of old pieces of wood. If the scaffolding is there, it certainly is not obvious.

The myosin has a shape reminiscent of the letter "r," but with the rodlike shaft much elongated. The rods bind together to form myosin filaments, each filament having several hundred protruding hands. The hand is composed of two globular molecules that between them grasp the actin site. In the power stroke, the hand binds to an adjacent actin site, and then bends down, pulling the myosin filament along the actin cable, causing the actin and myosin to slide by each other. The visible result of this phenomenon is the bulge of the bi- and triceps when we flex our muscles. As the fibers slide along and the muscle contracts, by nature, the muscle must bulge out to contain the contracted fibers in the shortened muscle length. Each stroke moves the myosin approximately ten billionths of a meter. Each hand can cycle some five times per second.

I say the hand of the myosin bends down, but that hand is not like ours, having a supple wrist and distinct skeletal parts. Here, with myosin, the hinge is one biomolecular unit, a molecule that is flexible. It actually bends upon command, the result of just-right electrical charges pulling among the individual atoms. When working in unison with thousands of other myosin heads, the pull can have a pair of biceps raise a three-hundred-pound weight, and yet not have the force at any one of the heads exceed the strength of the actin molecular bonds, which would tear the cable.

Each muscle cell consists of thousands of fibers, each of which contracts upon stimulation by a nerve. Transverse tubules carry the signal to the individual fibers. In synchronous action, some myosin hands grasp the actin and pull, while others are releasing their grip and reaching further along the actin cable. On a nano-scale, it's one hand over the other, pulling along a rope. The combined effect is a continuous and smooth flow of muscular power. Once the nerve signal ceases, the calcium is pumped out; the muscle relaxes and extends, ready for its next task.

Calcium ions flood in; ATP, manufactured from glucose, is on site to power the myosin; neurotransmitters bridge the synaptic gap. All these materials and a myriad of others I have not mentioned must be on site and on time. And in a healthy body they are. Not by chance, surely.

To move the needed materials from the place of their manufacture to the location at which they are to be utilized, our bodies are provided with a type of muscle quite different from those just described, a muscle not much larger than a molecule. It is the motor protein.

Most materials produced by a cell are made in the cell body. Within these compressed dimensions transport by diffusion is adequately fast. In 50 milliseconds, diffusion can spread a product about 10 microns, a typical path within the cell. But in a nerve cell, with its elongated axon, the point of need may be a meter distant from the point of production. If a particular protein required at the terminal were to travel from cell body to terminal by diffusion, a year would not suffice to have it complete the journey. Clearly some means of active transport is essential. Nature has provided both the vehicle and the pathways, the vehicle in the form of motor proteins and the pathways as microtubules that run the length of the cells.

Motor proteins, approximately two billionths of a meter across (almost ten times smaller than the actin/myosin muscle units), move along the microtubule tracks in pairs. The lead motor protein locks onto the microtubule thrusting forward, and in doing so pulls free the trailing motor protein, which then seems to leapfrog to the lead position. This action is repeated as the pair travels along the microtubule. On a molecular scale, these two proteins appear to be walking on a rail. Sites on their backs provide a bond for the produce that is to be transported. It is a micro-world of traffic. As with muscle motion, ATP is the power source that allows the motor proteins to leap one past the other.

At a transport rate of 20 to 40 centimeters per day, two days and more is required to move Golgi apparatus housing neurotransmitters from cell body to axon terminal. It seems the cell must plan for events several days in advance. The neurotransmitter packaged within the Golgi apparatus somehow finds and mounts a motor protein headed not by chance down the very track that will bring it to the axon terminal two days later.

Our cells are a nonstop marvel. Transport in all directions satisfies the needs for the two thousand proteins manufactured every second of every day, seven days a week. No nighttime snooze and no Sabbath rest here.

Muscle distribution within our bodies is filled with cleverness. Hold your hand up and bend your fingers. Notice that the muscles that allow you to cup your hand by bending your fingers down are not located in your fingers. Make a fist and feel the inner, smooth side of your arm just below the elbow. Feel those muscles flex. They are connected via tendons to your fingers and give the pull that shapes your fist. By having the muscles located on the arm rather than at the fingers, the fingers remain slim enough to do fine work such as holding a stick or typing a page. But when you pull your fingers down, there is another joint in the line of action, the wrist. Why isn't that pulled down along with the fingers? Now feel the outer, hairy side of your arm just below the elbow. Feel the other muscles at work there. They get the command to apply just the correct force to hold the wrist steady when your brain says bend fingers only, and they allow the wrist to bend when the cranial message is: wrist also in action. But we never think of it because it's all controlled at the less-than-conscious level.

There's a brain of which we are conscious and one we are not. Just as there is a world we perceive and one we do not. Both are real. And with careful thinking we can realize the presence of both.

# ACKNOWLEDGMENTS

Prior to the start of my work on *The Hidden Face of God*, I had the vague feeling that some commonality might pervade all existence. With the research for, and the writing of, this book the extent and manifestation of that universal oneness has become for me a part of my daily experience. Many persons contributed to this realization, some through sharing a brief moment of inspiration, some in many hours of conversation, some by their writings. I am greatly indebted to all. Among these are ordained theologians, persons with advanced academic degrees, winners of prizes in science and philosophy. I list here only the names, without titles.

First my wife, Barbara Sofer Schroeder, and our children, Hanna, Yael, Hadas, Joshua, and Avraham, provided plenty of food for thought in our many dinner table discussions.

As with *The Science of God*, Debra Harris and Beth Elon brilliantly directed the manuscript to Bruce Nichols, editor at The Free Press. Bruce's discerning editing helped rid the wheat of the considerable chaff I had initially included. Edith Lewis helped put the finishing touches on the style. As in the past, I am indebted to Helen Rees and then Marc Jaffe and Michelle Rapkin who, prior to the publishing of *Genesis and the Big Bang*, realized the potential for interest in a rigorous approach to the science/Bible debate.

Sincere thanks to Noah Weinberg, Yaakov Weinberg of blessed memory, Dennis Turner, Avraham Rosenthal, Shmuel Silinsky, Motty Berger, Sam Veffer, Nadine Shenkar, Ari Kahn, Zola Levitt, Sandra Levitt, Paul Joshua, Benji and Leah

Schreiber, Cedric Levy, Peggy Ketz, Helen Stone L'or, Mordechai Geduld, Aryeh Gallin, Susan Roth, Ilana Attia, Michael Behe, Lee Spetner, Michael Denton, Moshe Schatz, Michael and Karen Rosenberg, Sharon Goldstein, Nancy Sylvor, Naomi Geffen, Barry Bank, Marty Poenie, Jonathan and Elaine Sacks, David Lapin, Beril Wein, D. Homer Buck, Michael Corey, Phil Rosenbaum, Yigal Bloch, Barbara S. Goldstein.

The books and articles that provided information and insights were *Human Physiology* by D. Moffett, S. Moffett, and C. Schauf (Mosby Publishing, St. Louis, Mo., 1992); *Essential Cell Biology* by B. Alberts et al. (Garland Publishing, New York, 1998); *Essentials of Genetics* by W. Klug and M. Cummings (Prentice Hall, Upper Saddle River, N.J., 1999); *Mapping the Mind* by R. Carter (Weidenfeld & Nicolson, London, 1998); *Tour of a Living Cell* by C. de Duve (Scientific American Books, New York, 1984); "The Origin of the Universe," eds. H. Branover and I. Attia, *B'or Ha'Torah*, Number 11, 1999; *The Fifth Miracle* by Paul Davies (Simon & Schuster, New York, 1999); "Facing Up to the Problem of Consciousness" by D. J. Chalmers, *Journal of Consciousness Studies*, 1995; *The Natural History of Creation* by M. Corey (University Press of America, New York, 1995); *Incredible Voyage* (Book Division of National Geographic, Washington, D.C., 1998); *God and the Big Bang* by D. Matt (Jewish Lights Publishing Co., Woodstock, Vt., 1996); *Biochemistry* by L. Stryer (W. H. Freeman, New York, 1995).

# INDEX

That's just the way of it." Fedora ended the discussion.

"Well, I'm not like one of the colored women. I just can't go dropping me a baby in the morning and go back to the field right after dinner." Tillie shifted in her chair while Fedora gave her a "watch your mouth" look.

Eula took the cooling pot of peaches to the table to Jenny who began ladling them into the washed-out Mason jars.

"That puts me in mind of that colored wench on the Bredge place." Belle's choice of words always did border on the bawdy, and Eula wondered again why her younger brother had chosen a woman just a step away from white trash to marry. "I saw the wench at the store in Clarksville last planting time. She had these two springy-haired yella' children with her and her belly swoll out to here, again." To watch Belle's hand draw pictures of the woman in the air, the observer would have believed the poor creature's stomach was bigger than a bushel basket. "Now, she is as dark as dirt and her man is as black as midnight. Where did those yella'-skinned pickaninnies come from?" Belle shook her head for emphasis.

"Well, if I was forced to say, I'd specify that it was between that white sharecropper Jim Bredge hired on three years back and that squatter family down by the railroad station." The deliberate sound of Cora Lee clucking her tongue did not escape Eula's ears.

"Nobody's forcing you to say nothin', Cora Lee." Belle poured the sealing wax over a jar of just-filled plums. "Besides, I don't think it's that white trash squatter family. That man's got about eight of his own young 'uns crammed into that one-room shed already."

"That man's almost sixty." Tillie sat with her hand rubbing her belly.

"Age just makes them all the more randy." Jenny shot a look at Tillie that brought titters from Belle.

"By my reckoning, I'd speculate on that sharecropper." Belle swiped the outside of a Mason jar with a wet cloth. "Don't he crop the forty acres right next to that nigger woman with her own high-yella kids?"

Eula flinched when she heard Belle's "nigger" epithet flung into the kitchen. Mother Thornton had taught her girls that a lady used that word only when strongly provoked. It was a term reserved for men. Only low-class white women uttered it away from the sanctity of their own homes.

Coming out of the mouth of an in-law, the sound of it felt like fingers rubbing the wrong way against a blackboard.

"Why do you all think it has to be shiftless white trash? Hettie, on Papa's side-forty, has two yellow-skinned girls already, and her husband is as black as that stove Aunt Eula's standing at." Tillie spoke with a "case closed" attitude so like Fedora's. "I believe those women sell themselves when they get over to Clarksville."

"What do you know about a woman selling herself, missy?" Cora Lee asked. "If the truth be told, I'd say Jim Bredge hisself wasn't out of the woods."

A sudden quiet, deep as a pond in January, wrapped itself over the kitchen. Only the sound of the bubbling fruit in the kettles broke the silence. Eula held her spoon suspended in midair, afraid to lay it against anything, lest the sound rock the room like a rifle shot. She was quite aware that Fedora had not uttered a word in over five minutes. With her back still to the women, Eula heard someone clear her throat.

"Lord, here we are going on and on about babies. Eula please pay us no mind. We get to talkin' silly sometimes." The sound of Cora's nervous giggling brought Eula up from the kettle.

Before she turned toward the group, Eula set the muscles of her face into a mask and sealed them in the steam of the kettles. Every woman in that room, except Tillie, knew who fathered Hettie's three children, and it wasn't her long-gone stove-black husband, or the white trash squatter, or Jim Bredge. But no self-respecting white woman would ever dare utter the name of the father in the presence of his wife, not even when she was sitting no more than two feet away.

"I'm sorry, Aunt Eula," Tillie waved a hand around to include the other offending women in the room. "I forgot that you and Uncle Alex did have a baby once."

With her masklike face in place, Eula managed to give her niece a slight nod of her head. Fedora, concentrating on apple peels more than Eula had ever seen, lifted her head slightly as she sidled her eyes in Tillie's direction. Eula watched the younger woman's own eyes grow wider as she grabbed at the waist of her dress.

"If there is a baby in here, Aunt Eula, do you think I might lose it like you did?" Fedora's arm moved with such speed that Eula swore she only saw a white blur as her sister-in-law whacked Tillie near the elbow with an unpeeled apple.

"Will you stop your nonsense? You ain't about to lose Wiley George's baby." Fedora glowered at her daughter.

"I don't really know the story, Aunt Eula." Stubborn Tillie rubbed at her arm but kept her eyes on Eula. "I only know that you and Uncle Alex had a baby that died. Was it the whooping cough?"

The scraping sound of Fedora's chair as the woman jumped to her feet brought Eula's eyes hard around to her sister-in-law. Faster than a blink of an eye, Fedora reached across the table and snatched the spoon from the cooling pot of plums. Wielding the implement like a sword, she struck her daughter hard on the shoulder.

"I taught you better." Fedora's face had gone red. "You leave your aunt be. Not everybody wants to talk about what pains them like you do girl." Fedora held the spoon ready for a follow-up blow as Tillie shifted her hand to her shoulder.

"I ain't pryin', Ma. I just want to know what happened to Aunt Eula's baby."

The other women worked at a pace that Eula hadn't seen all day, but she knew every ear was turned up as keen as a hunting hound to hear her answer.

"She died abornin'." Eula's lips felt like they were moving through hardened clay. She tried to clear her throat. "Do we have another case of Mason jars?" Sweeping past Belle and Cora Lee, Eula stepped as slowly as she thought seemly out onto the back porch and the pump.

She bit into her lip. As she primed the pump handle, she wanted to shout out. Damn Jenny and Belle and Cora Lee. Why did their silly conversation have to drag her into it? Babies, yellow babies, coming babies, dead babies. Was it better to hurt her or Fedora? As the water started its trickle out of the pump, she cupped her shaking hand to gather a fistful and splash it over her face.

# CHAPTER SIX

"Henry, run these sticks over to the barn." Without looking at her son, Annalaura took two crooked branches off the pile of stripped tree limbs accumulating from Lottie's run between Doug and the mound at Annalaura's feet.

The little girl, drenched in sweat, dropped eight more sticks atop the slow-growing pile before she collapsed between two rows of tobacco.

"Momma, I'm hungry." Little Henry plopped on the dirt across from his sister and let the sticks in his hand fall in different directions.

Annalaura reached down for the water jug on the ground, unscrewed its top, and offered it to Henry, while she glowered at Lottie.

"Girl, get up from there. Don't you know that it's September and we've got to get these sticks sharpened for spearing the tobacco? Mr. McNaughton wants his tobacco now."

"I don't care 'bout no ole tobacco. Momma, I'm hot." Lottie crawled over to Henry, who held the water bottle to his lips.

Before her daughter could lay her hands on it, Annalaura slapped the child's wrist.

"You'd better care 'bout some old tobacco, and you'd better be glad you're feelin' some heat on your back right now. Store it up. If we don't get this harvest in, come winter, you'll feel nothin' but cold snow on your behind." Annalaura looked up at the sky. She was grateful that early September had brought cooler weather, although, with the work they all were doing, she knew the fires of Hades couldn't feel much hotter. "Henry," she commanded, "give your sister some water, and get those sticks over to the barn so Cleveland can sharpen them for the staking."

Annalaura took a menacing step toward her son. The child scrambled to his feet, grabbed at the sticks, and ran them to the barn. Lottie, her back to her mother, climbed to her feet and started a slow walk back toward Doug. Before the child disappeared between the rows of still-growing tobacco, she called out to Annalaura in a singsong.

"I'm still hot and I'm still hungry. Hot and hungry." The child took off running. Annalaura had no intention of chasing her daughter. A flash of anger fought with a shade of regret for just an instant as mother looked after her fleeing child.

John Welles had left them in this predicament. She was working her children worse than what Aunt Becky told her the overseers had done in the days of slavery. With her sleeves rolled up past her elbows, Annalaura brushed a bare arm across her eyes. Only some of the dampness clinging there was from her sweat. She swallowed hard. Blubbering was not going to help any of them. John had left her without a word, either angry or peaceful, and that was that. There was no one else in this world to save her children from certain disaster other than herself. She clamped down on her lower lip. Maybe that pain would stop her from feeling sorry for herself. After all, the Lord had answered part of her prayers.

She looked up at a sky streaked in white clouds. None of them bore any resemblance to rain. Two weeks earlier, she had prayed mightily for rain that never came. Now she knew she had prayed for the wrong thing. It might have been all the hoeing and weed pulling that did it rather than her prayers, but the tobacco had miraculously stretched itself above the ground to a just passable height for spearing. This would be no bumper crop, but if she, Cleveland, Douglas, Lottie, and even little Henry worked fourteen-hour days, maybe they could bring in half of McNaughton's forty acres. If she had the time, she would drop to her knees and pray like she had Sunday last. She wouldn't pray for rain this time. Rain would be a disaster. All she wanted was another miracle.

Henry came walking in slow motion back to the pile of sticks that needed to be sharpened. Out of the corner of her eye, Annalaura saw the little boy's toes poking out way beyond the sole of his sliced shoe.

She raised up from stacking the sticks to look at her youngest. He was only three years old, but already, he walked like an old man. From this distance, she swore she could see frown lines etching themselves into her baby's forehead. As he approached his mother, the tot began to sway. Annalaura dropped the sticks in her hands, swept up her skirt, and caught the child before he toppled into a tobacco plant.

"Momma, can I have some soup?" Henry whispered.

Annalaura tried to gauge the heat coming from the sun. She knew it couldn't have been over eighty degrees—a fairly cool September day. It wasn't the heat that made her baby swoon. Picking up Henry, she cradled him in her arms and carried him to the barn. There, in the stifling air of a building smelling of cows and pigs, Cleveland sat surrounded by two stacks of sticks.

"How many more you reckon we need of these, Ma?" Cleveland pointed to the shorter pile as Annalaura laid Henry on a slim bundle of hay.

At least the cows had food for their supper.

"That pile there," Annalaura pointed to the already sharpened sticks, "is enough to spear about five acres. Cleveland, we've got to bring in at least twenty so we can stay on this place this winter."

Her boy looked at her with the certainty of impossibility on his face.

"I'll get Doug to help with the sticks while you get the rafters ready to string up the tobacco once it's speared." She lifted her head to the top of the barn.

McNaughton had built their lodgings to take up less than a third of the rafter space. The rest was to be used for the hanging and curing of the tobacco.

"Doug don't like to work in the barn. That dryin' tobacco gives him the wheezin' attack," Cleveland reminded her, though Annalaura needed no help remembering her second son gasping for breath when he had a bout of what John called the "asthma."

If she had any other way, she would spare Doug the frightening experience. Hell, if she had any other way, she would spare all her children this misery. She turned toward Henry who lay still and ashen with his eyes closed. She had to feed her children. Stepping outside of the barn she called to Lottie.

"Come on up for dinner." As she let the words come out of her mouth, she fumbled in her pocket for the four eggs boiled this morning.

She moved to the smoke house and the low-simmering pot on the back burner. Filled to the brim with water, it contained less than a cup of dandelion greens. Annalaura looked over at the spot between barn and smoke house where she had gardened so well last year. Except for the brown tops of two scraggly onions, the garden plot was bare. There had been no time to tend it and, worse, no seed money to start it. She didn't dare ask McNaughton to advance her the seed like most of the other owners would do, because she knew she wouldn't be able to pay the farmer back. As Lottie headed from the field with Doug trailing behind her, Annalaura checked the sun. She had fed all four of her children with the next to last cup of cornmeal way before sunup, and now it was close to two o'clock and their dinner would be a boiled egg and a pot of water seasoned with no more than the look of a green dandelion. She wiped her forearm over her eyes again. She refused to let them see her cry.

It took her children less than five minutes to gobble down their egg and skimpy cup of dandelion water. When the little ones turned their faces to her, she saw the futility of asking for more in their eyes. She allowed all four to rest for another twenty minutes, though Cleveland had tried to argue that he was ready to go back to the sticks. She insisted that he lie down on his cot. She put Henry on her lap, and with the hem of her skirt, wiped his dusty toes and tried to push the top of his shoes over them.

"Cleveland, watch after the others, I'm taking Henry with me to Aunt Becky's."

"But, Momma, the tobacco…" Cleveland tried to protest before Annalaura silenced him with a hand.

"I'll be back way befo' dark to work on the sticks." She saw the question in Lottie's eyes but silenced her daughter before it could be asked.

❧

If she'd had a buckboard, she could have made the ride to the Thornton back-forty in under fifteen minutes, but with walking and carrying an exhausted Henry, it had taken her close to an hour. He had fallen asleep and his three-year-old weight felt like thirty pounds of unbroken rock. Just as she spotted the cabin in the distance, she shifted the child to her other hip. She hoped the sight of him would keep Aunt Becky from asking too many questions. Annalaura loved her aunt, but the woman had the all-seeing eye of her Cherokee mother. She would know what the trouble was before Annalaura could get out a good word. Annalaura nudged Henry awake as she turned up the path to the old mud-chinked cabin where Aunt Becky had been born a slave on Thornton land sixty years earlier. After the War, she had married a man who called himself Murdock, though Becky never used his name, nor did she ever move away to live with him. Rebecca Murdock had always lived on Thornton land, and she had always been far more than an aunt to Annalaura.

Geneva Thornton Robbins had been just twenty-nine, and Annalaura four, when the galloping consumption separated mother from daughter. Steps from the gray, weathered front door of the old slave cabin with its iron ring for a knocker, Annalaura rummaged in her head for a memory of that day. Her mother's face was less than a blur these twenty-five years later. Not even her smell lingered in the adult Annalaura. What was clear as a pond after a springtime thaw was the remembrance of how the four-year-old had begged to be allowed to live with her pipe-smoking grandmother after her mother's death.

Grandma Charity's lap had always been inviting even when Annalaura's mother was alive. The woman with the strong nose, straight, still-black hair, and cinnamon-colored skin had rocked young Annalaura to sleep many a time. As a youngster, she nestled into her grandmother's ample bosom and let her drowsy mind take in the stories of the old woman's own girlhood.

Charity was just nine when the soldiers marched her entire village out of their North Carolina home. The woman had repeated the story so often that Annalaura could almost see the Cherokee cabins. Grandma had always insisted to her doubting granddaughter that the Indian cabins had been much finer than the ones the white men built for their slaves. The Cherokee had always lived in houses of wood that kept a body warm and snug in the winter with fireplaces that were built right in the center.

Annalaura set the half-asleep Henry on his feet as they paused outside Becky's front door. The unhappy-looking child set up a wail that sounded just like the one Charity used to imitate all the crying that occurred on that long walk out of North Carolina.

"That walk to Oklahoma was from sunup to way past sundown, with the soldiers on their tall horses poking their fire sticks into the backs of anyone who lagged." The old woman had always begun the tale the same way, as her bare and calloused feet set the old rocker in motion.

"That piece of a trail was just about washed away in Cherokee tears." Grandma Charity creaked the chair.

Henry stamped his feet in time to his wailing as he reached his impatient arms up to Annalaura. The cabin door creaked open on its rusty hinges, and Aunt Becky peeped her head out of the gloom. She looked her visitors up and down, and as usual, Annalaura could not read her impassive face.

"You feedin' that chile?" All Cherokee women had the knowledge of the herbs, and most had at least a smattering of the gift of second sight. Aunt Becky always told her that. Rebecca Thornton Murdock was only half-Cherokee and that should have slowed her down, but the speed with which the old woman could read anybody's intentions both annoyed and amazed Annalaura.

"Just had dinner." Even if Becky was close to knowing why she had come to her cabin, Annalaura couldn't just blurt out that her children were so close to starving that it seemed an actual fact.

Becky still held the door open only a crack. She turned that eye on Annalaura again. When the earthen-skinned woman finally did open the door, it was to grin a gap-toothed smile at Henry.

"Come on in here, chile. Auntie's got a treat fo' you." Becky slid a thin, calico-clad arm around the back of Henry's head and guided him into the interior.

Though the sun still showed promise of another good three hours of bright daylight, the kerosene lamp in the center of the old wood-blocked table that Annalaura remembered so well from her own girlhood remained the main source of light within the long room. The open-spaced cabin boasted one window, but Becky kept the lamp-oil-smudged window glass covered with particularly heavy burlap curtains. The heat from the lamp made the room even more suffocating than normal. Annalaura pulled out one of the two table chairs only to discover that a leg had loosened from its seat. Without a word of caution, Becky pulled out the other chair and pointed Henry in its direction. The old Cherokee disappeared into the semi-gloom of a corner where Annalaura knew the food safe, with its supplies, stood. After Rebecca twisted off a lid from a nearly empty jar, she reached for a spoon from the table. Scraping the sides and bottom, Becky came up with a tablespoon of peach preserves and held it out to Henry. The child beamed at the treat.

"Jest had dinner, did he?" Becky spoke without laying her eyes on Annalaura. "What you gonna feed him fo' supper?"

Annalaura knew her aunt was indicting her for not being able to feed her own children. "Times is a little bit hard these days, is all." She bent down to knock the wooden peg back into the chair leg.

"Uh huh." Becky had the weary sound of a woman who knew all about hard times.

Times had gotten even harder after Geneva died. The good-as-orphaned Annalaura had never understood why her mother's older sister had insisted upon keeping her at the cabin instead of letting her live with her grandmother. Her aunt would never say more than it had been a promise made to the dying Geneva. "Raise my girl. Don't let her ever go to Momma." But, when Annalaura got up to some size, she deviled the other colored children in the neighborhood until they whispered in her ear when they were sure her aunt wasn't around. "There's bad blood between yo' Grandma Charity and yo' Aunt Becky." If they knew more, Annalaura couldn't even beat it out of them.

"Come on over here, baby." Becky, her back bent from fifty-five years stooped over the tobacco plants, beckoned Henry who was playing with the spoon in his mouth.

The boy licked at the long-gone taste of peach preserves. The wraith of a woman walked the child back into the gloom of the cabin and pulled out a small square of corn bread from the food safe.

"Take that outside, baby, so you don't get crumbs on Aunt Becky's clean flo'." She left the door open a crack as Henry sat on the stoop stuffing the corn bread into his mouth.

Rebecca walked back over to Annalaura and took over the one sound chair. One bony arm reached out for her pipe. She put it in her mouth unlit.

"When you last heard from that sportin' man husban' of your'n?" Becky tapped at the bowl of her pipe. Her eyes had not yet settled on Annalaura.

"I reckon he's too busy to send word," Annalaura lied.

"Uh huh." Becky took a draw on the dry pipe.

Annalaura stiffened. Her aunt always had a way of dragging out the torment when she wanted to lay into Annalaura. "I 'spect he'll be back right before the end of the harvest."

"Thorntons brought in the last of their tobacco day befo' yestiddy." Becky finally let her eyes light on Annalaura's face. "Yours ain't barely ready to spear yet. You need a man to help."

Annalaura steeled herself against those Cherokee eyes that always made her squirm. "No good not keepin' yo' man satisfied in bed," Becky pronounced.

Annalaura felt the blood rush to her face.

"You need the conjure woman."

"I don't need no conjure woman." Annalaura wanted to take back the rise in her voice, but not the words. "I need a speck of food to feed my children until the tobacco is sold."

"A woman who can keep her man happy don't need to be searchin' fo' scraps of food." Becky took another draw on the cold pipe and took her time pointing it at Annalaura. "I'll say this fo' you. Keepin' a sportin' man under one set of covers ain't easy for no woman, 'specially one as strong-headed as you."

"I don't care if John Welles is satisfied or not. I need to feed my children." She caught a glimpse of Henry coming back through the partially opened door.

With an upward tilt of her head, Becky shooed Henry back outside.

"Now, you talkin' nothin' but foolishness girl. A man ain't good fo' much, but a woman needs one around to lessen this world's misery." Rebecca frowned at her. "What you need is some extra special strong herbs from the conjure woman."

"Aunt Becky, if I could kindly borrow some meal, a dab of flour, and maybe some jars of greens, I will pay you back double when my harvest money comes in. I don't need no conjure woman to make me up no love potions. A man will only double my misery." Annalaura looked back at her aunt, who held her cold pipe at arm's length like a rifle aimed at her niece's face.

She slowly let it circle in the air.

"I could never settle my mind on why a fast-thinkin' man like John Welles wanted to marry up with a fresh gal like you. You got a comely shape, right 'nough. God's truth, men can act the fool over a plumped-out behind and a big pushed-up bosom. But you got yo' shortcomin's. Ain't many a man wantin' a woman with a troublesome quick tongue." Becky clamped the pipe stem between her teeth and kept her lips drawn back. "Gal, you better bless yo'self twice that John Welles ain't minded layin' in a bed with a woman with a keen reckonin' head."

"I wouldn't want John Welles in my bed even if he was still around here," Annalaura flared. "I wanted to leave that man right after I had me Cleveland." She had heard John speak his, "I'm doin' this fo' our family, darlin'," at least two dozen times too many. Even if her quick-thinking husband won more than he lost at gambling, she still hated the thought that he could lose her babies' hard-come-by school money.

"What foolishness is you talkin' now, gal?" Surprise slipped out of Becky's mouth.

"John Welles hid my letter from Grandma Charity. The one she wrote tellin' me where she was and invitin' me to come live with her in Oklahoma."

Annalaura nearly jumped from her rickety chair as Becky slammed her hand on the wooden block of the table with such force that Annalaura wondered if the old woman had broken an arm.

"It ain't fittin' for a married woman to speak such talk. Letter or no letter, you got no business running off to Oklahoma." Becky's voice came out of her mouth like a snake spitting venom. "John Welles may have his gamblin' ways, but you married him. It's up to you to lay in his bed, lumpy as it may be, let him do what he's got to do, and act as happy as if you'd gotten your gold heaven crown right now." The sturdy table shook a second time as Becky pounded it.

"Hell, Aunt Becky, I can't lay in a bed with a man that is always laying up in some other woman's bed." Annalaura had barely gotten the words out of her mouth when she felt the stinging slap across her cheek.

Becky had knocked over the good chair as she reached across the table to deliver the blow.

"I learned you better than to cuss, gal. A good woman can always find 'nother word to say what she means. 'Sides, a man don't want to hear a decent woman sayin' a cuss word." Rebecca, who had been a woman of a fair size in her young days, had shrunk down to just a little bit above Annalaura's nose. But right now she towered over her like Goliath gloating over David. "Get that runnin' off foolishness out of yo' head. When that man does come back, you gonna keep him so tuckered out, he ain't never gonna stray no further than yo' front do' without him tellin' you first."

Becky moved to the corner safe and returned with a half jar of corn-meal and a tin of flour. She held out a bloodied parcel wrapped in tobacco leaves. Annalaura held her hand to her cheek.

"Take this here rabbit. It's just the front quarters. Ben Roy brought it to me this morning."

Annalaura knew that Becky lived alone in the cabin and had long ago passed her work usefulness on the farm. Still the Thorntons kept her fed and clothed with a few supplies every month.

"I'm right sorry, Aunt Becky." She held out her hand for the bloody package. "My babies haven't had meat since Independence Day. John only left us one slab of bacon and we just finished that up." She bit her tongue when the hard look threatened to come back into Becky's eyes for criticizing John.

"Here's some meal and flour. I'll fetch you some salt and baking soda, and I've got the last of the pole beans in the garden. You can have some of them fo' your children." Her voice had softened. "When he gets back here, yo' man don't want to hear that his woman couldn't keep his chil-dren fed. Now get on back home."

The walk back to the barn took even longer because Annalaura couldn't carry both the supplies and Henry. The boy took ample oppor-tunity to protest his fatigue by sitting down at every other rut in the much-rutted road. It was nearing dark when she guided her youngest up the ladder. Even before Henry reached the top rung, she knew something was wrong.

"Momma, Cleveland hurt his leg." Lottie shoved Henry aside before Annalaura's head had even cleared the opening.

"What do you mean, hurt his leg?" Before she could gaze in the direction of Cleveland's cot, Doug's desperate wheeze for breath caught her attention.

She dropped to the pallet on the floor where the young boy hung his face over a bowl of lukewarm water and frantically waved both hands to bring up the nonexistent steam.

"I heated the water myself, Momma." Lottie's note of pride was lost on Annalaura who had a sudden vision of her five-year-old aflame around the cantankerous stove McNaughton had rigged up in the smoke house.

"I told you not to go by that stove unless I was near." The sharpness in Annalaura's words was at counterpoint to the panic rapidly taking over her mind. She heard Cleveland groan a greeting to her.

"It's all right, Momma. Doug took sick before I had my fall. I was in the barn when Lottie was boilin' the water." Cleveland's wail sounded his hurt.

Another grunt came from the cot as Annalaura jumped to her feet and ran over to her eldest. The boy was laying in his alcove, the side of his face badly scraped.

"Lottie, get me some water from the basin there," she commanded the girl, as Henry ran to comply.

"Lottie already done give me a rag, Momma. It's my leg." Cleveland nodded his head in the direction of his left leg, and the effort brought forth another grunt.

She watched her rock of a son dissolve into tears.

"My Lord, Cleveland. What done happened?" The left pant leg of her son's one pair of work britches hung in bloody shreds around his knee.

The skin over the shinbone, between knee and calf, had doubled in size, and a deep scratch ran across the area. Annalaura grabbed the wet cloth from Henry and laid it gently across the wound. She knew the bone was broken. She only prayed it was a clean break.

"I was in the rafters, Momma. Gettin' things ready for tomorrow, so we could start to hang the tobacco. One of the beams broke and I fell. I'm sorry, Momma." Cleveland's voice dissolved into a howl that chilled Annalaura's heart.

"Don't you fret none 'bout that tobacco. Aunt Becky's heard word that yo' papa might be comin' back just in time." The lie came quick off her tongue as she took a second wet rag from Lottie and patted Cleveland's face with it.

She wanted to take her firstborn into her arms, hold him close, and take away all the grown-up worries that had no business on the shoulders of a twelve-year-old. Instead, she hugged Lottie and smoothed her daughter's hair.

"You done a real good job. Now go fetch me that broom handle in the corner. Henry, get the sacking off the bed. Doug, once I get Cleveland's leg tied up, I'm gonna heat you up another big bowl of water to get that ol' wheezin' gone." She spoke as though she had everything under control.

Only those older than twelve would know that the truth was the exact opposite.

# CHAPTER SEVEN

Alex hadn't needed to look at the calendar hanging on Eula's pantry door to tell him that two weeks had passed since he last visited the mid-forty. He had trotted the gray down the lane that bisected his acres every day since he spoke to the nigger woman on the place in mid-August, but he hadn't wanted to look too close at the tobacco.

As he slowed his horse this morning, what he had to do was clear. No lone woman, with four pickaninnies way too young to be of any use, could bring in what a full-grown man couldn't. He'd have to shoo her off the place and do what he could to get at least some of his acres in before first frost. Fifteen hundred dollars was a lot to lose. As he snapped the reins to the right to guide the gray up the path toward the barn on the mid-forty, memories of the woman's bare thighs fought to push aside his good sense. He shifted his weight in the saddle hoping against everything reasonable that the woman's husband hadn't yet returned home. Slowing the horse to a leisurely pace, he let his eyes drift over to the fields. With the barn still some two hundred yards up the path, he stopped the gray and slid to the ground.

Alex walked over to the tobacco stalks lining the path. Damn, if at least some of them didn't look just about passable for spearing. Remounting, he scanned the acres on the other side. Sure, the tops of the stalks were uneven, and without much of a look, he could see that all the plants hadn't reached harvest-ready yet, but the woman had done more than a middling good job. McNaughton galloped the horse to the barn and dismounted.

The September sky dawned a mild pink with the promise of good working weather. A wren chirped a greeting in this second hour past dawn. Pushing up the wide brim of his straw hat, Alex looked over the fields again. Strange, he didn't see the woman or her children. They should have been hard at work in the fields by now. He smelled the hogs at their slop and peered over toward the other side of the barn. Although he couldn't catch a good look inside the dark interior, he heard the cows grunting as they chewed their hay. They had been milked, for sure. Three scrawny chickens sauntered in front of him. He was certain that they, too, had already laid, and their eggs taken. The woman had been up, but where was she now?

He spotted the opened door of the smoke house and walked over to it. Inside, Alex felt the heat from the coal-fired oven. The whiff of baking biscuits greeted him. Walking closer, he looked into a cast-iron kettle and saw nothing but water dotted with blobs of grease. Where was breakfast? McNaughton looked around the walls of the smoke house. Even though it was September, and summer supplies were bound to be low, still there should have been at least one slab of bacon left on the hanging hook. And, where were the woman's winter preserves?

Alex stepped outside to glance at the garden plot between smoke house and barn. Last year he remembered the abundance of vegetables and noted to himself that his newest tenants had the gumption to tend their own food and bring him in a bumper tobacco crop to boot. These were niggers worth keeping. Now, the plot showed only bare reddish-brown Tennessee dirt. Not even the withered top of one green onion was left standing.

The slurping of the sows struck his ears as he neared the barn. Shaking his head, Alex no longer had any doubt that the slowest moving of the creatures had long passed her piglet days. He'd slaughter her for the fall and advance some of her meat to the tenants against their tobacco shares. Then Alex remembered. The harvest on the mid-forty promised to be a mighty iffy thing at best this year. Just inside the barn, the two cows were busy at work on their short stack of feed. He reminded himself to bring in more hay for the coming fall. That, too, could be held against any profits the Welleses might claim. The ladder to the living quarters above lay in place. He moved toward it, but before he could call out his halloo, the sound of someone struggling for breath in the upstairs living space caught his attention. It was the sound of a child. As Alex moved closer, the woman's voice, marshaling her children into order, floated down through the square opening.

"Hey, you up there," Alex shouted out. "When you comin' to the fields?" With his booted foot on the bottom rung of the ladder, he looked up to see the woman's young girl looking down at him, her eyes wide in surprise. Before he could negotiate the first two ladder rungs, the swirl of a skirt thrust the child aside. Two worn work boots and a flash of two milk-chocolate colored ankles confronted him.

"I'm comin' right down, suh." The woman dropped to her knees in a thud as she bent over the opening to look down at him. The top of her shirtwaist lay open and both breasts fell hard against the cloth, almost overflowing it. Colored like first-tapped maple syrup stirred with a cinnamon stick, they were three shades lighter than her sun-drenched face.

Alex stopped his climb to stare. The woman scrambled to her feet, turned, and began to back down the ladder with her skirt pulled tight over her buttocks. He stepped off the ladder, his eyes never off the rounded shape above him. His hands gripped the side rails as his eyes told him that he wanted to reach up and touch that rounded firmness coming straight at him. This was nothing like the vanilla pudding flabbiness Eula presented him every fortnight.

"I'm fixin' my children breakfast. I'll be with yo' tobacco directly, suh." On the ground, the woman turned toward him, her eyes flitting like a hummingbird between his shirt pocket and his face.

"It's past sunup. Ain't you a bit late gettin' to my tobacco?" He had the uneasy feeling that there was more to this woman's late start than she was telling.

Had the husband come home? Was that why her shirt fell open just enough to offer him a teasing glimpse at the top swell of one breast? He felt the heat starting in his britches again. He tried to push the thought away, then reconsidered. Why couldn't he have business and pleasure, too?

"Yes, suh. It's just that my oldest boy had a fall yesterday." She bobbed her head toward one of the rafters.

Alex followed her glance and spotted the place where a beam had given way. He turned toward the small pile of spearing cuttings.

"You ain't got enough sticks to spear more than five acres. There's forty that need to be brought in." He looked back toward her.

The woman's eyes flickered up to his face.

"You surely are right, suh, but my middle boy will be down in a minute to sharpen some more." Her eyes drifted down to his shirt pocket. "By tomorrow dawn, I'll have that first acre in for you."

"Your middle boy—is he the one I heard wheezin' upstairs?" Alex sucked in a breath.

This woman was a good talker. She couldn't possibly bring in an acre of harvested tobacco by tomorrow, not with nearly two good hours of this day already gone.

"The wheezin' sickness only takes him when he's with dryin' tobacco. I'm gonna have him sit outside to do the sharpening." Up went her eyes to his face for just an instant. "He'll do just fine."

"Is this the boy you told me was no more than twelve years old?" He had to hand it to her. She could move her words around with surprising quickness.

"He strong like a mule, suh, and my girl will fetch and carry the sticks. I'll do the spearin' myself." The eyes met his for the whisper of a second before he watched her drop them to the fourth button on his shirt.

"Woman, are you tellin' me that you and your three picka…" he remembered her admonition of two weeks earlier "…children are going to bring in an acre of tobacco by nightfall? What about tomorrow? Where is your man, anyway?" It was his turn to give her the hard look.

Quick as a flighty bird, she brushed past him and stepped out into the breaking day. Alex walked to catch up with her.

"No, suh. I don't reckon I can get an acre in by nightfall, but I can surely do it befo' midnight." This time the eyes met his and lingered long enough for him to look into their velvety bronze softness.

This woman had pools in her deeper than well water. What would it feel like to dip into those depths?

"My children and I will work from sunup to sundown spearin' the tobacco. After I put them to bed, I will sharpen the sticks 'til my oldest boy can do the job in two or three days." The eyes were up again, searching his.

He stared back at her until a look of remembrance crossed her face and she quickly dropped her eyes to his boot tops.

Alex moved away from her to survey the fields a third time. When he turned back, the woman had stepped inside the smoke house. He followed.

"How do you expect to work your children like grown-up field hands if you ain't feedin' them?"

The woman bent over the open oven door, a tin pan of four hot biscuits in her cloth-shielded hand. Surprised at the speed with which she straightened to her full height, Alex paused.

"I will get yo' tobacco in for you, Mr. McNaughton, suh." She drew out his name.

This woman had more sass in her than was good for any female, black or white. A good slap would remind her of her place, but he hesitated. Holding the hot tin in front of her, she moved to the door of the smoke house. Alex stopped her.

"I don't believe you can bring in my tobacco for me."

The right side of his body brushed her left shoulder. His left arm stretched across the door frame directly in front of her chest, his long shirtsleeve no more than inches away from her shirtwaist with its open top button.

"I've put on a hired man to help you," Alex made the decision as he spoke, "but the cost will be high."

She looked up at him, the cooling tin of biscuits in her hand, puzzlement on her face. "What can you give me to pay for the extra help?" He watched the woman's slow shake of her head as her eyes blinked the dawning understanding of what he was asking.

Her mouth, with those full kissable lips, opened and closed twice.

"Sir, please forgive my forgetfulness." The words came out slow. "You asked about my husband. He has been delayed in Kentucky—his auntie and all—but he left me a message for you." Each word came out as exact as a tobacco-weighing scale.

Did she think that sounding like a proper-talking Nashville colored schoolteacher was going to keep him away? He didn't appreciate uppity-talkin' niggers.

"My husband, John Welles…" she let the name linger on her lips, "wants to beg yo' pardon about the delay. He knows I can bring in twenty acres all by myself in the next two weeks. He also know he owe you forty and would like to make up for it in the winter."

No nigger, male or female, had ever gathered the nerve to try and bargain with him or any other white man. He should take a stick to this woman and beat her back to remembering her place. Instead, he let his hand slip down the front of her dress, stopping at the second buttonhole just at the rise of her breast. He listened to her soft intake of breath and waited while she failed to let it out. Her eyes remained on his face. He knew she dare not push his hand away.

"Suh, my husban' may not be home for a few days but he comin' back. I am a married woman."

Alex made sure his nod was neither a yes nor a no. This woman had more than crossed the line of disrespect.

"Your get needs some bacon and fresh meat. Maybe some greens and preserves. I'll be back tonight with all of that." He knew she was deserving of a good thrashing if for nothing more than raising her eyes to him.

But instead of laying into her, he wanted to drop his hand down into her shirt and discover what firmness awaited him there. Hell, he wanted to park his manhood within her right then and there, but in those eyes swimming with life, he didn't yet see compliance. He'd never forced a woman, and he especially didn't want to force this one. She stood stock still. As close as he was, he couldn't be sure that she had let out a breath.

"A few chickens wouldn't hurt none either," he added as he watched the biscuit tin jiggle in her hand for the flash of a second.

Alex did some quick reckoning, and the warning signs jumped into place. A woman who could think quick like this one was a dangerous thing. But the longer he stood close to her, the more certain he was that this was one danger he wanted to embrace. He steadied himself for patience. With a mind like hers, she would soon see that her options with him were none. He gave her a slow nod to make sure his meaning had registered before he turned and headed for the gray.

"I'll be back tonight," Alex repeated as he reached for the reins. Behind him, he heard her fight a grunt in her throat.

"I thank you kindly, suh, but no."

Did the wench say no? Why, even his wife had never hinted at refusing him. Not that he cared enough to press his case with Eula Mae. He turned back toward the woman, damning the hardening in his pants. Her eyes never left the dirty and scuffed toes of her own shoes. He took a step toward her. He could order this woman to do anything he wanted. Alex knew it and she knew it. He could have her on her knees with his manhood in her mouth within one minute if he wanted. Alex took another step closer and fumbled louder than he needed with his belt buckle. Though she fought hard to conceal it, he saw her cringe as she heard the rough, scraping sound of leather against metal.

Alex stepped close enough to brush her stomach with the leather end of his belt. He dropped it just below her belly button and let it slowly search for the outline of her drawers. He could feel his lungs searching for more air, but only by a close look at her chest could he see the woman flinch. Her eyes remained on the ground, her breasts poked full against the thin, much-washed fabric of her shirt. Alex suspected she hadn't taken a breath in many long seconds. He wanted to throw her to the ground and hike her skirt to her hips. He wanted to feel for the drawers he was certain weren't there. He moved his chest against her breast.

"Suh," her voice was no more than a whisper as she tried to push the cooling biscuit pan between herself and Alex, "my husban' would like to know if you would consider just one mo' thing."

The eyes were still respectfully down, but she had just turned the key that she knew would push him back, if even for an instant. This woman had more smarts than any colored needed. She was worth far more than a quick roll around on the smoke house floor. The throbbing in his pants moved to his chest. He wanted this woman right now, but he thrilled at the excitement that would come when she fell into his arms all on her own.

"Depends on what the one more thing is." His breath brushed her ear, but she held both her body and eyes steady.

"John Welles," she drew the name out in the warming morning air as though her husband deserved all the respect of a preacher, black or white. "John Welles know he wronged you." The eyes came up level with the lower part of the barn.

Alex watched her chest rise and fall as she finally took in little gulps of air.

"You been most generous to let us farm your land." The eyes moved to the mid-planks of the barn as she eased her body slowly around toward him, the pan of biscuits sliding across his shirt.

One breast brushed his chest as she pivoted. He laid his hand on her arm, stroked it from elbow to wrist, and flicked the pan of biscuits to the ground. He felt the recoil and quick recovery of her body.

"Last year, you let us keep 'most half of what we brought in to you." Without so much as a look at the fallen bread, she laid her now empty hand across her stomach like it was a metal shield.

Alex dismissed the move and did quick mental calculations. Of the three thousand dollars brought in last year, technically the Welleses were entitled to forty percent according to the terms of their original deal.

"Less advances I made to you, of course." He blurted the words out as he watched her resume her slow turn until she faced him directly.

Alex laid an arm back across the door frame. Her right shoulder touched his wrist. Her eyes stayed at pocket level.

"That was most generous of you, suh." The woman's eyes lifted to his face with all the speed of a snail on a tomato vine. "This year, my husband would like to ask if you would take two out of every three parts of what we bring in to you?" She had succeeded in making her face resemble an innocent angel, but she hadn't succeeded in masking her good sense.

This woman could do figures in her head as good or better than Eula who had all of her eighth grade education.

"Less advances." He spoke again, wanting to bite his tongue. A white man didn't bargain with a nigger, especially a woman.

But, then, if the miracle came and she brought in fifteen hundred dollars this year, she would be entitled to only five hundred instead of the contracted six. He had already advanced her that much. If he took the deal, he would get to keep the entire year's sale in exchange for letting her stay in the barn for the winter. Now, he had to convince her to give him a little something extra.

"Yes, suh. I would take it kindly if you would allow me seed money for my fall vegetables and for a new sow and maybe a dozen or so more chickens." Her eyes locked onto his.

Alex breathed hard. A young sow alone was worth almost sixty dollars. Where had a woman learned to figure like a man?

"Woman," his voice was hoarse, "I've told you before. I've got me a man who wants to farm this place right now. I know your man will be back, but if the God's truth be told, even you don't know when. You and those kids need food so you can work the harvest proper. Without it, there's no reason for me to let you stay, now is there?" Though he had dropped in his own bargaining chip, he had also just given a colored woman a power over her fate that he hadn't given even to Eula Mae.

# CHAPTER EIGHT

"John Welles, John Welles. Open this do'. I know you is in there." If the sound of the afternoon train rumbling down from Chicago hadn't been enough to wake him from his sleep, the rusty-razor voice of his landlady more than finished the job.

John put the rolled-up rag he used for a pillow over his head to block out the sound. He had no need to hide his eyes from the light in the crowded, windowless storeroom. The thin stretch of daylight that did make its way miraculously under Miz Sarah Lou Brown's storeroom door was all that he ever saw of brightness when he finally was lucky enough to fall asleep on the pee-stained mattress after his thirteen-hour workday.

"It's the first Friday in September, and you ain't gittin' out of here this day 'til you pays me my rent." Miz Brown's fist pounding on the door matched the voice in grating on his nerves.

John peeked out from the makeshift pillow. The little dab of daylight that had been there disappeared, covered over by Miz Brown's more than ample body. He knew she would either stand there all day or put a shoulder to the door and break through it like a cardboard oatmeal box. John rolled to his hands and knees and pushed his tired body to standing. Without the sun to guide him, he depended upon the three p.m. train heading to Florida from Chicago to give him his time bearings. Wearing only his summer drawers, John stepped over a crate of corn and two tins of lard to reach down to the floorboards he had first pried open right after he moved in to the two-story clapboard house. Squatting on the floor, he pulled the loose board up and lifted out the blue bandanna that held his money. Fumbling in the darkness, he retrieved four quarters, seven dimes, and sixteen nickels. Damn. He had hoped to have enough in change so that he wouldn't have to break one of his silver dollars. They were too hard to come by. For the flash of a second, Annalaura's disapproving face flicked into his mind. He pushed the vision away. He loved that woman more than he could show, but she had more mouth on her than he had time to deal.

"John Welles, I got no time to fool with you. I've got me people lined up 'round the block to take this here place. Country boys comin' into Nashville, fifteen, twenty a day. I don't need none of yo' foolishness." Sarah Lou's voice just about shook the storeroom door off the hinges.

That the old biddy was right riled John the most. So many men from out in the country got the same idea in their heads. No colored was ever going to amount to anything if he kept sharecropping for the white man.

"Let me get my britches on Miz Brown, ma'am, and I'll hand yo' money right to you." Even if he could have afforded better, there were too few rooms to rent for all the colored who were pouring in to the city.

Too few places to lay his head, too few jobs to make the rent, that's what a country man could expect if he decided to risk everything on Nashville. John pulled on his overalls and reached for his shirt as he walked over to the door, only stubbing his toe once on a forgotten box that sat near his mattress.

"Here's yo' money, Miz Brown, the whole of it." He opened the door wide enough for the space to frame his six-foot-tall body. John laid two silver dollars and the four quarters into his landlady's hand.

Her fingers snapped closed over the money as she tried to look around him and into the room.

"How much they pay you down there at that smithy shop?" Her wig, that looked more like the tail off a red squirrel, had been put on even more crooked than usual today.

"I've got me 'nuff money to keep the rent up. You don't need to worry none, Miz Brown."

"You don't work nothin' but fo' hours a day down at the smithy shop, but you out every night." The plump face frowned up worse than a prune. "I told you I don't allow no gamblin' peoples to live in my place. I runs a respectable boardinghouse. Why, I got two colored schoolteachers living right here. They..."

"Miz Brown, ma'am. I works six mornings a week, six o'clock to ten o'clock, cleanin' up horse sh...horse dumpin's...at the blacksmith shop. He pays me fifty cents a day. That's my room and board, and yo' rent right there."

This woman, who liked to set herself front and center in the second-best church pew at the Nashville colored Baptist church every Sunday, didn't need to know much more than that. John could see that her frown had only deepened. To keep her face from caving upon itself, he decided to give just a little bit more.

"I gets me a little extra money by running grocery deliveries." He doled out a hint of one of his patented smiles.

She didn't need to know the full truth. Six days a week, from five in the afternoon until one o'clock in the morning, John Welles ran hams, chickens, sides of beef, and just about every other fancy food a colored man could dream of to a certain Nashville address. At the same time, he also ran gin, bourbon, whiskey, and good branch water to Miz Zeola's whorehouse. For each of his eight- or nine-hour days, he got paid seventy-five cents. Altogether, he earned seven dollars and fifty cents a week.

"You just keep on bringin' me my rent every Friday, and you and me will get along just fine." Miz Brown almost had a smile on her face as she turned toward the three wooden steps that led to the back door and her kitchen.

Seven-plus dollars a week was a hell of a lot more than John ever earned in Lawnover. By the time that cracker McNaughton got through with his "advances" last year, John had only a three-hundred-dollar share to last him the whole of the next year. And this despite the bumper crop he and Annalaura had brought McNaughton. John and Annalaura had gone over the figuring together, but his wife had been the one to ask, in her most respectful way, for an accounting. McNaughton, thinking the two of them too dumb to understand figures, had rattled off bloated costs for rent, milk from the two cows, meat from one pig, garden seed, clothes, and a little starting food. His "advances" came to almost nine hundred dollars. Hell, the man on the back-forty had only brought in half the tobacco he and Annalaura had gathered, and that family's share had been two hundred and fifty dollars. That was when John knew he had to leave.

Closing the door behind him, he slipped in the big padlock that Miz Brown hated, and turned the key. Welles didn't want that woman snooping around. She wouldn't bother his money, there was way too little of it for her anyway, but she might put her hands on the pistol that he kept hidden behind a chink in the wall up near the ceiling.

Now that Miz Brown had awakened him with her rent nonsense, he decided to head off early to his brothel job. In the two months he'd worked there, he'd taken every opportunity to get in the face of the owner. Miz Zeola's whorehouse was just about the best in Nashville that serviced the workingman colored. Oh, John knew about the fanciest brothel in town with its curly-haired, high-yella gals, but they only serviced the colored doctors, lawyers, businessmen, and it was said in the quarter, more than a few of the richest white faces showed up there, too. But it was Miz Zeola who had the market cornered for the workingman with a decent paying job. And she treated her customers just fine.

To make up for the lack of looks in her women, Miz Zeola saw to it that her male guests had an extra fine dinner for just a fraction of the cost at a regular restaurant. She laid out her tables with white cloths, and in the summertime, even put sweet-smelling night-blooming jasmine in glass jars right there in the center. And it wasn't like Miz Zeola's girls were mud-duck ugly. While most of them didn't have the bright color and light eyes, her brown-skinned girls were more than fair-to-middling pretty. The madam trained them to keep a man in ecstasy longer than any other whorehouse in all of Nashville. John was sure of that by all those pleased moans and grunts he heard coming from the rooms upstairs when he went to the pantry door to pick up the dirty dishes. John had known better than to sample his employer's wares—that had gotten many a country boy fired right quick—so he wasn't sure what the women looked like under their fancy dresses, but he guessed their stuff couldn't have been any finer than his Annalaura's.

A ride on the horse-drawn trolley cost five cents. Since a nickel was hard come by, John opted to walk to Miz Zeola's. The twelve blocks gave him plenty of time to think about his wife. She hadn't been his pick right off, of course. At twenty-two, he was still too wild. He hadn't reached full-grown manhood when he first discovered the effect he had on women. Working the tobacco like he did had given him the muscles to fill out his tall frame. His older brother, who sheltered him after the death of their parents, always told him that he was far better with words than was safe for a black man. And when that first married woman started batting her eyes in his direction when he turned on his smile, he went straight home and got out the broken bit of that looking glass his sister-in-law used for her own primping. He practiced showing those teeth for hours.

When he used his "Yes, ma'ams," and "No, ma'ams," together with his well-practiced smile, along with those words that came easy to him, just about every colored woman in Lawnover fell all over herself trying to make nice. He was more than happy to accommodate, though he had never been a fool. John Welles wasn't about to get shot for messing with another man's woman. He poured on the charm for the ladies but always knew where to stop the dime. Trouble was, it didn't taken him long to go through most of the eligible women, old or young, in Lawnover. Soon, he stepped on over to Clarksville and the "sportin' houses." Now, there was action. The whorehouses in that town were about as rough as the back end of a barn compared to Miz Zeola's, but those old rusty country girls had taught him many a new trick in the back room of those juke joints. And then he'd met Annalaura.

Met wasn't exactly the right word since he'd watched her grow up. She was seventeen and still living on the Thornton place with her Aunt Becky when he finally took serious notice of her. There she stood in her Sunday best with those buttons across the front of her dress ready to pop. She was the only woman in the room who wasn't falling all over him.

He'd taken her a long way since those days. After they married, he'd even shared a few of his whorehouse secrets with Annalaura. Of course, he had sense enough not to tell her where he'd learned those things, but that was easy since she'd come to the marriage bed with no idea of what to do with a man. At first, he thought he could talk her into anything. He smiled at the remembrance of her frowned-up face when he reminded her that all wives were expected to do what their husbands wanted in bed. She'd gone along with most of it, but hardheaded Annalaura had drawn a quick and deep line in the dirt over some of what he asked. Neither the devil nor her husband could make her cross over it. The grin on his face was wide as he reached the brothel's screened-in back porch.

He knocked on the big oak kitchen door at Miz Zeola's. Most folks who had heavy front doors with fancy carvings and curlicues all over them had knotty pine, skinny-as-a-stick back doors. Not Miz Zeola. She always kept her kitchen door locked and insisted that it be made of two-inch solid oak with strong brass hinges. She let everybody know that she didn't want anybody coming through her doors, either front or back, without her knowing exactly who he or she might be.

"Yokel, ain't you early?" Big Red, Miz Zeola's head cook, was already elbow deep in flour and lard making the crusts for the pies when he undid the lock.

By the smell of it, at least two peach cobblers, heavy with cinnamon and nutmeg, were baking away. Big Red mounded up a pile of dough and slapped it on his wooden work board as he turned to stir the big kettle of cut-up sweet potatoes, sending their syrupy scent throughout the kitchen. On the eye next to the kettle stood a second pot full of turnip greens mixed with ham hocks bubbling their heavy promise into the air.

"Thought I'd get me an early start today." John didn't take kindly to Big Red.

All three hundred pounds of the red-boned man sneered down at Nashville newcomers, especially those who had sharecropped for the white farmers. Those, he thought, were too dumb to ever make it at big city living. To Big Red, John Welles was just one of about five "yokels" working for Miz Zeola this summer.

"Uh huh. You one slick country boy all right." Big Red gave him a quick look before he started rattling off the supplies needed for that night's supper.

Red never gave him a written list, and John was convinced the cook could neither read nor write.

"Six loaves of light bread, fo' hams, a peck of sweet potatoes, and half a bushel of black-eyed peas," John repeated the list back to Big Red in his most conciliatory voice. He had no time for an argument with this man.

When he first arrived in July, John watched the other yokels go the rounds with Big Red only to be fired in rapid order. Welles had spent his first month in Nashville eating only one meal a day trying to stretch the money he had taken from Annalaura until he could get himself a job. Losing both job and money after a few days to a big red-skinned cook who had no real power did not interest him. John would shuffle like Massa's best nigger until he could get his chance with Miz Zeola.

"Ain't I jest told you 'bout the liquor? We needs twelve mo' bottles of bourbon whiskey, three jars of good branch water, and fifteen mo' of gin." Red turned from the greens and glared at John.

"Ain't no need of you comin' 'round here early to get in yo' good licks with Miz Zeola. She eat country boys like you fo' breakfast." Spittle from Big Red's mouth found its way into the greens pot. "I see you flashin' them teeth at every woman come 'round here. I hear them words, smooth as rum, that you po' over Miz Zeola's head. I'm here to tell you, it ain't gonna work." Big Red gave the side of the greens kettle a loud whack with the wooden spoon.

"I hears every word you say, and I surely will take it to heart. But I think you gots me wrong. I'm headin' back to my family as soon as I put a little something away for the winter." John spread his lips but made sure his white-on-white teeth didn't show.

"Every country boy 'round here is runnin' from the tobacca. Ain't September harvest time out in tobacca country?" Red's eyes were narrow slits when he turned toward John.

Welles saw more of the Indian in the cook than he did in his wife.

"That it surely is, but my woman got kinfolk to help bring in the harvest." He let his words drip just a hint of apology as he lied to Big Red.

It wasn't that he hadn't thought of Annalaura in the three months he'd been gone. She was never really off of his mind. Of all the women he'd ever met, she was the cream on the milk. Sometimes, he'd fix his mouth to tell her how much she meant to him, but every husband knows that sweet-talkin' words can be the ruination of a good wife. When he left her after the two of them put the seed in the ground in late May, he had every intention of getting back to her bed way before the September harvest. But that thirty-two-dollar pot he won playing poker up in Clarksville seemed like a message straight from the Lord. It had Nashville written all over it. And there was no time to dance with Annalaura. Even though his wife had more sense than all the other females put together, still she was a woman and would never understand that it hurt a man to the middle of his soul to do no better by his family than to have them live in a white man's barn sleeping with the hogs.

"You jest make sure you don't git too big for them country overalls of your'n." Big Red liked to guffaw at his own jokes.

John had wasted enough time with this man who acted enormously satisfied with his role as chief cook in a whorehouse. John had grander plans. He hadn't slipped into the smoke house and pried open the locked metal box where Annalaura kept their savings for naught. He counted out over ten dollars that his wife had saved for school shoes for Cleveland and Doug. He had promised his sons that they could both go to school right after harvest. Doug, in particular, had been excited. Though his second boy was coming up fast on ten, the child hadn't had more than a year's schooling all put together. Yet, Doug could both read and write a little.

John was determined to do better by all his children. Even little Lottie would have her chance at school. He took only eight of the dollars and left Annalaura a little bit over two. A woman as clever as his wife would find a way to make it stretch. Of the two slabs of bacon hanging in the smoke house, John took only one though he did take most of what was left of last fall's preserves. He knew he was leaving his family in a tight spot, but he had every intention of being back in Montgomery County no later than mid-August. He had no idea that Nashville was going to hold him this long, but he was too close to satisfaction to go back home with nothing. For sure, he'd be back by Christmastime. By December, he would send both his boys off to school in fine style and he would put a fancy yellow, ruffled pinafore on Annalaura.

"Thank you for the words and the list. I'm jest gonna check to see if we have enough clean dinner dishes in the dining room safe." Before Big Red could mount a protest, John, his deferential half-smile in place, backed out through the swinging door that divided kitchen and pantry from the dining room.

This time of day, and with any luck, he might find Miz Zeola herself. He had wasted enough time with a colored man who was going nowhere in this world. John hadn't left the best thing in his life for nothing.

Miz Zeola's dining room cabinet was built into the wall, but you wouldn't know it because of all the mahogany surrounding the massive piece on the sides and even up to its nine-foot-high top. The safe was almost the length of the twenty-eight-foot room. Behind its four sets of leaded glass doors, Miz Zeola must have kept six different china patterns of twelve settings each, not to mention the matching crystal and silver. Though it wasn't part of John's job to check on the china, it was the only way an outside man could ever get into the main-floor rooms of the bordello. It was the job of the hired girls to keep the contents of the safe clean. They did the polishing of the silver 'sticks that sat four thick and stood three feet tall on the dining room table that sat twelve. At the entry, with the kitchen swinging door at his back, John scraped his boots clean on the rag rug Miz Zeola kept there for just that purpose. He wanted no telltale signs of dirt when he tipped across the burgundy-red carpet to peek into the small private hideaways Miz Zeola had set up for her best customers.

On the opposite side of the room from the china safe, the madam had carved out three little rooms, each no bigger than six feet by seven. Their doors were papered in the same wall covering as the rest of the room. The brass doorknobs held flowers that seemed to melt into the wallpaper. Three mirrors, each no bigger than two feet square, separated the doors. Miz Zeola wanted no outsider looking into her business. To the uninitiated, the doors looked like part of an elaborately paneled room. Inside, small tables were set for two, and a little kerosene lamp gave out the only light in the windowless rooms. If a person took a notion to peer close enough in the dimness, the outlines of a settee draped in satin sheeting could be seen. John smiled at the finery.

After he had plucked the money he needed from Nashville, he would drape satin sheets over a big four-poster bed and wallow all over it with Annalaura's tight, fine body wrapped in his arms. What his wife's face may have lacked in great beauty, that body of hers more than made up, with those curves and dips that made a man just about holler to the skies at their perfection. Of course, none of this could he tell her. Women didn't need to get their heads puffed out. A swish of air startled him out of his thoughts.

"You gittin' yo' eye fill of my rooms? Lessen you can pay, I don't want no country boy in this part of the house. What you doin' in here any-way?" The clock had not yet sounded five, and Miz Zeola hadn't quite finished her evening toilette.

Her usual perfume, which she declared came straight from New Orleans and always preceded her arrival by at least two rooms, had yet to be applied. John had neither smelled nor heard her approach and turned quickly around, closing the private door behind him. He whipped on his best sheepish, got-to-forgive-the-boy smile.

"I am mightily sorry, Miz Zeola. I had a little bit of extra time, and I jest wanted to make sure everything was at the ready." He kept his eyes on the dressing-gown belt around his employer's considerable middle while he let her absorb his words. He had learned, early on, that women loved it when a man gave them a chance to talk.

"If you got that much extra time, then maybe I needs to cut back on yo' hours." The satin of her purple dressing robe rustled as she stepped toward him.

He felt her eyes climb all over him more than once. He was used to that from women. Zeola was no different.

"Where you from again, country boy?" Those words on anybody's lips, male or female, were beginning to wear on his nerves almost as bad as when the white Lawnover farmers called him "boy," though he was thirty-four years old.

"Montgomery County, ma'am." He lifted his eyes to face her, leaving only the slightest trace of a smile.

"John, ain't it? How long you been with me?" She turned her face into the feathered boa at her dressing-gown collar and fluffed it with her long scarlet-painted fingernails.

"Yes'm, it's John. John Welles. And I'm pleased to say I've been working for you fo' almost three months." He let the smile flash for just an instant before he dropped his eyes back to the loosely knotted belt.

"Let me tell you one damn thing, John Welles. If you want to keep on workin' fo' me, you will stay where I put you." She poked one plump hand at him, as the curling rags in her hair wagged with each word she punched into the air. "You will come and go where and when I say you can, and you will not step foot into any place I say you cain't."

John snapped his eyes to her face. He couldn't recall when a woman had laid into him like this unless it was Annalaura when she fumed over his gambling ways.

"You got a woman?" She shot the question at him like she was accusing him of stealing her candlesticks right off her table.

"Yes'm." John had to think quick like he sometimes had to do with his wife.

"Chil'ren?"

"Yes'm." He let that answer hang in the air while he sized up this woman.

"But they ain't here in Nashville, is they?" She folded her arms over her big bosom.

"No'm. They's with her family in Montgomery County." What business of this woman's was his family?

"John Welles, yo' country boy games ain't gonna work on me. I knows you is in Nashville to make yo'self a killin' so you can get back to that fine gal you left on some white man's farm." Reaching into the pocket of her gown, she dragged out a long cigarette holder. Plunging a hand deep into the other pocket, she pulled out a cigarette and stuck it in the holder. "For all yo' shuckin' ways, you ain't stupid, John Welles, tho' you would have me believe you was close to simple-minded." She walked to the safe, pulled open a drawer, and withdrew a box of matches.

"No, ma'am. I don't reckon I am stupid. I'm in Nashville to earn enough money to bring up my family." He raised his head and eased his back up straight to give himself more than a head advantage over the stocky woman. Still, Zeola had a way of making a man feel just a little bit smaller than he actually was.

"Tell that pretty story to somebody else. Nashville runs thin real fast on country boys. Livin's too quick here. What you boys really want is big money in lightning time so's you can git back home and buy yo' own place." Zeola turned those hard-as-glass eyes on him, the just lit cigarette dangling from its holder in her hand. "Now, tell me, John Welles, if I ain't spoke the truth?"

John didn't even think Annalaura could read him this well, and she was damn good.

"I'm not braggin' on myself, Miz Zeola, ma'am, but I surely will make me enough money to buy my own farm." He'd given his wife credit for being able to see inside him like no other because of her Cherokee grandmother. "I already got my eye on twenty acres down near Lawnover. Me and my wife could make that work real well."

Everybody knew some Indian women had the second sight. That was why he'd taken off for Nashville without a word to Annalaura. If he'd stayed to argue the point, she would have known he was itching to go and would have tried to talk him out of leaving. She was hardheaded that way. She'd never understood that a wife was supposed to shut up and let the husband do what he knew to be right for the family, even if it meant a few sacrifices here and there. Even a blind man with one leg could see that 'cropping was never going to make a way for a colored man. When she saw all that money he was going to lay at her feet when he returned home, maybe that hard head of hers might soften into forgiveness.

Zeola circled him, looking him up and down just the way she did when a new girl came in for a job. She wouldn't take the applicant if she was too homely, too sickly, or looked too broke down. She preferred scared and ruined country girls who had been done in by no more than two men. John wondered what price she was setting on him?

"You 'surely will' make you enough money here in Nashville, is it? You know how many country boys tell me the same thing? They cain't last six months in this town." As she spoke, her hips swayed in time to her words. "What makes you think you won't be back in yo' Lawnover right after first snowfall?"

"Miz Zeola, I am truly sorry to hear 'bout them other men, but ma'am, ain't none of them me." He could feel himself looking down upon her from his full six feet. "Beggin' yo' pardon, ma'am, this job may not last as long as I'd like, but there's other ways fo' a man to earn good money here in Nashville." He kept his smile inside.

"Uh huh. Some of them ways will get you into the white man's jail quicker than I can yell po-lice." Her eyes had squinted down so far that a body couldn't tell if their color was gray or brown. "Believe me, John Welles, you don't want no parts of a Nashville jail." The mound of purple shimmered as she wagged her head. "A man with a strong back but a little mind may wish fo' Nashville all he wants, but you is right, wishin' ain't gettin'."

"And a man with a strong back, strong mind, and quick hands is a natural in Nashville." He put his full stare into her eyes.

She gave it back to him and then some.

"A natural, is it, John Welles? How much gamblin' you done?" The red-painted fingers splayed themselves under two of her considerable chins.

"It was gamblin' money that got me this far." This was the chance he'd waited for, but he knew he had to let this woman think she was drawing the truth out of him.

"I don't allow no lyin', no cheatin', and no liftin' of my money. Country boy, believe me, I'll know if one nickel of my money is missin'." The eyes opened a bit. She pointed one finger at his temple.

"Ma'am?" John faced her, puzzled.

"I needs me a country boy who's smart enough to act dumb so as to be taken fo' honest at my poker games. I want you to hold the pot. If even one dime is gone, it won't take me to kill you, the players in the game will do it fo' me. And more than one of my gentleman callers is handy with both a knife and a pistol. Do you take my meanin'?"

"Yes ma'am. My head figurin' is right good." John's heart picked up a beat. A thousand dollars would buy him a farm in Lawnover and all the stock that went with it.

"My rules is simple. You play it straight at my tables, take account of every nickel, and I gives you five dollars flat up out of every pot. To see if you can work it, I'll let you sit in on two of my weekday games. That'll give you ten extra dollars a week. Do right by me and I might sweeten that deal considerable." Zeola strutted to the sideboard, opened one of the doors, and pulled out a bottle of bourbon and a shot glass.

"Monday's yo' slowest day, Miz Zeola. I'd be pleased to start then." He thought he saw her flash a quick gold-toothed smile at him.

"Two mo' things. You treat my high rollers like they was Gawd Almighty, and once a week, I'll have one of my girls service you for free. You can have Sally. She close to thirty-five and I'm gonna have to let her go soon, but she'll show you enough new stuff to keep you satisfied fo' a mighty long time. Now git on back to yo' regular job." With the back of her hand, she waved him off.

# CHAPTER NINE

Eula slapped the pork chop on the floured board, gave the pepper holder a shake, reached her hand into the salt jar, and spread a pinch over the meat. She flipped the chop over and repeated the process before she dropped it into the hot grease sizzling in her skillet. She counted the chops as she blew a stray strand of hair out of her eyes. Three in the skillet cooking and six on the platter already done.

"Oh," she jumped back as a particularly large splatter of grease landed on her bare forearm. She dabbed at the burning place with a dry kitchen rag, draped the cloth near the stove edge, and hurried to lift the heavy lid off the potatoes bubbling in their own skillet. She reached into her apron pocket and pulled out two green onions. With the knife she'd yanked from her spice rack, Eula sliced the green tops into the potatoes.

Eula chanced a glimpse at her husband. Alex, sitting at the table with a cup of coffee, gave no notice that he'd paid attention to any of her hurried activities. Coffee at night? Fresh-cooked pork chops at supper? This was more like breakfast or dinner rather than the last meal of the day. True, Alex had spent a heavy week finishing up the harvest. All but the mid-forty were in the barns and hanging on the drying poles. It was still just the first week of September. Maybe that accounted for her husband having her cook up enough food to feed the entire Lawnover Joseph-the-Shepherd Baptist Church on a Friday night.

"How much of that chicken we got left from dinner?" Alex stood and walked over to the stove to stare at the skillet.

In twenty years, her husband had never watched her cook, nor questioned her portions. The surprise of it all had just about taken every word out of Eula's mouth.

"We got a whole one left. I'm going to warm it for breakfast," she managed.

Alex barely escaped a second grease splatter.

"Where's the butcher paper?" He paced from the stove to the table to the porch door and back again.

Eula watched him walk the same path a second time and forgot to turn over a pork chop. Was her husband walking in circles?

"Butcher paper?" She ventured a tentative response.

"Yeah. The butcher paper." More than a trace of annoyance shot out of Alex's mouth.

Startled at a husband who almost never raised his voice to her, she neglected to remove one of the chops from the skillet.

"It's rolled up in a corner in the pantry." She watched Alex brush past her and head into the back room off the kitchen.

Eula inventoried the day to see what might have addled her husband into such a frenzy. Breakfast had been as usual and dinner had been hearty enough. Surely, he couldn't be this hungry. There had been nothing untoward with the chores. Yes, Alex had milked the cows when, technically, that was a wife's job. In fact, he had done the milking the last two mornings. But she hadn't paid much mind to that. In their marriage, the two of them had worked out most things in a way that didn't require talking. Each could just see what outside chores needed to be done and head straight to it. Whichever one happened to be in the barn at milking time just did the milking. Alex liked it that way.

But this Friday night puzzled her. Perhaps the lagging harvest on the mid-forty worried her husband more than he let on, though Eula prided herself on being able to read Alex better and faster than he could read himself. She bit down on her lip as her husband came back into the kitchen with a torn-off strip of butcher paper in his hands and three jars of her peach preserves. Laying them all on the table, Alex walked over to the food safe and removed tomorrow's breakfast chicken. The smoky smell of burning cloth finally told Eula she had dropped her dry kitchen rag too close to the fire.

"My Lord," she shouted as she began to beat out the flame with her hands.

"Here." Alex reached her in two strides and poured his coffee over the rag. "How much longer for those chops?"

Eula pulled at the collar of her dress. It was too tight in the heat of tonight's kitchen.

"They'll be ready by the time I heat up your pole beans and corn bread. Do you want buttermilk or sweet milk for supper?" When had her husband last acted like this?

"I'll take the pole beans in a jar. Where's the corn bread?" He had moved back to the kitchen table, the four cooked pork chops dripping grease over his hands. He scrunched the brown paper over the meat.

Eula stood stock-still. In the back of her mind, somewhere, the scent of burning meat registered, but, for the life of her, she couldn't match up the smell with anything she was supposed to do about it.

"Take with you? You're takin' the food out of here? I thought you wanted a big supper tonight—the harvest being over and all." The grease popping all around her sounded like firecrackers on the Fourth of July.

Only the sight of Alex's shocked face brought her back to herself. Realizing too late that she had walked dangerously close to the line of wifely impropriety, she turned to the skillet and pulled out the last of the singed pork chops. Her head down, she scurried to the safe and reached for a clean rag. She busied herself with scrubbing the grease splatters off the stove and her wall as though her life depended upon it. Eula had stunned her husband and herself with her questions.

"I'm droppin' some food off for that old woman over on the Thornton place." Alex had slowed his pace as he moved to the stove to retrieve the remaining pork chops. He turned to the china cabinet, reached to the top shelf, and pulled down her biggest bowl, with its blue sprig of flowers and thin strip of silver around the base.

Eula stood openmouthed. That oval-shaped piece of china was her only "silver" piece and her pride. In twenty years, she had never actually put anything in her favorite wedding gift. A hundred questions marched themselves into her mind. She worked her mouth hard to get them out, but her mind worked even harder to keep them in. Her mouth won the battle.

"Ben Roy's place? Old woman? Surely, you're not talkin' about Rebecca usin' my weddin' present bowl?" Eula wanted to drive her teeth through her tongue when she saw the quick flash of anger on her husband's face.

"Yeah. Rebecca. Ain't you always tellin' me she's a Thornton responsibility? Well, you're a Thornton. I don't want your brother thinkin' we're too hard up to do our share." He walked into the pantry and returned with two big jars of pole beans and a Mason jar full of honey.

"Is it the mid-forty?" Eula's mind reeled. She didn't dare chance a look at Alex.

Her husband appreciated a woman who had her own mind, but he appreciated the female even more when that woman kept it to herself and didn't bother him with it. Despite all the warnings shooting through her head, Eula had to sort this one out.

"I thought that colored man you hired was goin' to start on Monday. Two weeks or more 'til a cold snap." Though she held her head down, the flashes from Alex's eyes pierced right down to her belly. If she didn't look at him, she could finish her words. "That should give him time aplenty to get in most of the forty."

Alex's face flamed the color of over-ripe cherries.

"I don't want to hear you fret over the mid-forty again. That's all taken care of." His voice sounded like the hiss of a snake in her ears.

"Name's Rebecca Murdock, you know. She don't ever use that name, but she was a married woman once. Married a colored man from outside of Lawnover. Still likes to call herself a Thornton even though she's not really one. Just one of our colored from before the War." Eula prayed that what Alex called "mindless woman's chatter" would guide his own mind away from her wifely slip.

She could bear his annoyance much easier than his anger. Scooping up the corn bread, he added it to the other wrapped parcels and headed out the kitchen door.

"You reckon she's still up this late at night?" Eula fought hard to make her voice sound everyday.

Alex was already at the porch door when he stopped and turned back to her. Behind him the last of the September sunlight had gone thirty minutes earlier, and the moon had already gained supremacy in the sky.

"I told you I'm takin' some of this food over to that nigger woman on your brother's place, then I'm takin' the rest to the tavern for a poker game. Don't stay awake." The spring in the screen door was a strong one, and when Alex let it close on its own, the sound of it snapping back into place reminded Eula of a rifle shot.

Moving quickly to the kitchen door, she called after him. "Won't Ben Roy be there? Fedora usually packs up the food for those poker games when Bobby Lee's busy." She tried to put the picture of Alex bringing food to the Lawnover store into her mind. It wouldn't fit.

Alex reached the barn and his horse. He gave no sign that he'd heard her call. Eula turned around to stare at the shambles of her kitchen. It wasn't that her husband had never gone out after supper before, nor that he had never gone to play poker at Bobby Lee's General Store, but he had never been the one to bring the food. Pouring the hot grease into a tin can, Eula caught herself talking out loud.

"Like Alex always says, this is just women's silliness. Some foolishness I can't even put a name to." Eula scrubbed at the pork chop skillet hard enough to scrape a knuckle. "Alex and Ben Roy just got into it again. Taking food to old Becky must be Alex's way of getting back at my brother." Hearing her own words did not soothe her worries. She walked over to the safe and pulled out her journal. Alex had surely made a dent in her supplies tonight, but, like always, she would find a way to manage.

Climbing into bed, Eula touched her husband's empty pillow. Alex wasn't much of a gambling man. She hoped he and Ben Roy could work out whatever their fuss was about. With the moon at half-mast, a thought that the problem lay somewhere on the mid-forty fought its way into Eula's mind. She pushed it back into the darkness where it belonged. Just woman's foolishness. She and her husband had had other iffy harvests, and they always made out just fine. As she turned down the wick on the bedside lamp, that churning feeling in her stomach that wouldn't settle down told her that sleep would come hard this night.

# CHAPTER TEN

The moon had already launched its transit across the sky when Alex saw the lantern light shining through the small window at the top of the barn. It was the first thing he spotted as he turned off the lane and onto the path leading toward the living quarters on the mid-forty. Somehow, he had supposed the place would already be in darkness. He gave the gray a little kick in the side. The horse reached the barn in a few quick strides. There, standing just inside the partially opened barn door and wrapped in a frayed quilt, was the woman.

Though it was nearly nine o'clock, and the September evenings were cooling down, Alex didn't believe it was quilt weather just yet. Her boots from the morning peeked from underneath the heavy cover. What was she wearing underneath that quilt? Breasts, hips, and those bare thighs—pictures of her nakedness rumbled through his mind. He didn't bother to push back his grin as he scrambled down from the horse. Alex kept his eyes on the woman as he fumbled with the parcels slung across the saddle. The warmth moving up his body heightened his anticipation as he walked toward her. She stood, unmoving, with her head bowed. Alex had almost reached her when the ruckus from the living quarters dashed down upon him.

Loud crying from the woman's youngest child competed heavily with shouts from the shrill voice of her girl, and the running feet of he knew not who. Alex stopped several yards from the woman and pointed upward. Didn't she know to be ready for him?

"Why ain't your children asleep?" He didn't want an audience of children.

"They'll be asleep soon enough, suh."

He had to strain to catch her words and wondered why. She certainly spoke her wants loud enough this morning. Now, it was too late for her to back out of their bargain. Shifting the pouch to his shoulder, Alex reached toward the woman. His hands reached to loosen the tight grip she kept on the edge of the quilt. Just as his hand touched hers, she turned and moved toward the barn, the quilt sliding halfway down her back. Alex saw that she still wore the same dress from the morning.

As he stepped inside, the lure of fresh-cut tobacco that should have signaled an early September greeting at his entrance was noticeably absent. As his eyes adjusted to the gloom, he watched the woman walk past the cows bedded for the night. All was quiet on the lower floor of the barn. The chickens had tucked in their wings for sleep. Outside in their pens, the three hogs had gone silent. He watched his tenant gather the quilt over one arm as she stopped near a small stack of spearing sticks. Behind her, the last few bales of hay lay mounded in a short pile topped with the pitchfork. Bits of straw covered the barn floor. He looked at the woman with her head still dropped. Did she want to tumble him right there? His back was too stiff to roll around on the hard boards of a barn floor.

The light shining down from the small opening of the living quarters framed the woman. Headed toward the ladder, he made out the curves of her body, even under that shapeless dress she wore. Moving after her, he watched the sway of those rounded hips, her skirt hitched just above her knees.

"Momma?" A boy of about nine stared at Alex as he cleared the top rung of the ladder. Dropping the quilt to the floor, the woman stooped to pick up her suddenly quiet toddler. The young girl stood barefoot near an alcove that held an older child with his leg tied to a broomstick. The girl's mouth lay open, and Alex could see that she still had most of her baby teeth. The boy in the alcove pushed himself to sitting as he grabbed his sister by the arm and yanked her toward him. The boy's shocked face matched the girl's.

"Evenin'." Alex gave a half nod in the general direction of the children as he laid open the leather pouch on the small wooden table.

He wanted them fed and out of the way as soon as possible. Hell, he'd send them down to sleep with the cows if this took much longer. Pulling out the jars of spring beans, he noticed the two barrels lined up at the table. Then he remembered he had given two of his own cast-off chairs to the tenants on the back-forty when he furnished their quarters. He made a mental note to look for real chairs for the woman. Alex's hand smeared with grease as he unwrapped the pork chops. He thought he heard a quick intake of breath from the woman when he pulled out the bowl of smothered potatoes. He laid out one of the jars of peach preserves along with several squares of Eula's corn bread. As he looked around for plates, Alex caught the eye of the toddler in the woman's arms. The child's arm suddenly jutted toward a pork chop. Squirming to be released, the toddler ducked his head inches from the rough table edge.

"Henry." Embarrassment shadowed the woman's voice. "You ain't been asked yet."

The boy gave a quick look at Alex and buried his head in his mother's shoulder, setting up a soft whine. The woman bounced the child gently in her arms.

"Where's your plates?" Alex asked.

"Lottie, go get the tins." The woman nodded toward her girl.

The little girl started to move but her older brother's grip held tight on her arm.

"Momma, can I be asked, too?" The girl wrestled her arm free as her eyes darted from the food-laden table to Alex and back again.

"Lottie, ain't I taught you to wait 'til you're spoken to?" The woman's rebuke pleased Alex.

She was raising this get right.

"And, drop your head." The woman's audible whisper landed on the girl, who ran to a corner of the room and retrieved four tin plates, which she placed on one of the crates since the table was full.

The child scurried behind her mother's skirt and dropped her head. Alex gave a halfhearted try at suppressing his smile. If the truth be told, this get had some pleasing ways, but they were definitely pos-ing a problem. He stepped back to look for hanging hooks on the wall only to brush into the woman's middle boy. Stepping quickly to avoid knocking over the child, Alex caught his heel in a large opening in the floorboard.

"Momma, that man don't know to stay away from the knotholes." Little Henry's tears turned to howls of glee as he pointed at Alex.

The woman looked as though she wanted to squeeze herself down through the two-inch round hole and disappear. Instead, she pushed her son's head deep into her shoulder. Alex could hear the child's muffled pleas for air. The woman kept her hand tight on the back of the boy's head. Wordlessly, she bobbed her apologies toward Alex. He backed up and started to walk toward the wall opposite the oldest boy's alcove of a bed.

"Suh, be careful. There's holes all over these floors." The middle boy pointed a finger at the next opening less than eighteen inches from where Alex stood.

"Doug." The woman's rebuke to her child was mixed with sharpness, embarrassment, and clear confusion.

Alex followed Doug's pointing hand and saw no fewer than four knotholes that could definitely cause damage to a small child's foot. He turned to look at Henry, his buried face finally released by his mother though she held her hand over his mouth. The little boy shook his head vehemently trying to push away his mother's hand.

"Why don't you feed your get and tell me where the clothesline is." Alex had barely gotten the words out of his mouth when Lottie shot past him and grabbed a pork chop.

It took her mother one stride to reach the girl and remove the chop before it reached the child's mouth.

"Lottie Welles, you got mo' manners than that. You turn yo'self around and thank Mr. McNaughton for his kindness. You, too, Doug and Henry." As her mother delivered the message, the little girl turned and dropped a quick curtsy to Alex and mouthed a barely heard thank-you.

The toddler finally succeeded in freeing himself from his mother's arms and grinned his thanks.

"Thank you, suh," came from Doug, who sidled up to the table but waited until his mother started to fill the plates.

The woman turned to Alex.

"The hooks is in the wall on either side of the sleepin' corner, and the line is hangin' from the far one." She spoke to his shirt pocket while her two youngest children shifted their eagerness from one foot to the other.

Despite the obvious hunger of her children, the woman did not hurry her words. She waited until Alex gave her a nod before she returned to ladling food onto the plates.

As the woman busied herself, Alex maneuvered his way around the porous floor. A fresh dropped cow pie shot its fragrance up to him through one of the openings as he neared the sleeping alcove. He didn't remember the floorboards being this full of knots when he hurriedly built the tenant quarters four years ago. Of course, the wood had been salvage. Holding his breath, he spotted the thin wire clothesline hanging from one hook, its mass bunched up on the floor near the space the woman called the "sleepin' corner." Alex frowned when he looked at what passed for her bed space. Hadn't he given his tenants a proper bed? This was an area barely six feet long and four feet wide. For his six-foot frame, this would be a tight fit. Worse, instead of a bed, a thin mattress lay on top of a built-in box of wood. The mattress was bare.

"Where's the sheets?" Alex spoke out quickly without remembering that the children were within earshot.

He grimaced at his error as he glanced across the room to the opposite alcove and the woman's oldest son.

"Sorry, suh. I had to shred it up yesterday when my Cleveland fell from the rafters. I think his leg is broke. I used the sheet to tie the leg to the broomstick." She pointed a filled food tin at Cleveland.

Watching for more knotholes, Alex walked to the woman and took the tin from her. The quiet in the room felt steep. He handed the plate to the woman's eldest while he stared at the broomstick-trussed leg. He ran a hand over the skinny extremity, careful not to increase the pain. He stopped at mid-shin, bent down, and placed his other hand behind the boy's leg at mid-calf. With a quick jerk, he moved his hands in opposing directions.

"Uhh." The child yelped. Tears splashed his cheeks, the plate of food trembling in his hand.

Before Alex could reach for one of the torn sheet strips, the woman was at her son's side. The look of a mother tigress protecting her cub flashed out of her face. Alex spoke up as he retied the strips of cloth.

"Break ain't bad. Just had to set it proper. You did a good job tying it up. Have the boy eat his supper." Alex moved to his feet, grazing his head on the low sloping roof.

The woman lifted a hand to touch the forming lump, and just as fast, withdrew it.

"I thank you kindly, suh." It was the softest tone he'd heard from her yet.

With the clothesline strung, Alex retrieved the dropped quilt and hung it across the wire to fashion a semi-private bedroom. He sat on the formless mattress and made another note to himself to find a better cast-off. Behind the makeshift screen, he heard the woman stack the tin plates and put away the extra food. Alex took off his boots and willed her to hurry.

"Doug. You sleep with Cleveland tonight. Lottie will take yo' place with Henry on the pallet." As much as the woman's arrangements signaled progress to Alex, the protests of her children foretold more delay.

Behind the quilt, Alex unbuttoned his shirt.

"Why cain't I sleep with you like always, Momma?" It was the whine of little Lottie.

"You'll sleep where I tell you, and I don't want another word about it. I need Doug to sleep with Cleveland in case he needs help in the night."

Behind the quilt, Alex heard the woman drop to her knees.

"Shh. Henry's already asleep. Momma will give you a goodnight kiss and you go on to sleep yo'self."

Alex strained his ears to catch the sound of the woman's lips on flesh. He undid the top button of his trousers when he heard her footsteps coming toward him. He watched the quilt slide back a few inches.

"I thank you most kindly fo' the food, suh. My little ones is tired from their long day. They'll be asleep in a minute, but my Cleveland is in some pain." Only her face showed through the quilt. Her eyes rested on the bottom of the windowsill next to Alex. "If it pleases you, suh, now that they've got some food in their bellies, I will have my children asleep right at dark tomorrow night. Would you care to come back then?" Her eyes sidled from the sill to the top of his head.

Despite himself, Alex had to admire this woman. Here she was still trying to bargain with him when the deal was long closed. He reached for a small metal flask in his pants pocket. Unscrewing the cap, he leveled it toward her.

"Give him this. It'll cut the pain." He didn't bother to suppress the little smile that kept coming to his lips. His excitement for this woman was growing by the minute.

"Suh?" She looked puzzled, confused.

"It's whiskey. Give him a tablespoon and take one for yourself." Alex pushed the flask at her.

"My babies don't take no strong drink, suh, and neither do I." Her eyes went directly to his own. Her head bobbed her no thanks.

Alex's smile broadened. "It will cut the pain. Give it to him." Alex moved toward her. She took a half step backward, almost pulling down the quilt. He pressed the flask into her hand.

Her eyes looked at him, wide apart. Alex smiled as he reached for the second button of his trousers. The woman stumbled away from the quilt. Alex nearly laughed out loud. Sitting on the corn husk mattress, he removed his boots.

Thank goodness the living quarters were small enough for him to almost make out the sound of sleeping children despite the muffling of the quilt. Unlike adults, they didn't snore and it was hard to figure out their sleep-breathing. He was certain the two youngest had been out for nearly twenty minutes and the middle boy's—Doug was his name— asthma-wracked breathing showed that he had soon followed. Alex just wasn't sure about the one the woman called Cleveland. He'd heard her administer the whiskey and the boy's reaction to the burn as it went down his throat. That had been nearly fifteen minutes ago. Behind the quilt, Alex took off his shirt.

Finally, he heard the woman stirring. He listened to her footsteps as she walked around the room and paused. He guessed she was making sure each child was asleep. He pulled his trousers over his feet. The woman's footsteps slowed as she neared the quilt. She stopped right on the other side. He heard a long-drawn intake of breath as she eased herself around to his side of the blanket. Wearing only his summer drawers, Alex stood to greet her. Her hand went to the makeshift partition, almost pulling it down again. He steadied it as he pulled her toward him.

"It'll be cold tonight without that blanket." The woman pointed to the askew quilt. "Sorry, suh, I don't have another." She held her crossed hands over her chest.

Her gaze must have been on some faraway star she spotted out of the small window above the sleeping alcove, because they sure weren't on him. She nodded toward the temporary wall.

"If you want, it might be best to do this tomorrow. I can borrow a quilt from my Aunt Becky by then." She couldn't quite disguise the note of desperate hopefulness that fleshed out her words.

Alex sucked in his lips to dampen his smile. She was running out of excuses, and Lord knows, she'd tried just about every one. Soon, she'd concede defeat and come to him willingly. He put his hand under her chin and lifted her face to his. She lowered her eyes.

"What do they call you, woman?" He turned her chin to look at the planes of her face. Even with her wide-set eyes held nearly closed, he liked their look. But they came in second to those pouty lips. He ran his thumb from her lower lip to chin. Alex's breath came in hot spurts.

"Annalaura. Annalaura Welles. My husband is John Welles."

Just when he thought she was ready to give in to him on his own terms, she fired back with her little reminder. No husband foolish enough to leave such a woman was coming back tonight, and tonight was all he needed.

"No, no. Anna was my momma's name, and it ain't my momma I want to think about tonight. I'll call you Laura." He pulled her closer, feeling the crush of her breasts against his chest.

Alex slid his body sideways against their firmness. When he pushed her back, his voice was almost a whisper in his own ears.

"And, tonight, Laura, I can guarantee that you won't get cold without that quilt." He bent to kiss her only to find her lips pressed tight together like he was trying to feed her castor oil.

Her whole body trembled under his touch. He could feel those shaking shoulders move to push him away, tense up in mid-motion, and stop. Her body was fighting him. He released her to reach down for the whiskey flask.

"Take a swig of this." He removed the cap and held the metal bottle in front of her.

She nodded her no, blinked the remembrance of her place, and answered in a voice that held almost no sound.

"Thank you no, suh."

Alex took a drink from the container and put it to her lips.

"You need a drink." He tipped the flask to her closed mouth.

She barely parted her lips. He was certain that almost none of the liquid made its way into her throat. Alex pulled her back toward him and let his lips glide slowly over hers. He felt her body go limp. Suddenly afraid that she might pass out, he slipped an arm around her back. He'd never heard of a colored woman fainting. He'd always been told that colored women were strong enough to take anything. Alex pulled back from her and watched her face closely. Laura's eyes were now tightly closed. He put a hand over the spout of the flask and poured a bit more of the liquid over his hand. Though she clenched her jaw as tight as a wood vise, he managed to part her lips with his fingers. She shook even more as he took her into his arms again. Alex tapped his liquor-soaked finger over her lower lip, rubbing little circles over its fleshy softness as he moved his hand from left to right. Even in the dim light, Laura's lip glistened like a juicy plum as he leaned over and drew it into his own mouth. The taste of spiced honey and ripe, fresh-picked strawberries filled his throat and went straight to his head faster than any whiskey he'd ever drunk. The scent of her was like a sandalwood platter he had once smelled on the back of a peddler's cart. It was from Africa, the peddler declared, and cost five dollars. Alex hadn't spent the money. Kissing her again, he sucked in her upper lip, drawing it out long and slow into his own mouth. He let his tongue probe inside Laura's mouth, but there was no answering response. In his embrace, he let his hands run down the sleeve of her blouse. She held her arms stiff at her side.

"Put your arms around me," he whispered in his half-voice.

Slow like a herky-jerky at a carnival, Laura reached her arms to his shoulders and stopped. They lay there like two leaves of untied tobacco. Shining through the window, the silver light from the moon laid a streak across the back of Laura's blouse. Alex stepped away and put a hand at the open band collar of her top. He felt the tension in her increase. He knew she wanted to slap his hand away. In the dimness he felt her hands on his shoulders open and close as though she were pleading with them to complete Alex's bidding. She was still fighting the battle to push him away, but soon, she would acknowledge their bargain.

Slowly he undid the four remaining fasteners on her blouse. Laying the garment open he was surprised to see a chemise underneath. It hadn't been there this morning. Fingering the thin cotton, he could tell it was her Sunday best. Had she put it on to please him, or to hide herself from him? With great care, he unfastened the buttons.

"We can lay our heads on this." He slipped off her blouse and tossed it on the skinny mattress.

The woman's eyes still hadn't opened. Alex slid his hand from her waist up between the open chemise as he pushed the garment off her body. Threads of silvery moonlight shone across her breasts. Alex struggled to catch his breath. Her nipples looked like lumps of nutmeg ready for the Christmas baking sitting atop a high-mounded cinnamon bun. His own fingers began to tremble as he encircled each nipple, letting his fingers explore in ever-widening circles. He tried to cup each breast, but the fullness overflowed even his large hands. He felt Laura's shaking increase with each breath she took, pushing her breasts even more firmly into his hands.

Alex sat heavily onto the corn husk mattress, his thoughts refusing to marshal themselves into coherent order. He managed to lay his hand against the waistband of her skirt. He fumbled for the button and let the garment drop to the floor. Underneath, Laura wore her summer drawers. He was certain they hadn't been there this morning, either. He let his hands slide to the side to undo their ties. The woman dropped one hand from her breasts to grab at the top of her falling drawers. Alex brushed her hand away as the pantaloons drifted slowly down on top of her skirt. He felt her body sway away from him, and he pulled her closer.

Laying his head on her bare stomach, he let his tongue slide from her belly button up to the beginning roundness of her breasts. With his hands on those firm hips, he pulled her down toward him and sucked one nipple into his mouth. A long ago memory of fresh-cut raw cane sugar flooded his throat, and he felt he was about to float away into a river of delirium. He slid his tongue across her chest and drew in the other nipple. He felt her knees soften under him, and Laura grabbed at his shoulders to steady herself. He glanced up at her face. The moonlight had captured the top half in a glow that reminded him of a sepia-toned Bible picture. He let his tongue linger as he traced it down her belly again. He moved his mouth lower, exploring her stomach until he reached her spirally haired triangle. Breaths were getting harder for him to take. He turned her to the mattress. She sat with a loud rustle of the husks. Her eyes, which had been closed tighter than a new-trussed pig for many long minutes, opened wide at the sound.

"My babies will wake up, suh. Maybe you could come back tomorrow?" The trembling in her body had reached her voice, and she could barely get out the words.

Alex pushed her gently down on the mattress. Her shaking rattled the corn husks even louder. She lay straight-legged on the bed with one arm wrapped over her breasts and the other stretching to cover that triangle of hair. He rolled on top of her and bent down to kiss her just as she turned her head slightly. His lips caught the corner of her mouth. He used the strength of his legs to spread her thighs. She held them together like an unheated piece of iron at the smithy shop.

"Bend your knees." His voice was almost as gone as hers. He felt her falter in her effort to respond. Alex slipped his hands under her hips. Her legs felt like lead as he guided them upward. He moved his fingers to the spirally haired triangle to explore the depths of the wetness within. The woman arched her back and almost convulsed for breath as he pushed his manhood into what felt like the golden portals of heaven.

The moon shining in his eyes stirred Alex from sleep. He immediately knew where he lay. The satisfaction in his body hadn't gone away. He knew she was still there. Alex lay on his side with his knees bent. The woman, Laura, lay with her body curved spoonlike inside his. Her skirt was thrown over them both from waist down. Her blouse, which had earlier covered them, had slipped to the floor. His hand lay across her belly holding her close to him, her bare bottom firm against his manhood. Alex let his hand travel up to one breast. He squeezed it again, still pleasured by the rounded firmness. When he took hold, he knew there was more than soft cotton mush in his hand. Laura stirred.

"I'd best be gettin' to the smoke house." Her voice was weak as she tried to scoot toward the edge of the mattress.

He pulled her back knowing she could feel his growing manhood.

"You don't need the smoke house at this time of night." He let his hand drop back to the triangle.

Laura pointed at the window.

"Moon says it must be close to midnight. I reckon you'll want to be gettin' back home before much longer." She made another attempt to rise when Alex rolled her onto her back. She looked startled. "Suh, it's late." Her earlier fright had dissolved into alarm.

"I want you to put your arms around me and hold me like I'm the man who's goin' to take you north and pour sandalwood perfume all over you." Alex kissed her face, breasts, belly in quick succession.

Slowly, she laid her arms around his neck.

"Now squeeze tight," he commanded as he entered her for the second time that night.

The light from the moon now shining past its zenith made her skin glow copper like a new penny. He couldn't tell which he enjoyed more—the first or the second time. He knew he was more than up to a third go with Laura. He glanced over at her and was sure she hadn't slept at all. Her breathing was still too fast for that. He guessed she'd lain without moving the entire time he'd been asleep. Alex leaned over and kissed her on the ear.

"Honey, I do have to go now." He let his hand stroke down her breast one last time as he moved her to sitting.

She half turned toward him, a quizzical look on her face. If he had misspoken, it was understandable. He hadn't made love a second time since the earliest days of his marriage, and this had been so much sweeter. He reached for his drawers as Laura stepped into her skirt.

"You don't have to get up. I can find my way down the ladder in the dark." He touched her hand as she buttoned the skirt.

"I've got to go to the smoke house to wash out that pretty bowl you brought the potatoes in." She reached for her blouse.

He grabbed her wrists.

"Let me look at you just a minute longer." He pulled her between his bare thighs.

She laid her hands against his shoulders. This time, there was no trembling.

"Suh, thank you for the food fo' my children, but I reckon you don't want the sunshine to catch you here." Even her voice sounded stronger.

Alex released her as he stepped into his trousers. Laura started to move beyond the curtain.

"Wait," he whispered to her as he tossed the shirt he'd just retrieved back onto the floor. "I want to kiss you again."

Slowly she turned, but she didn't step toward him. He pulled her closer and began fumbling with the newly buttoned blouse. She pushed him away.

"Suh" was all she uttered as she walked beyond the quilt to the table and the bowl. Alex dropped to the mattress to put on his shirt and boots. Hearing her ready herself to descend the ladder, he pulled down the quilt.

"No need to hurry with that bowl. I'll be back in a couple of nights to pick it up." He didn't have to wait long for her response.

Laura stood with Eula's bowl in her hands, blinking non-understanding back at him. He walked over to the table and took the hand-painted container, with its streak of silver, from her hand, setting it on the table.

Alex gathered her in a tight embrace and kissed her lips. He released her to reach into his trouser pockets. He retrieved something and dropped it into the bowl. A thin shaft of moonlight caught the coin. Alex watched her face when the value of the silver dollar registered. It pleased him greatly to see that wide-eyed look of confusion on this most satisfying woman. As he headed down the ladder, Alex felt her stares into his back and sensed her confusion. He had just given her a whole day's wages for a working white man. As he reached the barn floor, he knew this woman was more than worth it.

# CHAPTER ELEVEN

John's new brown and tan houndstooth coat was stronger on looks than it was on warmth. Still, he had no cause to complain since he had given just one silver dollar for it to a man over from Davidson County who couldn't take Nashville anymore and lit out for home. A coat warm enough for the late November weather would cost John two dollars, and since it had to look good as well as offer warmth, the price could come closer to three. As John stepped through the back door of Zeola's, he didn't bother recounting his money situation. He already had the rent for December, and that wasn't due until a week from tomorrow.

"Close that do'. You born in a barn?" Big Red slid his bone-handled butcher knife sideways into the just-cooked Thanksgiving turkey.

Two pecan and two sweet potato pies cooled on the sideboard, and, by the smell of the cinnamon in the air, two apple pies were about to come out of the oven.

Reaching behind him, John absentmindedly pulled the door closed as he gave a quick nod to the cook. Big Red, with his no-manners self, was no longer of any interest to John.

"I sees you got yo'self a new checkered coat for the occasion." Big Red made short work of both turkey legs and now started on the breast.

With only a quick head shake in the cook's direction, John walked toward the connecting pantry door into the dining room.

"Well, slick, you ain't as good as you thinks you is. I sees she's got you sittin' the first pot. How many mens you think is gonna leave the family table and come on over to Zeola's on Thanksgivin' evenin'?" The knife in Big Red's hand swayed in the air like a leaf on a flimsy branch.

John stopped short and turned to the cook.

"Red, you save me that there juicy thigh and I'll give you fifty cents from tonight's pot." John walked through the swinging door with Big Red staring after him.

While the cook mostly talked nonsense, the thought of the amount of cash in the holiday pot had crossed John's mind. In early November, Zeola had moved him from holding two poker pots a week to four. Now, Monday through Thursday, John Welles was the pot man for all the early poker games at the whorehouse.

He opened the door into a dining room in mid-Thanksgiving meal. Not wanting to disturb the diners, John nodded a slight greeting in their direction and eased himself around behind the chairs holding six of Miz Zeola's girls. Miz Zeola sat at the head of the table with the elderly and widowed Mr. Jackson who, according to the girls, couldn't really lay a woman—though on more than one occasion, he had just about died trying. This was the early shift. Zeola fed her girls and got them ready to take in any walk-off-the-streets. She figured that on a holiday, some lone man might hanker for female company. Zeola didn't count on getting more than a dollar a girl for these quickies. Mr. Jackson was a charity case. He paid three dollars once a week, come summer or winter, just to spend two hours staring at one of Zeola's buck-naked, less-than-prized girls.

The real money didn't usually come until after eight o'clock, when the wives of all the regulars had fallen asleep, dog-tired from their Thanksgiving efforts. Miz Zeola included a whiskey-laced pecan pie, bought her ladies new finery, and gave her regulars a dollar off the usual price, all in the spirit of the holiday. The liquor flowed free on Thanksgiving and Christmas. Zeola figured if she could just get the men to come in, relax 'em with good booze and cheap women, they would be more than happy to sit in at a high-stakes poker game.

John just hoped her plan would work now that he was holding the pot. He'd heard tales that on some past holidays, there'd been so many players lined up that Zeola had to run three or four games instead of the usual two. It was hinted that for her best customers, Zeola even held the pot herself.

"I hope you had a bite to eat already, John Welles." Zeola unthreaded the watch chain from her thickly powdered bosom and pulled out the timepiece. "I got two men sittin' in the room already and two mo' just stepped into the house." She tilted her head and squinted one eye to get a better look at the watch. "Them last two gonna play first and lay later."

Everybody at the table, including Mr. Jackson, who almost popped out his false teeth, let out a howl.

"Big Red will be savin' me a plate, Miz Zeola. I'm gonna see that you have a bang-up first pot to help you celebrate yo' Thanksgivin'." John turned down the wattage of his smile as he left the dining room and walked across the empty hall into the parlor.

Miz Zeola's parlor was the biggest room in the house. She'd had some colored carpenter come over and put up four floor-to-ceiling pillars, although they did no earthly good at holding up anything. In front of each one, she'd set two velvet-cushioned chairs, but what made the space special, were the tall potted plants she placed on the side opposite the chairs. She fixed them in such a way that the big leaves just about covered whoever was sitting there. With the candlelight scattered just so, and those bushy plants, Zeola hoped that nervous newcomers, and the Nashville law that sometimes sniffed around on official business, would be too busy looking to see who was hiding behind those palms to pay much notice to the two doors that looked like skinny five-foot-tall amateur painted pictures. They weren't as well hidden as the little rooms off the dining area, but it would take somebody who'd stepped through them before to know where on the picture frame to push at the hidden handle to get them to open into the poker rooms.

Of course, the piano man was also part of Madame Zeola's plan. This afternoon, the parlor was empty except for A.C. playing the blues down low. Miz Zeola always patted herself on the back at her find of the piano man and paid him well to stay at her establishment. In that rough voice of his, he could sing as well as play. The only thing Zeola didn't like about A.C.'s playing was his choice of the blues. She absolutely banished it from her parlor during business hours. She only wanted to hear that new music coming up from New Orleans, jazz she called it. The madam said it made her customers want to tap their feet and move their bodies and that's what she wanted—men to keep moving their dollars out of their pockets into hers.

A.C. looked up as John walked toward the concealed doors but didn't break a chord in his rendition of the Memphis blues. As good as the piano man was, John sided with Miz Zeola. Tonight he didn't want his poker players crying into their bourbons-and-branch over some woman gone to another man. He wanted their minds fixed on straights, full houses, and aces over queens.

"Game's 'bout to start." John leaned over the top of the upright. "Can you give us a little bit of that jazz music?"

A.C. stopped and pulled down his lit cigarette from the top of the piano. He narrowed his already small eyes at John and went back to playing the blues. John gave a half-smile and pushed open the picture-frame concealed door and walked into the smaller of the two smoke-filled poker rooms.

He took a quick survey of the occupants while nodding his polite good evenings. Two men looked up from the liquor Miz Zeola had so generously poured for their wait.

"You here to play?" The dark-skinned speaker, dressed in a plaid shirt and striped pants, had obviously spruced himself up for the evening.

"Miz Zeola's wants to know if you needs yo' drinks freshened? Two mo' players will be here directly. I'm the pot man." John took in the look of the other player without letting him see his eyes shift in the man's direction.

While the fellow had on a starched white collar and the beginnings of a suit, the second player didn't look much more prosperous than the first. Since John's promotion three weeks ago, his cut of the cash put in the gamblers' pot had gone from a straight five dollars to a dollar per player and ten cents on every dollar in the pot. The trouble was he was still assigned to the small games where a big pot might be forty dollars with a top of six players. That gave him ten dollars a game, but in the three weeks since Zeola had changed the rules, games like that had been hard to come by. Still, his newfound wealth was enough to move him out of the storeroom and up to the second-best room on the top floor of Miz Brown's rooming house.

Without waiting for an answer, John walked over to the small round table set up as a makeshift bar and picked up the opened bottle of bourbon and the branch water sitting nearby. He splashed the dark liquid into the men's glasses, three quarters full. He topped each off with just a few drops of the branch. Both players nodded their approval, and the one in the starched collar slapped John on the back, just as the trick door pulled open.

"Whoo whee, look at that purty checkered suit the pot man's got on." Pete, one of Zeola's long-standing regulars, stepped into the room.

When he wasn't building fake columns in the whorehouse parlor, Pete did carpenter work for the white folks who lived on the outskirts of Nashville. Pistol Pete, as he ordered the girls to call him, came in about every two weeks for servicing and a poker game. He was a loud talker but a small-time player. John nodded his good evenings to the newcomers.

"I brought me a friend from Memphis. Git him one of them drinks, pot man, and don't skimp on the liquor." Pete pulled out one of Zeola's straight-backed chairs as he pointed to the new man.

John reached down deep to pull out his best meek expression as he lifted two glasses off the liquor table and poured each almost to the top with bourbon.

"Pot man, here, is from the country. Way out in the country, the girls tell me." Pete swigged down half the glass in one gulp.

John guessed this was not his first drink of the evening.

"This here's Bubba." Pete started to slur his words.

"Pleased to meet you, Mr. . . . . er." John held out a hand to the newcomer who reached to shake it to loud laughter from Pistol Pete.

"Whoo whee. That's how you know they is from the country. Manners fallin' all over theyself." Pete finished off the glass and handed it to John.

The first two players exchanged nervous looks as Bubba sat down. John was certain he saw the starched-collared one pat at the pocket where he kept his watch.

"Let's get this game on." Bubba's liquor glass remained untouched.

"Name's Johnson. Who's dealin? Anybody but Pete."

All eyes at the table swung toward Bubba Johnson. With his thick November coat and porkpie hat, Mr. Johnson looked like Miz Zeola had mistakenly sent him to the wrong poker game. This one looked and sounded like he meant business. Maybe John would get a decent cut tonight, after all.

"It's up to you all." John remembered the drill.

Zeola had insisted that her pot men give the appearance of being absolutely neutral. That's what made a man good at the job, she offered. The players had to trust the man guarding the dollars they'd anted up for the play. And a good pot man could show no favorites. He not only had to look well fed, but he had to be dressed in Nashville's latest to show the players that their money didn't mean a thing to him.

"How long did it take you to get that hayseed out of yo' head? Even a conk in yo' nappy hair ain't gonna make you look like you from the city." Pistol Pete's meanness grew with each passing minute. "A country boy is a country boy and they ain't no two ways 'bout that."

The man with the starched collar gave out a quick, nervous laugh then, just as fast, washed it away with a hardy swallow of bourbon. John watched Bubba cut his eyes at Pete.

"Like a lot of folks, I started out in the country." John pulled out the fifth chair and sat down, setting the bottles next to him. "And I ain't learned it all, that's for sure, but I do know my way around a big city poker table, and that's all that matters tonight." John peeled the tape off the card box and pushed the unopened carton to the center of the table.

Mr. Plaid Shirt, sitting next to John, took in a deep breath. Bubba slapped his hand on the table as he squinted approval at John.

"I can deal," the plaid-shirted man responded as he pulled the cards from the carton and began to shuffle.

The starched-collared one was the first to lay his dollar on the table. The dealer added a second as did Bubba. Pistol Pete tossed in two. John made sure that nobody could see it, but his smile almost broke out wide all over his face. This Thanksgiving could be his first truly thankful one in a long time.

⚬⚬⚬

"That's it for me." The fellow in the starched collar pushed back from the table and pulled out his pocket watch from his makeshift suit. "It's 'bout nine o'clock. I been here for close to fo' hours. I'd best be headin' home."

"I'm goin' upstairs and lay me a woman. Pot man, I'm comin' later to get my money back. Come on, Bubba." Pistol Pete knocked over his chair as he stood and glowered at the plaid-shirted dealer. "Pot man, this ho house got any new gals fresh from the country?"

Bubba leaned over to John, ignoring his friend. "See if you can get me into a bigger game when I come back downstairs." Bubba spoke low as he reached into his pocket and pulled out a silver dollar. He slid the coin across the table toward John.

Nodding his thanks, John smelled the perfume before the wearer pushed open the hidden door.

"Evenin' gentlemen." Zeola, and her favorite fragrance, almost overpowered the small room.

Even Pistol Pete, with all of his senses drowned in alcohol, reeled backward.

"Woman, I wants yo' best gal tonight and I wants her fo' two hours." Pete scowled at Zeola who didn't break her smile as she nudged him in the side.

"All my girls never stop talkin' 'bout Mr. Pistol Pete. I got one set aside special just fo' you." She leaned in close to his ear, but her whisper came out as loud as a rooster crowing at dawn. "And I'm gonna give her to you fo' two hours for only three dollars, and I guarantee she won't disappoint." Dismissing Pete, Zeola swirled her chiffons toward Bubba. "I do hope you enjoyed the evenin', Mr. Johnson." She held out her hand.

"I'll enjoy it even more in about forty-five minutes." He reached into his jacket pocket and pulled out a billfold. He laid a five-dollar gold piece in Zeola's palm. "I wants yo' next game." He winked at her as he helped a very drunk Pistol Pete through the door.

"Well, who's the big pot winner, tonight?" Zeola stuffed the gold piece into her bosom.

"That would be me, ma'am." The plaid-shirted dealer finally lifted his eyes from the pot.

"Ain't you the one." Zeola sounded impressed as she turned to John. "Pot man, how big is it?"

"Sixty-six dollars and fifty cents, Miz Zeola." John had already run the figures through his head. He would be getting over eleven dollars, counting Bubba's tip.

Zeola made quick work of counting up the pot, but she slowly and carefully placed each of the coins and bills into the winner's hands. With the two remaining players watching, she fanned out six one-dollar bills and laid one five-dollar gold piece into John's hands.

"That extra forty-five cents is from the house cut." She turned to the plaid-suited fellow. "Since this is yo' first time here, I want you to enjoy Miz Zeola's hospitality. I'll charge you only a dollar for one of my best girls." She reached over to the bourbon bottle and filled up the winner's glass and handed it to him as she walked him to the door.

John clutched his twelve dollars. If the pots continued to be this good, he could be home by Christmas with enough money to make the day very merry indeed for Annalaura and the children.

"Don't put that money in yo' pocket just yet, John Welles." Zeola closed the door behind the last two players. She kept both hands behind her back, holding the knob shut tight.

"How much of that money you sending home to yo' wife?" Her fake eyelashes swept her face like big black cobwebs as she reached a hand toward him.

Startled, John swung his eyes toward her.

"Yeah, I know you country boys leave yo' women at home when you come over here to Nashville." She turned her palm upward and rubbed her fingers together.

"Beggin' yo' pardon, ma'am, but I know how to take care of my family." John pulled out his special grin, the one he always used when it was especially important to put the charm on a woman getting too big for her drawers.

Zeola dropped her hand and walked a slow circle around him. At his back, he could feel her eyes boring into him.

"Looky here, Johnny-boy, I don't need you makin' eyes at me. I been in this business since befo' you ever laid yo' first woman." She stood flat-footed in front of him. "Now, I likes you well enough to see that you don't mess up."

"No, ma'am. I ain't got no intentions of messing up." John clamped down on his tongue to control himself. He wasn't used to not speaking his mind with a woman unless it was Annalaura. He knew an argument with her was useless.

"You ain't my first country boy, you know. I seen a lot of them come and go. Some get so full of theyselves that they go outta here feet first with a Texas jacknife stuck in their gut. Others try to shuck and jive me, and I show them the do' fast." Zeola waved her chiffon sleeves in the air. The scent of her perfume attached itself to John's new coat. "You may think it, but I'm not meddling in yo' business. But, if you don't know that a good-lookin' country boy—who thinks he's the smartest thing to ever hit Nashville—and his money is soon parted, then you ain't nothin' but a fool."

"Yes'm. I do understand but there's no need for you to worry 'bout my family. I'm gonna take care of them just fine." His grin came so automatically when he tried to fool the women that he now had to work hard to keep Zeola from seeing it.

"I'm proud to hear it, but I'm takin' five dollars a week outta yo' pots to save for yo' woman whether you likes it or not." Her hands went to her hips, and all of her chins waggled when she shook her head for emphasis.

John seethed as he let the smallest of smiles cross his face. He'd done without a mother since he was six years old. With more than twenty dollars already stashed in the floorboard box back at Miz Brown's, he knew what he was doing.

"'Tending no offense, Miz Zeola, but I'm used to taking care of my wife myself."

"Man, how long you been in this town? You been at my place fo' months. How long you think any woman on a cracker's farm, sharecroppin', can take care of herself on her own?"

"She ain't on her own, Miz Zeola. She's got her people to go to." It wasn't quite a lie.

"You done good tonight." Zeola nodded.

"Now I'm gonna cut yo' rate on my Sally 'cause she done got herself in a family way. You can have her fo' a dollar, but it betta be after business hours."

John swallowed twice before he trusted his voice. "That's most kind of you, Miz Zeola, but ain't Sally too old to have a baby?"

The idea of paying for a woman had never set well with John. He had always been surrounded by women who had been more than willing to give him anything he wanted for free, and now Zeola wanted him to pay for one of her girls as some kind of bonus for his loyalty. He knew he should think of it as just another part of goin'-along-to-get-along just like at the sporting houses, but this one didn't set right with him. He held nothing against the whore, but he wanted no part of another man's leavings, business or no business.

"If you got yo' eyes set on one of my chambermaids, you can get yo'self ready to step on out my back do' and don't never look back." Zeola's face looked like the aftermath of the battle of Clarksville.

"No, ma'am. I don't want yo' hired girls, nor Sally neither. I don't needs to pay fo' no woman." The words had slipped out before he could snap them back.

The darkest, most dreary winter's day couldn't compare to Miz Zeola's face. "Too good to pay, is you? Just stay away from my chambermaids."

"Miz Zeola, ma'am, I'd feel real bad showin' disrespect to a woman 'bout to have a baby. But, if you say she wouldn't mind, it would pleasure me to take you up on yo' kind offer of a dollar fo' her time." He hoped he'd found the right angle to bend his neck in apology.

"Hmm. Don't let Sally be none of yo' worry. She ain't but six months anyway." In a flurry of chiffon, she huffed through the door.

Alone, John pulled out his new, squared-off handkerchief. He mopped his forehead before he began tidying the game room. He had risked too much to ever slip like that again. Zeola was nosy, and he had already told this woman more of his business than she needed to know, but he was learning quick that she was not a woman to wink at and forget. John had to have her on his side. The idea bounced in his head that his wife and his boss were both women who took more than a shuck and a grin to satisfy.

At the rate he was going, he would have plenty of money to make a big splash next month at Christmas. But, a big splash would not be enough to ward off Annalaura's anger. John needed far more than toys, pinafores, and satin sheets to warm up his wife. That made her all the more priceless. He had to get enough money to buy a place of his own, and he needed Zeola to hand him those late-night pots. Big Red had said the second game pots sometimes got up to two hundred dollars with eight men playing two decks. That was the kind of money he had to have to make all this worthwhile.

John put the dirty shot glasses in the basin Zeola kept under the little table. The room had no mirrors—bad for poker playing—but he made an attempt to pat his hair in place. Cut short, John knew he didn't need a greasy-style conk to make him attractive to the ladies. After he'd fulfilled his obligations to Sally, so she could rave about him to Zeola, he would head back to the boardinghouse and his two new housemates. One or even both of the colored schoolteachers could help him pass the time away from his wife. When he got home in the next couple of months, he knew not to tell Annalaura about this part of his adventure. In that one way, she was just like all other women. A man gone from home a long time on family business got lonely. Paying for a woman was no substitute. The companionship of the schoolteachers would make the time away from Annalaura tolerable. Even if his wife could never accept it, a man had to be a man, and what the wife didn't know couldn't hurt her none.

# CHAPTER TWELVE

Eula tugged at her shawl to see if it would fit around her a second time to ward off the mid-January cold in the smoke house. Even the closed door didn't stop the wind that had whistled at her window most of the night from creeping into the room through the unplastered walls of the outbuilding. With fingers frigid from the weather, she held up the lantern to the shelf she reserved for her canned vegetables. Normally, she checked her stock every Monday, but not until after breakfast. Today, she'd gotten an early start because Wiley George planned to drop off Tillie for a few hours while he and Alex went over to Lawnover. As she held the light higher to read the labels on her Mason jars, the frown in her forehead deepened.

With her lamp held close, she counted the jars of canned tomatoes a third time, but the number still came up thirty-five. She lowered the lamp to the worktable where her journal lay open to the page marked "vegetables." She traced one rapidly numbing finger across the rows. There it was—tomatoes—and the number beside it read fifty. Since Alex enjoyed her tomato preserves and she included them on his plate at least twice a week, she was quite certain of the accuracy of her count. Fifteen Mason jars of tomatoes were missing. She drew her fingers down the other columns. Turnips, collard greens, carrots, pole beans, and butter beans all came up short in the recount. There were no errors. Eula pulled her fringed shawl closer as she lowered the lamp to the fruit shelf below. Mason jars of peaches, pears, plums, cherries, and strawberries were missing with no accounting. Even more perplexing than the vacant spaces on her shelves was the clear evidence that the shortage was increasing each week. She glanced over at the hams and hog sides hanging on the hooks. Their numbers were also down, but Alex had explained their decrease. Back in December, she watched him box up a particularly large package of just-cured bacon and two whole hams that he said were destined for his sister in Kentucky. Eula remembered being so stunned by this unusual generosity that she forgot to conceal the surprise in her voice. When she mentioned that his sister had always appeared too well-off to need food from them, Alex flashed a quick torrent of angry words that doubly baffled her.

Holding the lamp in front of her, Eula put the closed journal in the crook of her arm and stepped out of the smoke house, latching the door behind her. She glanced around to see if the sun was preparing its appearance. The ground was still hard with frost and the night smells still lingered in the darkness. With caution, she stepped up the gravel path to her porch door and into the kitchen. Sitting at the table, Eula opened her account book and carefully lined out the old numbers. In her square handwriting, she wrote in the new. The missing jars both puzzled and annoyed her, though she was keenly aware that annoyance had to be banished from her mind as unworthy of a good wife.

Her irritation stemmed from her pride. Eula didn't need Reverend Hawkins to remind her that pride was a sin almost as bad as bearing false witness or coveting thy neighbor's ox. But, if pride hadn't been strictly forbidden to the righteous, Eula knew she could hold her head higher than any other white woman in all of Lawnover when it came to farm managing. That the food stocks were dwindling and she couldn't explain the loss gnawed at the one talent where she knew she was the best. Still, such a boast was prideful and she tried to push the sinful thought out of her mind. Eula tapped her pencil on one of the open journal pages. The assault on her pride did nag at her, and tonight on her knees, she would have to pray extra hard for forgiveness. But another worry, far more serious than missing tomatoes, clouded her mind, and she wasn't quite sure how she should address that particular subject to the Lord.

Dropping her pencil on the table, Eula stood and walked to the corner behind the stove and lifted the poker. She jabbed it into the coal box to stir up the fire in the cook stove. The kitchen remained too cold for her to remove her shawl. Most mornings when Alex got up first, he stoked the fire, but not this morning. She took a corner of her shawl and rubbed clear a little spot on the kitchen window. A pale white glimmer of a January day shone back at her, and still no sign of Alex. Eula dropped to her chair and wrapped her arms under the shawl as she started a slow rock. Maybe she should offer up a prayer right now and get these unreligious thoughts out of her head.

The wind had been fierce last night, and sleep had been hard to come by. Though she never, on purpose, touched Alex unless he asked, Eula didn't have to feel the cold sheets on his side of the bed to know that her husband had been away all night. And she didn't need the whistling wind to remind her that it wasn't the first time her husband had stayed away from their bed. Some years, his nights away had been as many as four or five. Those times, he had casually mentioned, were spent playing cards with Ben Roy at the Lawnover tavern. Last evening, Alex had made no mention of a game nor of the tavern.

Returning the poker to its corner, Eula went back to the table and stared at the columns and numbers in her journal. She needed no written record to recall that her husband hadn't been close to her more than three times since September. She required no jog to her memory to wonder if something, or someone, other than a game of cards was the cause.

The first time Alex passed over their Friday together, she paid scant attention. He hadn't joined her at bedtime, and he had been beside her in the morning. When that second September Friday slid by, she admitted to a fleeting moment of curiosity, but that passed when she attributed her husband's lovemaking absence to the hard work and long hours he'd spent bringing in that troublesome mid-forty. After he returned to her bed right after full moon in October, she dismissed her earlier fretting. Her relief was short-lived. After his two quick visits in November, the last of which she had carefully maneuvered herself, there had been nothing. That was when her impure thoughts began.

Momma Thornton had never said it outright, but Eula had pieced together enough information in her growing-up years to know that men, long married, sometimes acted like Alex. When she was fifteen, she caught her mother in a back bedroom rouging her cheeks and puffing a white powder all over her face and down the front of her open shirtwaist. She had never seen the frown lines etched so deep in her mother's face before, nor had those eyes looked as though they'd seen more than they wanted.

Eula's mother had promised her a new set of hair ribbons if she wouldn't tell her father that his wife's newfound attractiveness was anything but natural, glowing beauty. The next morning, her mother came out of her room smiling, with the rouge long gone from her cheeks. Momma had winked at her and whispered, "It's a wife's job to keep her man happy." Eula got red, blue, and green ribbons the very next time Momma went to town. And she learned that a less than vigilant wife could cause a husband to stray.

After twenty years of marriage, had she forgotten that early lesson? Though she loathed the thought of rubbing her face in white powder and painting on red cheeks and lips, she knew a wife's responsibilities. It had taken careful planning, since a trip to the Lawnover store was out of the question. A description of everything she bought there would have gone straight to Alex's ear. Instead, she managed to catch a ride to Clarksville with Ben Roy when Fedora was over visiting Tillie. While her impatient brother took care of his own business, she hurried over to the mercantile store. She bought a bar of a sweet-smelling soap—the clerk called it lavender. She wasn't at all sure how Alex would take to soap that smelled like a flower rather than good old homemade rendered tallow. In case the lavender didn't work, she bought a bottle of rose water. Back home, she splashed half the bottle of roses into the galvanized tin wash tub and scrubbed her skin raw with soap that smelled more like the bark of the tree rather than the purple flower. Even so, it had worked. When Alex came in from the fields that Tuesday night, he took her to bed. That had been in late November and no amount of scrubbing and soaking had worked again.

The kitchen had not only warmed up, it had become fiery hot. Eula dropped her shawl from her shoulders and let it fall to the floor. She reached for her journal and used it as a fan. There were images flying around her head that no good wife should entertain. Sin and sinful thoughts were the province of Reverend Hawkins, but she would never dare discuss such things with him. Over the past few weeks, she'd nearly rubbed the print off some of the pages of her Bible looking for the precise chapter and verse that said distrust of a husband outranked pride as a major sin. With two hands, Eula batted at the air with the journal. God help her, despite weeks of effort, she could not banish the thought of Alex with another woman. If he had strayed, and she couldn't accept that he had, it must have been her fault.

Though Momma Thornton had been very strong on telling her eldest daughter about her shortcomings, she had never been much at sharing talk about the bed habits of men. Eula got most of her information on the ways of men and women at quilting bees, canning parties, and even at church socials when she sometimes overheard snatches of conversation. Fedora had hinted that husbands lost "the feeling" after a number of years of marriage. And her sister-in-law had made it clear to the ladies that it wasn't the fault of the wife. Eula wished it were so. If she could ask Fedora how a husband lost such a thing and what a wife could do to help him retrieve it, she would do everything within her power to keep Alex home with her.

Alone in her kitchen with dawn finally breaking full and pale this January morning, Eula failed in her efforts to put flesh and bone to the other woman. If such a person existed, who could she be? Surely, no one Eula knew. None of the women at the ladies' quilting bee or the church socials had ever paid Alex more than a passing glance in twenty years, despite his good looks. Eula slammed her journal to the table.

As she stood to start breakfast, she was certain Alex was out riding the acres. Perhaps a fence had blown down in last night's wind. Still, the missing food, her husband's decreasing interest in her as a wife, and the early risings had all happened about the same time. Gripping her pencil, she suspended it over the margin of her journal. Pressing the pencil down hard to stop the wavering lines her shaking hand might make, she wrote down question marks next to the columns of missing foodstuff. In the margin she lettered, "Must ask Alex," and changed her mind. As she drew her pencil over and over the words, a gnawing sensation, worse than a mouse at the corn barrel, grew in her belly. She didn't know how or why, but her stomach told her that somehow the mid-forty was part of the answer to her quandary.

# CHAPTER THIRTEEN

It was the sound of another blast of wind rattling the barn window rather than the sliver of pale light creeping through the shutter that finally forced Alex to open his eyes. Even without looking, he knew the ground outside would be frosty this January morning, but under two quilts with Laura's soft body spooned inside his, he felt bathed in warmth. He awakened with his arm over the roundness of her belly. He let his hand slide even lower. She stirred and he leaned to kiss her ear. She swiveled her head toward him and peered at the slender shaft of light from above.

"It's morning," she murmured as she turned back and started to scoot to the edge of the sheet-covered mattress.

He moved his hand to her waist and held her even tighter. In the last four months he had worked on training himself to awaken just at daybreak. That way he could be in his own barn starting chores in twenty minutes. But when the weather turned cold and he put up the new shutter over the barn window, too little light entered the sleeping alcove he now shared with Laura to alert him to the breaking dawn.

"Morning can wait." He stroked her shoulder.

It was getting more and more difficult to leave her. She put her hand over his and broke his hold as she squirmed to face him.

"Daylight just about here. It's best you get back home." She pulled away and sat on the edge of the mattress as she reached between the two quilts for the new flannel nightgown he had given her.

He let his fingers run down her bare back. She didn't resist. In the past few months, he was almost sure he detected a softening in her behavior toward his embraces and kisses. Now, she followed his lead without being instructed, and sometimes, he could almost swear, he felt her lips responding to his. He knew she no longer trembled when he caressed her. More than once he heard her sharp intake of breath when he thrust inside her. He was almost certain her moan came from pleasure. Alex knew it pleased him to wait several seconds for her reaction, and when he sensed her letting go of her resistance, he plunged even deeper. Then, there were the times he felt her fingernails gently rake his back. He was convinced that Laura was fighting, and losing, the battle against the passion she didn't want him to see.

"I'll put your coffee on." She slipped the gown over her hips in the small cubicle Alex had constructed for their privacy.

Where once the quilt had been the only thing separating them from the children in the barn loft, he had built a wood-slatted wall with a narrow opening for a doorway, which he covered with a heavy burlap curtain.

"Watch yo' head when you get up." Laura pointed to the sloped roof that allowed just enough headroom for her to stand upright at its highest point in the new little bedroom but caused Alex to stoop at all times. She pushed aside the curtain and walked out into the main room.

Despite her best efforts to get him up, Alex pulled the bottom quilt under his chin to compensate for the loss of Laura's body heat. It was getting harder and harder to be without her embrace.

After his first visit in September, he knew he'd soon return to the barn and Laura's arms, but he told himself he would come no more than one or two times a month. Before September ended, he was already riding up the lane three times a week and found the gray wanting to turn onto the path into the mid-forty even more often. Now it was a daily battle. He found himself on those acres at least once each day even if it was just to check on Laura to make sure that neither she nor the children wanted for anything.

When Alex first recognized his dilemma, he tried to solve the problem of his frequent nights away from his house, and Eula, by vowing to leave right after the lovemaking. That had always been difficult. And, though it had been slow in coming, Laura was beginning to respond. Alex could not leave her now.

This half hour before dawn had become his enemy. Most mornings he tried to wake up even earlier so he could make love to her before leaving for his own home. This morning's late awakening denied him that pleasure. He pushed both quilts away and sat on the side of the bed, his feet resting on one of the cotton rugs he'd supplied next to each of the sleeping areas. He didn't really need Laura's warning. He had spent so many nights in the little cubicle that he had become expert at judging the slope of the roof. He hadn't bumped his head since before Christmas. If he sat upright on the edge of the mattress he still had almost two inches of clearance. As Alex leaned over in the semi-darkness to retrieve his pants from the floor, the light from the low-burning kerosene lamp flickered a path from the kitchen table where it sat, to the floor and to his trousers. Following the path of light through the opened burlap curtain, he watched Laura busy herself at the little potbellied stove he'd installed. Vented to the outside by a stovepipe, and with a small brick hearth set on the repaired floorboards, the stove gave off just enough heat to warm water and coffee, and take the worst of the chill off Laura and the children. Slipping on his pants, he watched her stretch an arm for the coal shovel to replenish the fire. As she opened the door of the little stove, a welcoming burst of heat drifted into the chilly room.

Shirtless, Alex walked through the narrow doorway to finish buttoning his pants. As he slipped belt into buckle, he glimpsed Laura pulling her flannel gown tight around her middle to keep it away from the open flame. While his hands mechanically pulled his belt tighter, his eyes trained on Laura. Something was different about her movements this morning. He wasn't aware of how deep his frown had furrowed as he let his eyes roam over her. Replacing the shovel, she gave him only a quick look as she turned to reach for a rag to lift the hot coffeepot from the embers.

The slenderness of Laura's body made those places where it curved and poked up and out all the more appreciated and pronounced. For the past four months Alex had explored every contour of that body and had thrilled at every in and out and every up and down. This morning, for the first time, he saw a new fullness. Even in the gloom, only partially relieved by the kerosene lamp, he noted that her breasts still bounced under the flannel of her gown. That always pleased him. As the dimness in the main room gave way to the breaking dawn, and to the lamp that Laura had deliberately trimmed low, he could see that the roundness of her hips still filled out the lower half of her gown. But looking closely at her now, he wondered how he had missed this new plumpness. As he rummaged his mind over the past few weeks, he was certain he had felt the changes, but until right now he hadn't really given any thought to the cause of that little protuberance at Laura's belly. His eyes on her stomach, he walked toward her. Busy pouring coffee into a mug, she gave him another quick glance. He waited while she topped off the mug and handed it to him.

"Careful. It's hot." Laura turned the cup, handle-side, toward him.

He took it from her as he reached to turn up the kerosene lamp. She started to move away, stopped, and pointed to his mug.

"I can cool it if you want." Her voice sounded unsure, questioning.

Alex set the mug on the table as she reached for it.

"Leave it." He couldn't move his eyes away from her middle.

"What?" She sounded confused.

"Come back to bed." He moved his eyes to her face as the storms of protest began to mount.

Before she could voice her refusal, a look of dawning awareness that he wanted something more than lovemaking reflected out of her eyes. Alex took her arm and led her back through the bedroom opening. He turned and bent his head as he backed through the door frame and sat down on the mattress. He pulled her between his spread legs. She looked puzzled. Reaching behind her, she slid the burlap curtain closed while Alex kept his hands tight around her waist. Looking at her face, he pushed the gown up to her waist, showing her triangle and belly. She grabbed at the flannel with one hand as she rested the other on his shoulder. He let one hand glide gently over her stomach from just below her breasts to the triangle, then across her belly from side to side. He heard her take in a breath and hold it for several long seconds. He laid the side of his face against her belly and listened to her racing heart.

"Laurie, what's this?" He watched her hand fly to her stomach before she could win her usual battle to restrain it. He could hear little gurgling sounds in her throat as she tried to answer.

"I'm getting a little fat on your food, is all." Her voice sounded strained and false.

This time he let his fingertips circle her stomach. He could feel the trembling in her whole body. He took the palm of his hand and pushed in on the swelling. She trembled.

"Laurie, is that a baby in there?" Alex fought the huskiness in his voice.

The slim shaft of light from the shuttered window above them came in brighter as the fullness of dawn broke. The light caught the lower half of Laura's face in its white grip. He watched her mouth open and close, her lips turn up to a smile and then down to a frown. He heard little bubbling sounds in her throat and watched her swallow two, three times. He had never known her not to have a quick answer for everything.

"You would tell me if there's a baby in there, wouldn't you?" Before her movements registered fully in his head, she dropped to her knees, forcing her body between his legs.

Faster than he could respond, she lifted the flannel gown over her head, showing her nakedness. As she leaned forward brushing her breasts against his pant leg, he saw for the first time, a playful smile on her face. Her hands reached for his belt buckle. As she unbuttoned his trousers, that funny little smile that seemed to fade and revive itself only to repeat the cycle within the blink of his eyes was accompanied by a sassy little twitch of her shoulders. Too surprised to react, Alex followed her lead as she lifted her body and leaned her naked breasts hard into his bare chest and pushed him back onto the bed. With the light from the shuttered window glowing behind her like an old painting, he watched her climb onto the mattress beside him. While he lay there, she pulled his trousers over his hips and off his legs. Torn between surprise and extreme pleasure, Alex fought to keep his eyes open. He'd waited too long to miss even one delicious moment of Laura's real feelings for him.

Naked, he watched her face move down his chest to his own tuft of hair. His eyes slid closed against his will when he felt her wet kisses just above his manhood. An involuntary shudder rocked him as he sensed Laura's mouth moving up his body. Her legs straddled him, pushing her thighs against his body. His own heart pounded in his ears louder than any blacksmith's hammer on an anvil.

Alex forced his eyes open when her lips reached his chin. She stopped and raised her nutmeg-ripe breasts level with his eyes. With a gentle sway to the side, she pushed one nipple into his open mouth. Alex sucked on the honeyed taste, his eyes shut, his mind floating, his body quivering. He felt the kisses planted on his forehead, on his hair. He squinted open his eyes just as she took his two hands in hers and guided each to her breasts and she bent down again to find his lips. Her tongue searched inside his mouth and caressed his own. The taste was sweeter than Christmas oranges. The feel of silken threads covered his body and stroked every inch of him. Somewhere, rummaging in his mind, was the thought of a baby, but it faded away when Laura lifted her body just enough to guide his manhood into her with her own hand.

"At last," thundered out of him as his mind drank in Laura's passion.

# CHAPTER FOURTEEN

"They say some colored women eat the starch right out of the walls. Dig their nails right into it, pull it out, and pop it straight into their mouths." At seven months pregnant, Tillie squirmed on one of Eula's hard-bottomed kitchen chairs.

Standing at the washbasin rinsing the last of the breakfast dishes, Eula paid scant attention to her niece's chatter. As she reached for a drying towel, her mind pondered on that non-seeing look in Alex's eyes when he finally wandered into the kitchen hours after the sun woke up.

"What?" Eula forced her eyes to focus on her niece, but her head insisted upon returning to the recollection of Alex.

It had been more than an hour after the break of dawn when her husband walked through the back door. Except for the eggs and the last of the bacon, his breakfast had already been cooked. Eula cracked the just-laid eggs into a bowl, and she braced herself for some words, any words, from her husband that would tell her where he had passed the night. Already dressed for Tillie's upcoming visit, Eula stood at the stove, the fork pressing down hard on the bacon in the cast-iron skillet, making sure that it browned evenly just the way Alex liked it. Careful not to utter a word, she made certain that the look she carved on her face showed her husband neither approval nor disapproval, simply that she was ready to accept whatever answer he gave her.

The bacon sizzled its doneness. Eula kept her face toward him. But as she stood there waiting for him to break the cold human silence of the kitchen, she realized that she might be there all day rooted like a stalk of harvest-ready tobacco waiting for a hired hand to wander by and spear it. If the truth be told, she couldn't be sure that Alex had actually noticed her in the kitchen at all this morning.

To him, she might have been a great, gray blob that blended in with the gray of her kitchen walls. If he was inclined to pay any mind to her, it would have been when she set his coffee mug down and his hand brushed hers as he pushed the hot coffee aside. Surprised, she said nothing. As she set his breakfast before him, she gave him a close look out of the corner of her eye. There was a tiredness about her husband like none other she had seen. It was not like he had plowed five acres without a mule and every muscle in his body had ached him into immobility. No, his fatigue was covered all over with an air of supreme satisfaction.

As she watched him slice into his eggs, she noticed a little smile that kept coming and going across his face. She got the feeling that her husband wanted to laugh out loud like he had just brought in the biggest and best tobacco crop in all of Montgomery County.

He had eaten his breakfast as if he were the only person in the house. The first words she heard from him all morning were when Wiley George hallooed at the door right after Alex pushed away his empty breakfast plate. There had been no "good morning," no "hello," no "Eula, pass the salt." There had been nothing between them except that look on his face and in his eyes that definitely put him in some place other than in the McNaughton kitchen.

"You got to watch 'em when they're working in your house." Tillie labored to turn in her chair to face her aunt.

Eula jolted at the sound of her niece's voice and nearly jabbed the fork she was drying into her hand. Tillie's short grabs for air between every two or three words finally worked their way into Eula's ears and shook her out of her reverie.

"What are you talkin' about? What starch?" Eula sat down heavily, only half paying attention to Tillie, who looked miserable with discomfort.

Her mind started to drift back to the morning and Alex. Knowing that the pictures of him flooding her head would do her no good, she decided to hang on to every word Ben Roy's silliest child spoke this morning.

"Momma's new hired girl told me that her sister's husband's cousin does it. She says that the walls of the cousin's cabin are filled with big holes, all from eating the starch right out of them." Tillie grimaced as she grabbed at her stomach.

The gasps were becoming deeper and more frequent. A shaft of concern flickered in Eula's mind. Though she never wanted to remember those days, had she panted like Tillie right before she lost Alex's baby?

"Aunt Eula, I've had myself these pains for the last week. If I can't have my own hired girl, maybe the starch will help. The girl says it might."

All the old worries bubbled in Eula's mind. There had been no early pains, no warning, really. But, there had been the weight. She reckoned she had put on fifty pounds. She stared at Tillie's midsection. Sitting out front like a watermelon, the baby didn't seem to take up much more than twenty pounds. Maybe it had been Eula's extra weight.

"Did you speak to your momma or Wiley George?"

"I told Momma, and she says pains are natural." Tillie blushed and looked down at the floor. "'Course, I didn't tell Wiley George. But, Aunt Eula, I don't think pains are natural. If I don't eat me some starch, I could lose this baby." Tillie's wide-opened eyes stared at her.

"Hush, girl. You're not about to lose Wiley George's baby." At this moment, it was 1893 and Eula was a twenty-two-year-old mother-to-be. She closed her eyes to gather strength before she gave Tillie a reassuring nod.

"But, how do I know that? Did you know that your baby was gonna be born dead?" Tillie's hand flew to her mouth almost before the words squeaked out. "Aunt Eula, I didn't mean it. It's just that I'm so worried."

Moving like a woman twice her age, Eula's leaden feet dragged across the floor to the other side of the table and Tillie. She pressed the girl's head against her chest.

"No, I didn't have no warning that the baby wasn't going to live." Eula clung to Tillie. Her disembodied hand stroked Tillie's hair. "I didn't have no pains and my breathing came easy."

"I know you did all you could to have Uncle Alex's baby." Tillie hung on so tight that Eula couldn't catch her breath.

"That was twenty years ago." Eula fought one hand free and reached for the back of the chair to steady herself.

"I know Uncle Alex don't hold it against you. Even Momma says he don't." Tears spilled. "Twenty years? Didn't you and Uncle Alex try to have another baby?"

An eerie silence draped over Eula's kitchen as the sadness that she could never fully lock away—those precious moments waiting for her baby to take that first breath that never came—pushed out.

"Of course we did, sugar. The good Lord just didn't see fit to send us another." Eula dropped to the chair. She reached for her niece's hand. "I'll ask Alex to speak to Wiley George. He'll hire on your own girl."

After those first months, she and Alex had never mentioned the baby again. Even the cradle her husband had fashioned himself was covered with heavy burlap and stashed in the back of the pantry. She'd never before asked a real favor of Alex, but she knew that at the mention of a baby, he would do what she asked. She patted Tillie's hand. "By March, you will have you a fine, healthy child."

# CHAPTER FIFTEEN

"You can't go to school today." Annalaura stirred the egg in the cast-iron skillet with one of the new forks Alex had given her.

Declaring it done, she reached for the bread she had already sliced and covered with two strips of bacon. Hurriedly, she slid the scrambled egg onto the bread and reached for a second piece to lay on top. She paid scant attention to Cleveland sitting on the edge of his mattress, rustling the corn husks as he rocked on his bed in his little alcove.

"Why not, Momma? You're makin' lunch for Doug. You're lettin' him go to school today. Why can't I? You know my leg is as good as new." The petulance that should have shadowed her eldest boy's voice was absent— replaced with a knowing resignation.

Annalaura wrapped the sandwich in butcher paper and placed it in the pocket of Doug's oversize hand-me-down, but serviceable, corduroy coat. Sitting on one of their two new kitchen chairs, Doug finished lacing up his winter boots. With her mind refining every word of the pleas she was going to have to make, she barely noticed any of her children.

"'Cause I said so, that's all." Her voice snapped in the frosty air.

The silence from Cleveland's corner, when there should have been great protest, finally pushed aside her carefully practiced speech. She turned toward her boy. A sudden shaft of remorse brought her up short.

"'Sides, I need you to look after Lottie and Henry. You can go to school tomorrow." Cleveland sat in his nightshirt watching Doug slip into his new jacket. The middle boy patted the pocket where Laura had put his dinner as he bounded across the floor toward the ladder.

"You goin' somewhere, Momma?" Cleveland asked, moving to his feet to grab Henry's arm as the youngster skidded too close to the ladder opening in his attempt to follow Doug.

Lottie, sitting in front of the hearth of the potbellied stove, ignored the ruckus as she played with her Christmas porcelain-faced doll. Annalaura poked her tongue into her cheek while she watched Cleveland. A chill ran down her back in the mildly warm quarters. She feared her almost twelve-year-old knew more than he should.

"I'm goin' to Aunt Becky's. Nothin' for you to fret over. Just watch after the youngsters, and I'll be back in time to start supper. There's plenty to eat for dinner in the cupboard." She gathered the breakfast dishes into a basket and readied herself to take them down to the smoke house. "Have Lottie wash these." She looked at her daughter, who did not let on that she had heard a word her mother said.

As Annalaura turned to back down the ladder with the basket over one arm and a handful of skirt in the other, she hoped that Alex would not drop by before she could return. Better yet, she prayed that he wouldn't come at all. If she could hold him off just a while longer, this nightmare might be over. She had made a desperate effort this morning to keep him from knowing.

⸎

The usual trek took Annalaura almost an hour as she carefully picked her way down the frost-covered dirt lane. Water puddled in the deep ruts, some of them deceptively icy in the January morning, making the journey all the more treacherous. If it didn't also mean a broken neck, she would have welcomed a fall. Maybe that would be the best thing after all. As she walked those last few yards past Aunt Becky's dead, snow-covered garden, even the blue wool coat Alex had given her could not keep away the chill sweeping over her body.

Shaking from more than the weather on this white sky day, she tapped at the door of the old cabin on Ben Roy Thornton's place. Her first few knocks were soft and unsure. She knew what she had to face inside, and it was a coward's way not to get to it. She knocked harder.

With her ear to the door, she listened for Becky's slow shuffle.

"Aunt Becky, open the door. It's me. Annalaura." She fought to keep the fright out of her words.

Slowly, the door opened a crack, and Annalaura stumbled into a murky interior made no brighter by the low-burning fire in the old fireplace. Rebecca Murdock, at sixty-one, was still spry enough to step out of her niece's way. The smell of black-eyed peas, but without the soul-warming scent of ham hocks, came from the big iron kettle setting on the fireplace grate.

"Whoa, girl. What you doin' rushin' in here like the night riders is after you, and here it is broad daylight?" Becky shut the door, dipping the long, low-ceiling room into even more dimness.

The old woman waved an arm in the direction of her wrought-iron bed while she made her way toward the oak-wood table. Annalaura headed for the four-poster, the nicest piece of furniture in the room by far. Weary from more than her walk, she let herself sink into the bed's feathery softness. Across the room, Becky seemed to lose quick interest in her visitor as she picked up her pipe from the table and tore a sheet of paper. Her back to Annalaura, Becky stepped toward the fireplace and held the rolled-up paper to the flame. She brought the lit strip to the bowl of her cob pipe. Drawing in a heavy breath to ignite the tobacco, Becky walked back to her chair and finally trained her eyes on her niece.

Annalaura swallowed hard, worked her mouth to start the stream of words she knew she had to get out. Just as she started to speak, Becky raised a palm toward her. Annalaura remembered. It was the same "wait" sign she recalled Grandma Charity making when the old Cherokee wanted to ready herself for bad news. If waiting helped bring on the second sight, Annalaura wanted it to pass quickly. Her courage was rapidly waning.

"Aunt Becky, I need to see that conjure woman." Her mouth was so dry she could only hope that her aunt's hearing was still good enough to catch her words.

She hadn't given much thought to it before, and maybe it was because Rebecca was getting older, but the resemblance between her Aunt Becky and her Grandmother Charity was startling. Becky's nose had a decided bent to it, and her skin had settled into a clearly defined copper under-tone. Becky's hair, still mostly black, was twisted into one long braid that hung over her left shoulder. It was almost as straight as Grandma Charity's. While Annalaura's hands fumbled with a corner of the feather quilt, Becky took another long draw on her pipe, letting the smoke rise in the air. Becky's narrowed eyes looked at her without blinking. Anna-laura pulled the coat tighter. Her body shook with chills, but her face was almost as warm as the fire in Becky's fireplace. How much longer was this Cherokee second sight going to last? Perhaps Aunt Becky was going deaf. Maybe she hadn't heard after all. Could Annalaura get those words out again?

"I can pay her something, but I need to see her quick." Annalaura rocked on the feather bed.

"When's he comin' back?" The sound of the voice that had chewed and spit snuff for over forty years rattled off every wall in the one-room cabin, as Becky took a third draw on her pipe.

Though she turned her face away from the fireplace, Annalaura still felt its heat blasting her from chin to forehead. Was it true? Could these old Indian women really read minds?

"That's just it. I don't want him back. I've got to find me a way to keep him from knowing." She bit her lip.

She had let too much slip to a woman who already knew what she was thinking. Creak, creak went the old rocker. It was the same one Grandma Charity had sat in when she rocked Annalaura to sleep all those years ago. Becky, with those eyes narrowed to pinpoints, said nothing. Annalaura wrapped both arms around her middle and squeezed her eyes shut.

"Just tell me where I can find her, and I'll walk all the way there tonight." Annalaura let the pleading in her soul whisper in her voice.

"You got money to pay?" The sound of the old woman's rocker hadn't broken stride.

"Yes, ma'am, I do." Annalaura moved to the edge of the feather mat-tress as she tried to get the sound of the rocker out of her head.

"You takin' yo' babies' food money for this conjure woman?" It was a cross between a question and an accusation coming out of Becky's mouth.

Annalaura opened her eyes and trained them on the low-burning embers in the fireplace.

"No, Auntie, I got money to feed my babies and seed money to plant the summer vegetables come April." She gave her head a vigorous nod for emphasis.

She had to find the key that would unlock the secret of the conjure woman from her cantankerous aunt. Maybe it was money. If Becky believed she had enough to pay, perhaps she would not guard the woman's name like it was the secret to the Holy Grail. Becky took another draw on the pipe, and Annalaura watched the smoke encircle the woman's head. The quiet in the cabin was broken only by the snap of a log in the fireplace and the hiss of the peas boiling in the pot.

"Let me see, gal. You do have food fo' your chil'ren." Those eyes trained on her niece. "You got money fo' spring seed." Those eyes didn't blink. "You got you a warm new winter coat." Becky smacked her lips on the pipe stem, but her eyes bored into Annalaura. "You don't want 'him' back, whoever 'him' may be." The word came out of Rebecca's mouth like she was the hangin' judge pronouncing Annalaura "guilty." "And you tell me you needs a conjure woman." If there had been a wall clock, Becky would have marked off a good five minutes of puffing on her pipe as her rocker kept up its unceasing creaking. The old woman finally leveled the worn pipe like it was a firing pistol. She aimed it straight at her niece's stomach.

"Who's the daddy?"

Annalaura felt her body leap from the mattress and her feet propel her to the rocker without her willing them.

"I've got to see the conjure woman, Auntie, please help me." All the morning's planning on the exact words and tone to use to gain Becky's assistance left her. She stood at the rocker before her aunt, her whole body shaking.

"I said who's the daddy, and don't hover over me, gal. Set down." Becky's accusing voice had always sounded like an ax splitting a chunk of wood.

She stumbled into the chair by the table. Words tumbled over themselves trying to get out of her mouth.

"The hired man that helped me bring in the tobacco last fall." Anna-laura let the lie slide out on short bursts of air.

"Uh huh." Creak, creak went the rocker.

"Me and him got real close during those tough times." She prayed Becky would believe her long enough to give up the name.

"He the one give you that coat and the money for the conjure woman?" Creak, creak.

"Yes'm."

"He the one give you food fo' yo' chil'ren?" Creak, creak.

"Yes'm." She couldn't prod her brain into action fast enough to keep up with the circle thinking of a Cherokee woman.

"He the one tie up that gray hoss of his in yo' barn three, fo' nights outta the week since September?"

A tornado could have swept up the old cabin, but neither the pace of Becky's creaking rocker nor the puffs on her pipe would have changed one bit. The aroma from the meatless black-eyed peas melted into the gray pipe smoke and collided with the din in Annalaura's ears.

"He the one name of Alexander McNaughton?" Creak, creak.

Somewhere in Annalaura's head a long, low, keen moan crowded out the sound of the rocker. Some thing whirling in the core of her being forced her feet out from under her. A woman's voice that sounded like it might once have belonged to herself screamed at the roof beams.

"Gawd help me. I don't want it. I don't want it." Annalaura dropped to her knees with no control over her legs, but she felt no pain.

Like a puppet pulled by a string, her neck arched back, bringing her face up toward the rafters. She slapped her hands together in prayer. "Please, Lord Jesus, take this thing outta me. I don't want it." That other woman's voice had not dropped one octave. The tears flooding out of her own eyes stopped her from seeing anything except the soot-black cabin roof, but the touch of still-strong hands around her shoulders guided her to her feet. Some great force from behind propelled her to the chair. She didn't know how long she sat there listening to the strange woman's moans, but something hot and steamy touched her trembling lips. Work-worn, cinnamon-brown hands passed before her eyes as they came to rest over her own. Together, they tipped the container to Annalaura's lips.

"How far gone is you, chile?" Becky's voice sounded like Sister Muriel in the church choir singing "Sweet Chariot." But the funeral song "Precious Memories" better matched how she felt.

Again, Becky pushed the cup of thick black-eyed-pea soup to her mouth. Her aunt's hand pushed down on the back of her head, forcing her to take in sips of the soothing soup.

"I think he hit right off. September. I think I'm four months." The words came slow and hard between sobs. With the hot soup and her own shame, the room filled with heat. Annalaura shook her arms out of the coat and let it lie across her shoulders.

"Why didn't you come to me right off, girl?" Becky turned the pipe upside down on a pottery plate as she lifted Annalaura's head to face her. "I'd been able to fix you up some special herb tea then."

"I kept thinking it couldn't be so." The shaking in her body continued, but now it came inside a room that seemed warmer than the hottest mid-August day. "He don't have no other children. I thought, maybe, he couldn't have none. At least, I wouldn't have to worry 'bout that."

Only Becky's steadying hand kept her from splashing soup on the table. Sucking in her lips until they disappeared into her mouth, Annalaura dropped her head.

"I haven't had me no word from John since he took off in June." She took another gulp of the hot soup. Becky's hands were still over hers as she tipped the cup.

"And you ain't thought that I'd know what to do? That I wouldn't understand?" The chastising sound in Becky's voice was unmistakable.

The tears stopped on Annalaura's face as she remembered Johnny.

"I was 'shamed, Auntie." She whispered the words as her chin dropped down to rest on the blue serge coat. "I'm a married woman. Now, everyone in this whole county will know that I've been with another man." The tears started again and spilled down her cheeks as she finally faced her aunt, her hand covering her mouth. She spoke through barely split fingers. "Don't matter none that John is off cattin' around again. That he ain't never comin' back to me. Aunt Becky, I'm a married woman." That other woman's voice threatened to take over her own again.

"And what do you think you could have done about it?" Becky's voice was soft as she tightened her grip on Annalaura's arm.

Annalaura dropped her eyes again. She wanted to pull the collar up around her ears. She wanted to climb back inside that coat so deep that nobody in this world could ever look upon her again.

"You think yo' husband would go easier on you if he thought it was a colored man's chile?"

"Colored or not, it's my fault," she whispered into the coat. "If I had just told him no when he brought all that food. If I hadn't hitched up my dress that day. If I had just kept my mouth shut. If…" The sobs racked her body.

"If you'd let yo' babies go hungry, if you had yo'self on two crinolines and three chemises under yo' dress and that coat, if you'd looked cold and dead…" Becky's strong hands raised Annalaura's head from her chest. "Ain't none of them ifs would have made a damn bit of difference."

Annalaura's eyes widened. She had never heard her aunt speak a cuss word in all the years she had known Rebecca Thornton Murdock.

"But somethin' I did or said must have made him look at me. Made him think I wouldn't fuss too much if he came to my bed…if he touched my body," Annalaura stammered.

Becky sat back in the kitchen chair with the wobbly leg. Her hand tapped at the bowl of her pipe.

"If a Tennessee white man comes ridin' along and spots an apple orchard and decides he wants him an apple, ain't nothin' that apple can do to make him pick a different one. It can't hope that a breeze will come up and knock it to the ground. It can't pray that a bird will peck a piece out of its side so it won't look so good. A colored woman in Tennessee is just like that apple. Ain't never been a brown-skinned woman who had any say over what a Tennessee white man can do with her body." Becky's voice trailed off as if she had been suddenly transported to another world.

Annalaura clung to the mug as though it were her lifeline.

"I was only fifteen when Old Ben Thornton came sniffin' 'round me. It was slavery times, but when it comes to a white man wantin' a black woman, slavery or no, it's all the same in Tennessee." The pain in Aunt Becky's voice edged deeper, and for the first time, the agony of it filled Annalaura's soul.

"Cousin Johnny." She didn't have to work hard to let her tone sound reverential.

The name on her niece's lips seemed to shake Aunt Becky out of her growing morass.

"I'll give you the herbs, but I can't get you no conjure woman." Becky's voice sounded gruff.

Was it about Johnny?

"I don't want this child, Aunt Becky. The conjure woman can ream it out of me. I've heard of it workin' on others." She lowered her voice though the two were the only people in the lone cabin in the middle of forty acres.

"You hear tell of those 'others' dyin' from the conjure woman, did you?" Becky sat back in the chair, careful of its one shored-up leg, the pipe clenched between her teeth.

"I know some die, but I've already had me fo' babies. I'm strong enough for her to do her business." Strength was eking its slow way back into Annalaura as she nodded her head for emphasis.

Becky took a long draw on her pipe. She let the smoke curl in the space between them.

"You know 'bout yo momma and the conjure woman?" Aunt Becky rocked slightly in the broken chair.

"No'm." Annalaura feared her aunt would crash to the floor.

"You 'member yo' momma dying?" Becky kept the pipe between her teeth without taking another draw.

Annalaura shivered. "I can barely remember." She peeled back her own memory curtains.

"'Cose you can't. I shooed you out into the rain when I brought yo' momma back from the conjure woman." Becky let her eyes slowly drift across Annalaura's face.

Annalaura read in those eyes that her aunt believed she was speaking the truth.

"Yo' momma was about as fur gone as you. Too late for the herbs. Happened 'bout this time of year, too."

Annalaura tried to bring the mug to her dry throat. "Auntie, Momma died of the gallopin' consumption. Remember?" She let each word come out slow to bring her aunt gently back from the depths of time.

" 'Cose I remember, girl." Becky's eyes looked at her hard as glass. "It's you who don't remember." Aunt Becky tried to suck another draw on her pipe, but it had gone dead. She got up and relit it from the fireplace. She settled back into the rocker.

Annalaura swiveled her chair around to face Becky.

"This old cabin was here in slavery days." The old woman sounded of the long-ago. "It's where me and yo' momma was born. But, when yo' momma married yo' daddy, she moved off of Thornton land. Still, every Sunday she would come by here to visit yo' grandma Charity." She took a deep draw on the relit pipe. "Yo' grandma Charity had herself a man caller by the name of Jessie. Much younger than yo' grandma. I swear to goodness, that grandma of your'n was a high stepper."

Becky had slipped back in time. Suppose Annalaura couldn't bring her aunt back in time to help?

"I ain't gonna dwell on it." Becky slumped back in the chair. "Jessie forced yo' momma and she come up with a baby she didn't want to tell yo' daddy about."

"Forced her? Forced her to do wh…" Dawning shock stopped the word in Annalaura's mouth. "You mean he…like…you?"

"No. Not like us. Jessie wasn't white. He knew better'n to take what wasn't his. Yo' momma was married to yo' daddy just like you is married to John Welles. A black man should know to respect that. Yo' momma was powerful 'shamed and didn't tell nobody 'til it was too late. Just like you. Ain't none of us knowed it 'til the conjure woman come up and told us Geneva was near 'bout dead. I fetched me a wagon off of young Ben Thornton and lit out for the conjure woman's house. Got yo' momma back home."

Annalaura almost lost her perch on the chair. "Auntie, what are you saying?" The words cut her throat. "Momma died of the consumption."

"I tell you like it was. Yo' momma died befo' the rooster crowed that next mornin'. Died with Jessie's baby still half in her." Rebecca clamped her eyes shut.

"Momma died with a conjure woman? Died because of a baby?" Annalaura shook her head. "Daddy? What about my daddy? Did he know? Did he kill the man?"

"Jessie lit out of here even befo' yo' grandma left for Oklahoma, and don't nobody know where he went."

"I'll fix you some herbs that'll help yo' baby's color come in darker." Becky's regular voice returned. "Best when a white man's chile come out tan. Less trouble that way. Chile can be passed off as a throwback. My Johnny came out white." Aunt Becky drifted off into that other place where her Johnny still lived.

Annalaura could see by the rheumy stare in her aunt's eyes that she wasn't coming back this day.

Annalaura pushed herself up from the chair and headed for the door. "I thank you kindly for the soup."

"You got fo' chil'ren. I'm too old to raise another dead woman's young 'uns. No conjure woman." Aunt Becky shut her eyes and folded her arms across her chest. The creaking rocker sped up.

⁂

No conjure woman. That's what Aunt Becky had said. The sun had passed its height two hours ago. The sky threatened snow again. Annalaura made every step she took on the long walk home a slow one. Iced-over rutted ponds did not bother her this time. She needed every precious moment of this trip to plan out a way to keep Alex from knowing that this baby was his. There was no need to congratulate herself on her morning efforts. She had tried every lovemaking trick John had ever shown her, and some others that just popped into her head when she lay with Alex. This morning, every frisky move kept the man's mind off babies, but in a few weeks, all of Lawnover would know of her pregnancy. She would give it her best efforts, but she couldn't be sure if a repeat performance would keep Alex's mind occupied on her, and off his baby. John Welles was a different story.

She'd felt it in her marrow ever since August, but sometimes that little flicker of hope that John might find his way back to his children jumped into her head. But when her husband let Christmas pass without a sign, she knew she'd been right all along. John Welles was gone. Even so, word of the baby would get to him wherever he was, and as certain as colored would always tend tobacco, he would divorce her. Annalaura stopped by the side of the lane, standing no more than two feet away from an iced-over puddle. She waited to catch her breath from the flip-flopping of her heart. She clutched the coat to her chest. Despite his cattin' around with every flirting colored woman in Lawnover and his gambling ways, John had his loving side.

But every time she thought she could feel real love for John Welles, he would up and disappear for weeks at a time.

Hard as it was, she could manage the loss of her husband at her side because she never really had all of him anyway. But divorce was a cold, final thing. The Lawnover colored Baptist church would ban her as a Jezebel. Fresh tears began their frigid roll down her cheeks. She swiped them away with a mittened hand. There was no need to cry over a man who had left her over seven months ago, not when she had to put all her smarts into Alex McNaughton. His was the immediate threat.

What would Alex do if he learned this baby was his? Annalaura plodded past an oak tree. Would he treat her the way most Montgomery County white men did when they tired of their colored women? Would he just move on to his next one and drop by with an occasional silver dollar, or a ham, or a suit of hand-me-down clothes for his yellow child?

There were many such women and children in Lawnover, and if the baby didn't look too white, people could pretend that the father was some strange colored man from down county. If the woman was lucky enough, and the baby came out looking tan with dark eyes and frizzy hair, then maybe she could find a man who would take her to wife. Annalaura brightened for the first time that day. If Aunt Becky's herbs really worked, perhaps one day, she could find a toothless, old, widowed colored man to take her and her children into his house. The flash of encouragement that quickened her step along the lane faded as fast as it had come. Even such a man wouldn't want a woman carrying the sin of divorce.

Annalaura looked up the lane to see the barn in the distance. Each step felt as though she pulled fifty pounds of tobacco behind her. She had to make sure Alex tired of her.

In the beginning, she was sure her amateur ways would run him away in a few days, but he had stayed and she wondered why. She had steeled herself for quick and rough, even hurting, lovemaking. Instead, he had been considerate, tender, gentle, and especially passionate. His caresses were soft and warm. His kisses long and lingering. And when she didn't respond in kind, there had never been an explosive blow, or an angry word. He talked to her like he cared what she felt, and on some things, he even listened to her answers. And after lovemaking, he always held her close.

Sometimes, the kindness he showed to her and her children almost made her forget that she should hate this man. Still, if the state of Tennessee gave her the right to say no to him, she would use it. A woman ought to be able to say yea or nay to any man who wanted to come into her body. Yet, if Alex had been colored...well then, maybe her answer to him would be y... She pushed that thought deep into a box in her mind. There could be not even a little love with a white man.

Maybe Alex would treat her the way Ben Roy Thornton treated his Hettie. Together, those two had two-and-a-half yellow-skinned children living right there on the Thornton side-forty. Right under Fedora's nose. Of course, Ben Roy's wife, like all the other white wives in the county, had to pretend that those babies got dropped off at the side-forty by a passing peddler cart. Mr. Ben Roy had fed, housed, and clothed Hettie and his children for the past five years. Everybody in Lawnover, black or white, knew the parents of those children. The path leading to her barn was just a hundred yards away. She stopped again and jammed her hands into the pockets of the coat.

No matter how sweet his ways, it would be far worse for her if Alex treated her like Hettie. Being a white man's "other family" was almost as bad as being divorced. Annalaura started the final few yards to her turnoff thinking the words in her head so loud that she burst them out in the air.

"God, don't let me be Hettie. Don't make this man want to stay." With the path looming before her, she dropped her eyes to the ground. She had one more choice.

The hired man. Aunt Becky hadn't believed her, but if she could convince Alex that the tenant farmer who helped bring in the harvest in September was the father, he would leave her. And, with any luck, Alex might beat the baby out of her. She prayed that he wouldn't kill her only because her four children needed her alive. Turning onto the pathway, she kept her head down, trying to build her courage. For the sake of them all, she had to convince Alex McNaughton that he was not the father of this baby.

# CHAPTER SIXTEEN

Alex pitched the last of the hay bale to the cows for their afternoon feed. He glanced over at the remaining hay loosely piled under the ladder near the back wall. That should be enough to last the animals through the winter. He had already mucked out the barn, and that completed the last of Cleveland's afternoon chores. It was nearly three o'clock, and Doug had just climbed the ladder after coming home from school. Upstairs, Alex could hear the boy practicing his reading. Cleveland had spent the day looking after Lottie and little Henry. Laura wasn't home, and Cleveland had been close-lipped about his mother's whereabouts, telling Alex only that his mother had gone to her aunt's house. The youngster couldn't say why. Alex hadn't pressed, but he hoped that his taking over Cleveland's afternoon chores would loosen the boy's tongue. He had unfinished business with the mother.

The delicious memory of the morning had made him barely able to concentrate on his own morning chores. When he came in for breakfast he vaguely recalled Eula saying something to him but little of it registered. Whatever it was must have been fairly important to his wife, because she never bothered him with talk unless it was of some urgency. He did recall that her journal lay spread in front of her, and she may have asked him something about peach preserves and hams. He only knew that he rushed through his farm jobs with his mind nowhere on his duties. Alex couldn't focus on anything, or anybody, other than Laura. Her unexpected display of feelings for him, and that plumpness in her belly, crowded out everything else. He rushed back to the mid-forty right after he'd managed to rid himself of Wiley George in Lawnover.

Alex had returned to the mid-forty about one in the afternoon only to find Laura gone. He waited over two hours for her return. Where was she? His mind whirled back to the one thought that had driven him since morning. Could there really be a baby? Putting the pitchfork back against the wall, Alex walked to the doorway and looked down the path toward the lane. The wind had started to kick up and the sky threatened snow. If Laura was actually pregnant, he couldn't have her out in this. He took in several deep breaths. Could he believe what his senses were telling him?

He and Eula had made a baby, but just that once, and at twenty-four, he hadn't paid much attention to either Eula or what his senses told him about her changing body. They'd been married almost a year when she told him a baby was on the way. Back then, he thought that was the way of a marriage, and he gave neither Eula nor the coming baby any further thought until his child's birthday also became her death day.

The doctor hadn't said one way or another if Eula could ever have more children. She just hadn't. Mostly, he felt it was a weakness in his wife that kept him from becoming a father and there was no need to worry a good woman on her failing.

Although some men gave their women a lot of grief because of it, Eula deserved better. But, every now and again, as the years passed with no other babies from any of the four or five women he'd been with besides Eula, he fretted that maybe the fault did not all belong to his wife. Sometimes he thought that maybe there was a weakness in his family line. His own pa had sired only Alex and his sister, though he had lived with their mother more than thirty years. Was Laura about to change all that?

Standing in the shadows of the barn door opening, he watched her walk up the path long before she saw him. He had told her the coat was a cast-off from the church. In truth, he had bought it brand-new in Clarksville. Alex had purchased two, one blue and the other black. Neither had been very expensive, but both were warm. The black had tortoiseshell buttons, and he had given that one to Eula. He hadn't realized until he read her surprised face and heard her shocked comment that he had never given his wife a just-for-nothing gift in all of their married years. But it was the blue coat with its extra-thick lining that had really caught his eye. And today, with the snow clouds forming in the afternoon sky, Laura needed its warmth. As she passed the smoke house, it was no wonder she still hadn't seen him. Her head was bent down and the collar of the coat turned up almost to her eyes. As she closed in on the barn, he stepped out into the pale sun. He didn't dare risk taking her by surprise. Not now.

"Your Aunt Becky lives 'most a mile down the road. You shouldn't be walking that far in weather like this." Alex watched her stop in the path just beyond the smoke house.

The startled look on her face told him he had just brought her back from some distant place. Alex closed the space between them and touched her shoulder. He was surprised when she jumped.

"You're cold and it's about to snow." He began rubbing her shoulders to warm her.

Alex felt her body stiffen like their first days together. He frowned and pulled her to him in a tight embrace. Her arms hung at her sides.

"Is your Aunt Becky sick?"

She trembled in his arms again.

"No. Not sick." Her words muffled into his shoulder.

Alex led her into the barn and away from the drafty door. They stopped in front of the cows still chewing on their hay.

"Let's warm you up." He slid his hand to the middle button on her coat and slipped it open.

Alex laid his hand across her midsection. He felt her stiffen again. Annalaura's eyes stared at the barn floor. The beginnings of concern set up in his mind. Despite this morning's passion, Laura was keeping something from him. He rubbed his hand slowly around her belly.

"Are you sure there's no baby in there?"

Her head jerked up to face his before she turned around to stare at the feeding cows, but not before he glimpsed her look of panic.

"Laura?" He waited for an answer.

"No. Ain't no baby."

"Let's go upstairs." She was keeping the truth from him, but why? He pulled her toward the ladder. "I'm gonna get me a good look at your belly in the daylight. I'm gonna take off your clothes." Alex felt her feet dig into the barn floor like she was one of the planting mules.

"No," Laura almost shouted at him.

Startled, and with one hand on the ladder rail, he stopped and turned toward her. With those amber eyes opened wider than he'd ever seen them, he knew she couldn't miss the look of puzzlement on his face.

"What?" Alex hadn't meant his question to sound as harsh as it did.

Laura began a frantic shake of her head. "I mean…my children…you can't…sorry." Laura's skin was a beautiful chestnut brown, but it always had red undertones. Right now, her face looked like a ripe cherry. "Tonight. Please do it tonight. It's only afternoon. My children will see."

To his ears, her voice carried the sound of fear. A new sensation of worry rushed at him.

"I'll send them downstairs. Won't take but thirty minutes. I need a good look." He tightened his grip on her arm and started up the ladder.

"Mr. Alex, please wait. I'll tell you."

Laura hadn't addressed him with such formality since early October. In this fresh wave of surprise, he released her arm and she scurried around the ladder. He followed. Annalaura stood with her back to the pile of hay. Even though her head was down, Alex could see the tears falling. His own heart quickened.

"Tell it, then." He could barely catch his breath as he watched Laura slowly lift her face to his.

"Yes, suh. There is a baby." She stood with her eyes closed, her hands fisted at her side, just like those first days in September.

Alex wanted to smile his pleasure at news of a baby, but that strange look on Laura's face stopped any budding pride from building within him. He watched her swallow two, three times before her eyes slowly opened.

"Mr. Alex, suh, I has to tell you the truth."

He heard her words, but the pounding in his chest, loud as hailstones during a bad downpour, drowned out their meaning.

"If you've got a baby in there, Laura, why didn't you tell me before?" He placed his hands to her shoulders to steady her.

"Because I know you is goin' to kill me." She swayed under his hands.

Alex shook his head trying to let some understanding enter his mind. No words came from his mouth.

"This baby, suh, ain't yours. It's the hired man's from harvest time." Her eyes clamped shut.

"The hired man?" He could feel her tense for the first blow she knew was coming. "You talkin' about Isaiah?"

"Yes, suh. Isaiah Harris. He this baby's daddy."

Alex watched Laura reach down deep to get the breath to whisper out the name. Her hands lifted to her belly as though she thought he would send a blow in that direction first.

For long seconds, Alex couldn't find the words to speak. He remembered Isaiah Harris perfectly. The tan-skinned man had come to his place in early September to ask for a harvest job. He, Alex, Laura, Cleveland, Doug, and little Lottie had all worked fourteen-hour days to bring in the tobacco. Even Henry had helped by ferrying water from the pump outside the smoke house to the barn where Cleveland, Laura, and Lottie hung the tobacco. With Harris's help, the two men had been able to spear the tobacco on all forty acres. Even though it was far from prime, the crop had fetched a little over sixteen hundred dollars, not as much as last year but still a hundred dollars better than the tenants on his back-forty. Alex had thought it a miracle. All during those two weeks, he had kept close tabs on Isaiah Harris.

A man, any man, black or white, knows to mark out his territory. And, he had no doubt, Harris understood that Alex had marked out Laura as his own. Sometimes, the two men finished their spearing way after supper. Isaiah offered to sleep on the floor of the smoke house to get an even earlier start the next day. Alex would have none of it. He put Harris in the back of Eula's buckboard and drove him back to the main barn each and every night. He stashed the gray's saddle on the porch, and he seriously doubted that Isaiah Harris would walk all the way to the mid-forty at midnight only to come back again at four o'clock in the morning. No man, including Isaiah, would be fool enough to take another man's woman when she had been clearly spoken for. Laura was lying, but why?

"What makes you think Isaiah is the father? I was with you before he ever stepped foot on the mid-forty." He watched her eyes open and blink, although the shaft of sunlight coming through the barn door fell nowhere near the stack of hay.

"I...I just do." His head shook its confusion. "A woman knows these things."

Alex hadn't paid much attention to what Eula knew or didn't know, and maybe a woman did feel more when she was pregnant, but he doubted it. He pulled Laura into his arms and kissed her. He felt her body push into his as her lips responded for the wisp of an instant before she pulled away. He tightened his grip on her shoulders and he knew the truth.

"This baby is mine and ain't nothin' you say can make me think different." He pushed her down to the pile of hay and dropped beside her.

The rungs of the ladder lay angled above them. He saw her hand clutch at a piece of hay as he finished unbuttoning her coat. She turned her head to look at the bits of straw in her outstretched hand while he shifted her skirt to her waist. She brought a fistful of straw to his shoulder as he untied the string to her winter drawers and lowered them to just above her triangle. He put his face lightly to her stomach. Brushing her belly with his cheek, he was certain he felt something inside her move.

"Alex," she sounded almost out of breath as she let the straw trickle across his back. "If this baby frets you too much, I understands." Her voice grew stronger. "Sometimes a baby with the wrong woman can be bothersome to a man." Her breathing came easier and she let her eyes rest full on him. "It ain't too late if you don't want it. It would pain me to bring a world of trouble with yo' family down on yo' head." Her voice carried a strange mixture of fear and hope.

What did she mean? What possible difference could it make to Eula if he had an outside child with a colored woman as long as he looked after the farm and clothed and fed his wife? He could see Laura was just about bowed down with fear. He rubbed his hand over her belly as he raised up to look at her face. She looked as though she were willing him to say some particular words she wanted to hear. But what?

"Laurie, quit your worryin'." He brushed her forehead with his lips. He tasted the dampness of sweat, although the barn was just about cold enough to freeze the water in the pails by the cows. "I'm gonna take care of this baby. You don't need to worry about no midwife. I'll get the best colored doctor in all of Clarksville out here." Did he feel her shudder?

When she didn't respond, he stroked the side of her face with his fingertips. Didn't she understand that a man can't really call himself a man until he can show the world the living proof of his manhood?

"I'll send this child all the way to the eighth grade if you want." He leaned over to kiss her again.

Annalaura turned her head just as his lips approached her face, and he caught the corner of her mouth. Her breathing was coming in harder again. He reached down to pull off her drawers. Her hand stopped him.

"If you really want this chile, we can't be together so often." Her voice came soft, low, and filled with worry.

"What do you mean? Have I hurt you?" Twenty-one years peeled away in the flash of a second.

Had he been the cause of Eula's lost baby? His hand reflexively left Laura's stomach.

"No, no. You ain't hurt me." She turned her face back toward him, and a hint of alarm flashed quickly in those eyes. She dropped her head, and the cherry-red color returned to her cheeks. "You, ah, you been very good to me. It's just that you, ah, yo' love come on strong. Maybe too strong fo' the baby." She let her eyes flicker a second on him before she firmly placed them back on the pile of hay. In the fading afternoon light in the barn, they looked soft, even loving. "Don't come back tonight. Best if you come over no mo' than one or two times a week."

This time Alex thought he heard hope etched in her voice.

"If you think that'll be good for the baby." Before he could sort out her request, she sat up against the hay piled against the side wall and slipped her arms around his neck.

She pulled him close. He felt her fingers run down his neck and up into his hair as her tongue probed into his mouth. He pushed her back down onto the hay.

"After today, I won't bother you more than three times a week. I promise I will be careful."

"No more'n once a week," she whispered as she led his hand to the top of her winter drawers.

# CHAPTER SEVENTEEN

The sound reminded him of a woodpecker worrying a tree somewhere far off in the distance. Knock, knock. He wasn't ready to come out of his sleep just yet though there was nary a dream in it. The woodpecker kept up his work until a fantasy burst on him full grown. He was in the middle of the Clarksville battlefield and the Union cannon were bombarding the Clarksville rebels. With bullets flying from both directions, he didn't know which way to run.

"Welles, John Welles. Is you in there? Miz Zeola wants you." The knock thundered into his bedroom like a cannonball had fallen onto the pillow next to him.

He jumped bolt upright, flinging his arms up to ward off the next attack, only to come into hard contact with Savannah's nose. The grade-school teacher had lifted her head off the pillow beside him just as he reacted to Big Red's voice. He looked over at his lovemaking partner of a half hour ago. Her mouse-brown eyes stared big at the door. Her face looked like she expected the whole Confederate army to storm though and drag her off into slavery. Knock, knock. Big Red was getting even louder. John reached for a corner of the blanket on the double bed to cover himself.

"Oh, my Lawd." Ignoring the accidental blow, Savannah, the older of the two schoolteacher sisters living next door in Miz Brown's best room, snatched the blanket from his grasp, pulling it completely off the bed. She wrapped the heavy cotton coverlet around her ample body as she jumped to the floor.

Now naked in the room lit by late afternoon sun, John watched the woman turn in circles with his blanket snug around her as she sought a place to hide. One of her breasts escaped the swathing and drooped nearly to her waist. Savannah was five years younger than his Annalaura, but the schoolteacher had a body that looked more like a middle-aged Zeola than a young woman who had never birthed a baby.

"John Welles, Miz Zeola wants you and she wants you now. I ain't standin' here foolin' with you no mo'." Big Red's voice oozed its own particular charm as John motioned Savannah to stand behind the door.

"Hold on, Red. Let me git my britches on." John took his time donning his trousers.

He wished Savannah would stop bending and straightening her knees. Each time she did it, he could feel the floor shake. John pulled on the knob and opened the door into Miz Brown's second floor hallway. Big Red crowded out what sunlight shone through the hallway window at the front of the house. The cook's substantially sized feet, planted firmly on the floor, blocked out the garish red roses stitched into his landlady's carpet.

"My shift don't start 'til eight o'clock. Ain't but fo' now. What Zeola want?"

"It ain't fo' you to ask what Miz Zeola wants. It's fo' you to come when she calls." Big Red turned and rumbled down the staircase muttering all the way.

Before John shut the door, he peered over at the hallway window. Though he had the second-best room in the entire boardinghouse, and he paid a pretty four dollars a week for it too, his only view was of the brown clapboard siding of the house next door. Through the hall window he could see the limbs of the crape myrtle tree in the front yard bending under a gusty March wind. Tiny buds lumped out all over its branches, and he could see more brown dirt than snow on the ground surrounding it. As he looked through the second floor window, it was clear that spring was just about to make its full appearance. A jumble of images flooded his head. Pictures he didn't want to see. As Christmas gave way to Easter, sometimes he'd imagine Annalaura struggling to put food on the table. It was all he could do not to throw away Nashville and rush back to Lawnover. But if he gave up now, his family would always lead a life of nothing.

"Close the door," Savannah hissed. "You know I can't be seen in a man's room. The school board would fire me for sure." Savannah pounded on his arm as she moved her girth against the back of the door.

John slid his hand away just in time to avoid a broken finger in the quick-slamming door.

"Red ain't seen a thing. You got nothin' to worry 'bout, but you better get yo'self dressed and back to yo' room before yo' sister gets back." He watched her move toward the trail of clothes scattered along the thin carpet on the floor.

Savannah scurried into her chemise and tugged her winter drawers to just below her waist. She reached for, and then tossed on the rumpled bed, the corset he'd had so much trouble unlacing. As he grabbed at his own shirt, he took another look at the woman he'd been bedding ever since Sally got too big to entertain and Zeola sent her off to do the ironing for the other girls.

"Oh, my Lawd. I been on this job seven years. I sure don't want to lose it." Savannah pushed an arm through her white shirtwaist.

"It's Sunday. Ain't no school board members gonna be 'round to look in on you in yo' own house." John was pleased to be with Savannah though she was about the plainest woman he had ever bedded.

But if she hadn't been hovering over downright homely, she never would have risked her schoolmarm reputation to sleep with him. She was a smart woman, and at twenty-six, she knew she was already far gone into spinsterhood. She carried about a hundred extra pounds, mostly from chin to knee, Still, she was a kind-hearted woman, and she understood when he told her right up front that he couldn't do no serious courting. She nodded her head and admitted that her only marrying prospect was a sixty-year-old widowed Baptist deacon with four teenage children who told her plain that they didn't want another momma. Although he hadn't been Savannah's first, she wanted to find out what this lovemaking stuff was really all about before she got saddled with a man known to be tight with a nickel and quick with the hellfire and damnation, especially when it came to girl children and women.

"My sister won't be back for another hour." She stepped into her skirt and tugged hard at the too-tight waistband to get it fastened.

John sat on the edge of the bed to pull on his high-buttoned shoes. He looked over at Savannah as she tried to lift a leg to buckle her own shoe.

"I'll be on my way now." With one shoe still unfastened, and holding on to the offending corset with its strings dangling in all directions, Savannah cracked the door open a peep and looked both ways before she opened it barely enough to squeeze herself through.

"I'll see you next Sunday," he called after her back.

The schoolteacher didn't acknowledge him as she quietly closed the door behind her. On most Sunday afternoons, ever since January, he had slipped Savannah into his room while her younger sister was out courting. The plaid-shirted poker player from Miz Zeola's had taken a liking to the sister, though John thought he was far more interested in a teacher's first-of-the-month payday than he was in the charms of a rather ordinary looking woman. Still, it had kept Savannah free for nearly three hours every Sunday afternoon when the sister went off with her suitor to church socials and buggy rides chaperoned by the preacher's wife. That was just fine with John since Savannah didn't ask anything more of him than two hours in bed once a week. And being with a woman who didn't expect courting meant he was being true to Annalaura.

In all their married years, John had never been unfaithful to his wife. The women at the ho houses and the sportin' houses didn't count, and neither did Savannah. Just before he stepped out of the door, he patted at his pockets remembering to take some cash from his hidden savings box with him. What did Zeola want now?

⚭

The dining room was empty when he arrived some thirty minutes later. He walked right past Big Red, who, from the smell of it, was boiling cabbage to go along with the corned beef John spotted in the big kettle on the back of the stove. Red had mumbled and raved in turn, but John hadn't bothered to even glance in his direction as he pushed through the pantry door. Although no one sat at the dining room table, John knew to stand and wait until invited to sit. He walked over to the parlor. It, too, was empty though he noticed that the door to the main poker room was open. He went back to the dining room.

"I tole you to be here thirty minutes ago." Zeola stormed into the room in a flowered day wrapper that was untied.

John was sure that bunched-up bulge underneath her wrapper was her nightgown though the time showed close to five o'clock in the afternoon.

"Yes, ma'am. As soon as Big Red tole me, I walked here on the double." He let his face show concern he did not feel. What was so all important?

"Walk? Hell, man. Ain't you got a nickel fo' the trolley? I got me troubles here." Her nightcap sat askew atop her head, and half her curling rags were still tangled in her hair.

"Miz Zeola. Tell me what I can do to help you out." He watched her walk over to the safe and pull out the bourbon and a glass.

She shoved the branch water to the side as she poured the dark brown liquid nearly to the top. Zeola turned around and squinted one eye at him as she downed the liquor in two swallows.

"How much money you got saved in yo' poke, John Welles?" She belched.

What had gotten her into such a snit, he wondered? The woman hadn't even bothered to stick a wig on her head, let alone lay on that pound of rouge and powder she wore every night. Here she was, asking him about his personal business. Ever since November, and his four pots a week, he'd been able to save up nearly two hundred dollars to take back home. When he lit out the first of April, he would have close to three hundred if Zeola was true to her word and handed over the money she was holding for him. Of course, that wouldn't be enough to buy those twenty acres he had his eye on down in Lawnover, but he had no more time to wait.

"I gots me enough to take real good care of my family, Miz Zeola." The half smile on his face began to hurt as his jaw locked into place.

"I got no time to play with you, John Welles." She put the glass back to her lips, grimaced when she found it empty and stalked back to the safe. "Now, how much money you got saved?"

"I been able to save enough to keep my family fo' most of a year." He figured he was close to the truth. Three hundred dollars could keep a family alive if they had a place to stay and went hungry a little bit. He lifted his eyes to her face, but he was careful not to paste on a smile.

"You saved any mo' than two hundred dollars?" She pulled out the box of matches from the safe drawer.

John had to make this job last only another three weeks until April and his return home. If he was really careful or really lucky with the pots, he could bring home another forty-five dollars. He couldn't afford to anger Zeola right now.

"A little more than that, ma'am." He watched his boss dig into the pocket of her wrapper and pull out a cigarette.

She held it, unlit between her fingers. Why was she hell-bent on find-ing out how much money he had saved?

"Three hundred dollars won't let that woman of yours live in no kind of style." She bowed her head as she lit the cigarette, looking at him through the curling smoke.

"How long you plannin' on stayin' with me, John Welles?"

The question came at him like a sledgehammer.

"As long as you needs me, Miz Zeola." He watched her toss her head like she knew a lie when she heard one.

"I got a deal for you, but I ain't havin' you up and leave 'til I say so. You understand?" She pointed the cigarette at him as she slowly walked across the room and planted herself in front of the door leading to her parlor.

John could see the open door of the big pot room just behind her. "Ma'am, I'm here as long as you needs me."

"I tole you befo' not to mess with me. I knows you tryin' to lay up money to get back to tobacco country, and I knows it's plowin' time. You'll be outta here lickety-split as soon as you get five hundred dollars in yo' poke." She took two quick puffs on the cigarette, then let an ash fall on her fringed rug. She looked at him like she was daring him to call her a liar.

"I ain't got but a little over two hundred, Miz Zeola. And I ain't 'bout to leave you 'cause I . . ." He didn't know that a woman her size could flash around a table and four chairs to stand within six inches of him quicker than he could finish his words.

"I don't give a damn 'bout the lie you gettin' ready to tell me, country boy. I want to make you a deal." In addition to her no powder and no wig, Zeola hadn't taken the time to clean her teeth. The smell of onions from last night's smothered pork chops blasted at him.

"What you offerin', Miz Zeola?"

"The big pots." She took a step back, and John let out a breath.

"What you mean, the big pots? Alfred sits the big pots." He didn't like it when women talked in circles.

"I wants you to hold the big pots. Leastwise 'til I can get me some-body else. You take over the weekends, Fridays through Sundays. That's three of 'em." She sucked in on her cigarette.

"And Alfred? What pots will he sit?" It would be best to proceed slowly until Zeola could start talking like a woman with at least some of her wits about her.

"I see Red did keep that loud mouth of his shut and ain't tole you a thing. Alfred took off with Sally. He the father of Sally's baby. They both left me high and dry, and I got Bubba Johnson and seven other high rollers comin' in at ten o'clock." She jerked her head toward the secret door to the big gaming room off her parlor. "Take it myself, but I got me the police chief comin' in tonight. I got to make sure everythin' goes jest right. I gots my hands full. He wants that new young gal I took in last week, and she ain't nowhere near ready enough to entertain a white man." Zeola turned around in a swish of wrapper.

"Whoa, Miz Zeola." John held up a hand. For a woman used to being close-mouthed, Zeola was pouring a whole barrel of words over him. "You tellin' me that Alfred ain't sittin' pots no mo'? That he took off with Sally?" John flashed on all the nights he'd spent with the far gone preg-nant Sally knowing her baby wasn't his. "What kind of money we talkin' 'bout, Zeola?"

She stopped her puffing and her pacing. "Pots with Bubba Johnson playin' been runnin' to almost three hundred a night." Zeola's voice steadied.

"Three hundred dollars?" The figures he ran in his head were working wonders to calm him. "You say there's eight players tonight? What do I get fo' holdin' the big pots?"

"Twenty dollars out of every pot and a dollar a man at the table." Zeola was settling into her old self again.

"Make it forty and five dollars a man." John leveled his eyes at his boss.

Zeola let out a laugh so deep that it rolled right up her spacious belly to the front of her nightgown.

"Well, Johnny, I guess you ain't much of a country boy no mo'. You sure you wanna go back to plantin' tobacca?" Her laugh came so hard that he could barely catch her words.

"What 'bout my fo'ty dollars?" He had no intention of her joshing him off target.

As fast as her laughter came, it dried in her throat.

"You may not be a country boy no mo', but you ain't quite slick enough to git one over on Miz Zeola. I'll give you thirty out of every pot and two dollars a head. That'll give you over a hundred dollars a weekend just fo' yo'self. Take it or leave it." All of her earlier panic had evaporated.

"And I keeps my little pot games durin' the week." He made sure it sounded more like a foregone conclusion than the question he knew it should be.

A little smile flickered over Zeola's lips.

"You do both pot jobs." She motioned him to step across the hall and into her parlor. She pointed to the gaming room where the serious poker players sat.

As John moved to squeeze past her, she clamped her hand over his shoulder. He felt like he was being held in a red-hot blacksmith's vise.

"You give me two mo' months, and you will have enough money to buy them twenty acres in tobacca country. No mo' plantin' fo' the white man. Come this here June, the ground you put that seed in will be yo' own." She dug her nails deep into his shoulder.

He felt the blood oozing under his work shirt.

"You cross me, country boy, and Big Red will do mo' than wake you up from whorin' with that schoolteacher."

# CHAPTER EIGHTEEN

The Thornton farmhouse, sitting in the middle of Old Ben Thornton's six hundred and forty acres like it did, had always been the natural place to host the annual first day of planting dinner and prayer for the Thornton clan. Though the four Thornton boys each received unequal portions as their inheritance, with eldest son Ben Roy getting the biggest and the best, all the kin agreed to gather at the family home to share in the big meal and pray in a successful tobacco season. Farmers all over Montgomery County were doing much the same during mid-May. Throughout the county, farmers would group together on the first day of planting, stand side by side with their local preachers, and pray for warm skies and the proper amount of rain.

On that sun-filled, wind-free day when Ben Roy declared it planting time, Fedora had twenty-four hours to arrange for all the Thornton women to bring together the meal that would be served after the men had set their tenants to sowing seed. Because Reverend Hawkins had to cover fifteen farms in two days, he wouldn't get to the Thorntons 'til close to two o'clock this day.

While most of the Thornton women spent their time complaining at the short notice that came regular every year, Eula always had her stores in place. She could tell far better than her brother when the sky and the moisture in the earth teamed up to tell all farmers that the ground was ripe for seeding. When the first of May dawned, she went to her smoke house and got out one of the few remaining slabs of bacon to flavor the pole beans canned late last summer. Every May morning during those first couple of weeks, as she walked by the chicken coop, she set her eyes on the pullets that would make the best fryers and gave them extra feed. A day or two before she knew Fedora would come frantically knocking on her door, she began her cooking. She baked her chess pie and shoofly cakes, fried up her chicken, opened her Mason jars of beans, greens, and turnips, and when they were ready, put them all away in a cool place on the porch by the pump.

Sure enough, as the third week of May dawned, Fedora lashed her buckboard over to the McNaughtons and issued her breathless command to come join Cora Lee, Tillie, Belle, Jenny, and all the other Thornton women for the prayer dinner to wish for an easy plant, good growing weather, and a profitable harvest.

"My greens ain't nowhere big enough to pick, and I've got none canned." Belle, as disorganized as ever, made the same complaint Eula had heard for fifteen years now. She ignored her sister-in-law and continued to spread the red-checkered tablecloth over the last of the three tables Fedora had set out in the big yard behind the house. With close to thirty-five Thorntons showing up, Fedora always held the prayer dinners outside since Ben Roy had guaranteed perfect weather by his selection of the planting day.

"What did you bring then, Belle?" Cora Lee asked, as she spread out a flowered tablecloth on a second table set closer to the house.

"I had my girl fry up some chicken and make a sweet-potato pie." Belle stood to the side, acting, for all the Thornton world to see, like she was supervising the other women. Fedora's colored cook gave Belle a quick glare as the longtime servant started laying platters of smothered pork chops and preserved lima beans onto the first table, which was already set with its white cloth. Cora Lee stood up straight and whispered over her shoulder to Eula.

"I think your brother must have seen somethin' else in that one to marry her besides her cookin'." Cora took a platter of corn bread from Fedora's cook and set it on her table. Whatever her youngest brother saw in the ill-mannered Belle escaped Eula. Her table covered, she walked past Cora Lee without answering and headed into the kitchen to retrieve her own platter of chicken and bowl of greens. Let Belle and Cora Lee fight it out on their own.

"Cousin Eula, where do you want me to set this chess pie and cake?" Jenny balanced both in her hands as Eula entered Fedora's kitchen. She gestured to the red-checkered table just as she heard Belle's voice call through the open kitchen door.

"Eula, when you bring out your plates, would you grab some for my table, too?" Belle may have been an ongoing irritant for Cora Lee, but Eula had long ago gotten used to the ways of her sister-in-law.

If she hadn't married Eula's youngest brother, little Belle would have been the one scrubbing and cooking for others, and not the hired colored girl she liked to flaunt in front of Eula. Trying to balance the heavy platter of chicken, Eula struggled to slide the big bowl of greens next to it without spilling everything. Now, if she could just grab the pan of sweet potatoes, she could get out of the kitchen in just one food trip. Eula was smiling to herself as she started her careful way across the kitchen floor with her burden when she heard the baby's cry.

"Momma, he just won't take it. Can't you speak to Papa?" Tillie's shrill voice verged on the hysterical.

Eula stopped and turned as she carefully reset all the serving dishes back onto Fedora's table. She walked toward the back bedroom where the commotion was getting even louder.

"Hush up, Tillie," Fedora hissed at her daughter. "If you can just keep yourself still and stop that dratted screeching, the baby will take the tit."

Even though the window was open to capture the windless air, Fedora's forehead was beaded with sweat. She stood alternately jostling Tillie's two-month-old son in her arms and pushing the infant's mouth at Tillie's exposed breast.

The baby, Little Ben, was as red as the strawberries due to come in next month. The little arms and legs flailed in all directions.

"Fedora. Tillie. Can I get you all somethin' to help out with Little Ben?" Eula stood in the doorway.

Being around babies still upset her, but the frantic scene in the room compelled her to offer whatever help she could. Both women turned their heads in her direction. Tears streamed down Tillie's cheeks. Tillie squirmed in the rocker her mother wouldn't let her leave, her hair hanging in loosening strings from its pompadour. Every time Fedora tried to push Little Ben's mouth onto her nipple, Tillie flapped her hands hopelessly in the air and rocked away from the baby.

"It ain't workin'," Tillie screamed at Eula. "Please, Auntie, talk to my father. I need Hettie."

"Hush yo' mouth before I smack you one." Fedora crooked the baby in her arm as she turned to the bedside table in the back room and dipped her finger into a half empty jar of honey.

Apparently uncaring of the sticky trail dripping across her carpet and her bedcovering, Fedora walked over to Tillie and smeared the honey on her daughter's nipple. The girl reacted as though her mother had caught her breast in a wood vise.

"Noooo." Tillie looked like she could swoon at any moment.

Eula hurried over and stood between mother and daughter, her heart racing. "My Lord, is there somethin' wrong with Little Ben?" She stared at the squalling infant, too frightened to reach out a hand to comfort her great-nephew.

"He's starvin' to death is all, Aunt Eula. Starvin' to death because of my father." Tillie shoved her breast back under the cloth of her bodice.

"He's doin' no such thing. You're the mother. You can feed him." Fedora held the baby in her arms as he frantically nuzzled at his grandmother's clothed chest.

"He does look hungry, Fedora." The last thing Eula wanted was to challenge her sister-in-law.

"He's hungry because Hettie's not here to feed him." Tillie rocked the chair hard. Eula feared she might just pitch forward and fall flat on her face.

"He drinks Hettie's milk best. 'Sides, mine is 'bout dried up. I can't wet-nurse no baby. Momma didn't wet-nurse me." Tillie shot an accusing eye at Fedora. "Grandma Thornton didn't wet-nurse you, Aunt Eula. Only colored did that. Why should I have to wet-nurse this baby?" Tillie folded her arms tight under her breasts, making them poke out all the more.

"Because it's an emergency. There's plenty of milk still in there if you just let this baby get started." Fedora took her free hand and pushed on

Tillie's breast, setting up fresh howls from the young woman.

Eula knew better than to inquire on the whereabouts of Hettie.

"Emergency? Emergency?" Tillie's voice screamed out of her mouth. "You call having to cook for some family in Clarksville an emergency?"

"Your papa owed her to a friend. It's not for you nor me to question what Ben Roy says or does. He had to honor his word." There was a catch in Fedora's throat. "Fact is, Hettie ain't here today, and she ain't settin' foot in this house no matter how much you scream and shout. And somebody's got to feed this baby."

"Did Hettie send any milk with him?" Eula ventured.

A colored woman could be milked like a cow and the liquid dripped into an empty Mason jar, Eula knew. Tillie aimed a trembling finger toward the carpetbag on the bed. Fedora stood unmoving, staring at it. Eula crossed the short distance to the bed and pulled out the Mason jar full of Hettie's milk. Without warning, Fedora thrust Little Ben into Eula's arms and started toward the door. Eula hadn't held a baby in more than twenty years. Doc Starter wouldn't let her hold her daughter after the girl had been born dead. He wouldn't even hand the body to Alex. Eula dropped to the bed, fighting to control her own misery. The squealing child soothed enough to allow her to unscrew the lid to the jar. Unsure what to do next, she looked toward Tillie who shrugged her shoulders at the doorway. Fedora turned around.

"Dip your finger in and see if he will suck it off'n them. I ain't touchin' it." Fedora answered Eula's unspoken question.

Eula stared at the jar of mother's milk, and her own eyes clouded with tears.

"What's goin' on in here?" The sound of Belle's voice snapped Eula's head up. "Oh, come on, let me do that. You all act like you ain't never fed a baby milk from a wet nurse befo'." Belle pulled Little Ben out of Eula's arms.

Dipping her fingers into the jar, Belle poked them into the baby's mouth.

The child sucked on Belle's fingers from tip to knuckle.

"This little one's sure hungry." Belle looked at the Mason jar. "I'm surprised Hettie's got enough milk to fill that thing up. Fedora, ain't she nursin' for two?" Belle pulled out a square of clean cloth from the carpet-bag, dipped it into the Mason jar, and gave it to Little Ben.

Fedora remained silent.

"Colored women have lots of milk." A calming Tillie spoke up. "Her boy's four months. He eats after Little Ben's had his fill." Tillie buttoned her shirtwaist.

The baby's slurping could be heard throughout the room. Belle dipped the cloth into the milk while Little Ben fussed at its temporary absence.

"Fedora, don't that make three yella-skinned children Hettie's dropped?" Belle slipped the soaked cloth into the baby's mouth as Fedora turned in slow motion toward the door and moved her feet across the threshold.

Before she had gotten out of earshot, Eula's youngest sister-in-law turned on the next woman in the room.

"Eula, I do believe I saw that tenant of yours comin' out of the nigger church on Sunday." Belle soaked the cloth.

Every time Belle, who hadn't made it beyond the fourth grade, used that profane word, Eula winced. How could her baby brother have married a woman so unrefined?

"I'm not right sure what you believe, Belle. What tenant you talkin' about?" When Eula thought of the exchange in all the years to come, she would remember that Belle had led her into that trap like a fox treeing a rabbit.

"That nigg . . . , uh, colored woman on your mid-forty. What's her name?" Belle must have caught the frost in Eula's voice.

"You talkin' about Welles? That family tenant farmed our mid-forty last year?"

Glancing at Little Ben, Eula stood and started toward the door. Through the kitchen window, she could see Fedora standing outside, apart from the others.

"That ain't the way I hear it." The smirk in Belle's voice matched the one on her face.

"I heard her man's been gone for almost a year." Belle's voice sounded like it wanted to sing a church solo.

Standing in the doorway, Eula finally turned to her sister-in-law. What was the woman trying to say to her now?

"Nothin' to it. Alex hired a man to bring in the crop last fall, is all. Everthin' worked out just fine." She knew she should move her feet as fast as she could after the escaped Fedora. But, something in that curious look on Belle's face held her fast.

"I saw that Welles woman steppin' out of Bobby Lee's store with her belly 'bout as big as an August watermelon. And, the way she was dressed, I don't think I've ever seen such fancy on a nigger in all my life." Belle put Little Ben over her shoulder to burp.

Eula grabbed at her ear. She was certain that sudden roaring coming through the open window had distorted her hearing.

"What?" She could manage no other words.

"Your nigger tenant is 'bout to let loose another pickaninny on Thornton land. Tillie, come and get this baby."

Eula felt Tillie's groan. She could neither see nor hear anyone or anything else in the room.

"I wonder who the daddy is?" Nothing of a proper question lingered in Belle's voice.

By the grunts coming from Little Ben, Eula surmised that her sister-in-law had handed over the baby to Tillie.

"Aunt Belle, I swear you don't listen sometimes. Aunt Eula just told you 'bout the hired man. Of course he's the father. You know colored. They just rut around every chance they get." Tillie walked out of the room as Belle stood to smooth her crumpled dress.

"I don't believe that hired man was there long enough to give anybody a baby, do you, Eula Mae?" Belle swept past her.

Eula tried to make her feet move. There had to be some mistake. The woman...what was her name...couldn't be pregnant, not if her husband hadn't come home. Every imaginable thought poured into her head, but only one stuck to the inside of her brain like honey to a hive. She had to get home right now.

As she clicked the harness moving her horse faster, Eula's buckboard bounced on a rut in the lane. She could barely remember how she had gotten this far. She recalled stumbling out of Fedora's kitchen and mumbling something about forgetting her corn bread and she had to return home immediately to retrieve it. As she passed the stand of trees near her barn, she peered at the sky. The sun, a shade past its peak, told her it must be close to half past one. She had just missed the onslaught of men coming out of the fields and into the Thornton yard to begin the prayer dinner. She thought she spotted Reverend Hawkins as she made her hasty retreat, but she really couldn't be sure. She pulled the buckboard in front of the barn. Without unharnessing the animal, Eula led him to the horse trough filled with water, and ran into her own kitchen. Frantically, she reached for her journal.

Eula almost ripped the pages with her twitching fingers as they turned back to December, November, October, and finally September. All together the harvest had brought in over forty-five hundred dollars. Her books told her that after paying for food, supplies, and tenant needs for seven months, she and Alex should still have way over two thousand dollars. Grabbing one of her kitchen chairs and dragging it across her kitchen, parlor, and dining room floors to her bedroom, she stood it in front of the closet door. Gingerly stepping up onto its seat, she reached for the top shelf and began shifting boxes and bags until she felt the tin with her fingers. Dislodging a box of Confederate money left over from Alex's father that nearly spilled out over her head, she pulled down the tin. Eula wobbled off the chair, almost tipping it. She carried the box to the bed and dropped it down onto the feathered four-poster. Her hands worked to pry it open. The box was locked. Eula's mind rummaged over the whereabouts of the key. Remembering, she rushed to the closet and to Alex's funeral-and-wedding suit. There, in a breast pocket, she pulled out the key. Her hands shaking, she ran back to the bed and slipped the key into the lock. It turned easy as though it had been opened frequently in recent months.

Inside, Eula found, and tossed aside, the deeds for the house and land. She piled the sale papers for the animals on top of the bed. She scattered supply receipts across the floor. She dug her hands deep into the box until her fingers pulled out the envelope that should have been fat with cash. Her hands still trembling, Eula counted out one thousand four hundred fifty-five dollars and seventy-five cents. She shook her head.

That couldn't be right. Her journal records told her they should have over two thousand dollars. She recounted. One thousand four hundred fifty-five dollars and seventy-five cents. Just like the records on her preserves in the smoke house, there had been no mistake. Over five hundred dollars was missing.

A wave of dizziness came up sudden from her stomach. She felt like she did as an eight-year-old when she slid off a slippery rock and fell into the slow-flowing creek on the far side of Lawnover. She clutched at her middle and reached for one of the posters on her bed. She laid her body against it, clinging to it like it was the rope Old Roy threw to pull her out of the murky waters. It could not be true.

Belle was as much a liar as she was white trash. Every man and woman in Lawnover, black and white, knew that Eula's brother Ben Roy was the father of Hettie's three children. But Alexander? The father of a colored woman's child? Never. She and Alexander hadn't conceived a baby in twenty years. Now he was nearly forty-four, and there had never even been a whispered hint that he had achieved fatherhood. If he had sired children, the news would have been all over Lawnover in less than two days. No, her Alex couldn't be the father, because he was incapable of making a baby.

Alex had never reproached her for her inability to bear him a second child the way any other husband would. Deep in her mind, she thought it might be because he knew he was partially to blame. Still, if Belle was to be believed, the Welles woman was pregnant. How could that be so? For all of her treacherous ways, Eula's youngest sister-in-law was right about one thing. The hand Alex hired last September couldn't be the father either. Even on those three mornings when she had awakened during harvest time to find Alex's side of the bed empty, she had gone down to the barn to see Isaiah Harris carry his slop jar to the outhouse. He had been in her barn every one of the fourteen nights of the September harvest. Yet, if it wasn't Isaiah Harris, who was the father of the Welles woman's baby?

The wooziness increased, and Eula felt the bile rise to her throat. The grandfather clock in her parlor bonged two. She had never laid down in the middle of the afternoon because of sickness in all of her married years, not even when she was racked with the fever. But maybe, just this once, getting off her feet might help her put that imagination of hers to rest. Her father had always told her that imagination in a woman was not only unnecessary but a dangerous thing. Eula eased herself on top of the coverlet, not caring that her shoes carried the dirt and dust of Ben Roy's backyard on them. The room felt warm though the window was opened wide to let in the mild, new-rose-scented May afternoon. She put her hands to the side of her head to push out the thoughts that no decent woman should carry. If she was really a good wife to her husband, she should be able to come up with the real reason for the short supplies, the missing money, and Alex's peculiar behavior over the planting.

With barely a word, her husband had put another hired man on the mid-forty to set the tobacco seedlings. That would have been of no particular concern to Eula. All the farmers knew that it was difficult keeping tenants, black or white, for more than two years at a stretch. But Eula had to learn from Jenny that the man, an out-of-towner, was married with three small-sized children. Worse, Jenny had added that none of the man's family was staying on the mid-forty. Yet, the Welles woman and her get remained.

Instead of evicting her as he should when a tenant stopped producing, Alex had allowed the woman and her children to stay, even though their man had been gone for almost a year. Why would Alex let a woman who could do him no earthly farming good live in his barn, while a hard-working man with a family had to travel two miles to and from town every day to tend the tobacco? Maybe if she closed her eyes for just a moment, these devil-placed thoughts might fade away and make room for the true answer. Somehow she knew that even Reverend Hawkins couldn't help cleanse her mind. She had to pray directly to the Lord for forgiveness for thinking the unbelievable.

# CHAPTER NINETEEN

The two sixteen afternoon train from Nashville pulled into the Lawnover station right on time. The train, speeding to Chicago, stopped for only five minutes, and John, wearing his new cream-colored seersucker summer suit, scrambled to pull out his two valises from the overhead rack in the colored-only rail car. Kicking the box of gifts along in front of him, he moved toward the connecting carriage doors. John used his shoulder to open the heavy door and maneuver his belongings to the top of the stairs. The colored porter, standing on the platform at the bottom of the portable steps, stared up the station platform toward another white-coated porter two cars ahead. John's porter paid scant attention to his own departing passengers as they wrestled their boxes and string-tied bundles. John followed the man's stare, though he already knew what held the attention of his porter.

Two cars up the line from his own, steam hissed out of the under-carriage of the first of the whites-only train coaches. Colored porters there could expect tips, though they were seldom more than five or ten cents. Still, in times like these, every bit of change from four or five people at every stop could add up to substantial money. As John put one valise under an arm and juggled the second, he tried to lift up the gift box with his free hand.

"All aboard!" The disinterested porter called out, as a man dressed in an ill-fitting suit stood on the platform waiting for John to disembark so he could board.

For a fleeting moment, John wondered if the fellow was one of the fortunate ones heading for Chicago. The porter, finally turning back to his own disembarking passengers, reached out a hand to grab one of John's valises. On the ground, John lent the man one of his smiles as he fished in his pocket. It pleased him immensely to drop a fifty-cent piece in the porter's top pocket.

"Th-thank you, suh." The surprised train man could barely stutter out his appreciation.

Without a word, John winked and walked up the platform. With over a thousand dollars in his money belt, it felt better than good to show the world that John Welles was a man on the move. Judging by the time on the big round-faced clock on the platform, he figured he could rent a horse and buggy and be with Annalaura by three thirty. He nudged the gift box with his knee. The sturdy, rope-tied cardboard was stuffed full of presents for his family.

Every child had a complete outfit from underwear to shoes. Little Lottie even had bows for her hair. And, the doll, all made of cloth except for her head covered with real horsehair, would thrill his little girl. Henry had a fine wood-carved train. Cleveland would get a bone-handled knife. He knew Doug would take to the reading book he got for him. For Annalaura, he had gone wild.

Her pinafore was white with yellow trim over the ruffled parts that women liked. He had gotten her two dresses, one blue and one white to set it over. For the front of her pinafore, he bought a brooch that the man said was made of real shells from far-off California. And when he spread that money out on his bed, he could see her jumping in eagerness to let him scrub her all over with that sweet-smelling soap. He could already hear her moan her pleasure when he dabbed that rose water in all her secret places. Even Annalaura would grin and forgive him for being gone just a little longer than he wanted when she saw all that he was bringing home just for her.

John knew it was planting day even before the horse and buggy left Lawnover's wood-planked sidewalks. He could smell the new plowed earth, and the May air touched his face with just that right amount of warm and wet oozing out of it. The time was close to three thirty when he turned off the lane and up the path to the mid-forty barn. He reckoned Annalaura and the two older boys would be in the fields scattering seedlings right about now. He wondered if McNaughton had sense enough to put on some help for her.

He was surprised to see a tall, skinny boy of about ten standing just outside the smoke house door. The lad had apparently spotted the horse and buggy before John got a look at the child. There was something in the way the boy cocked his head to stare at the approaching buggy that told John the child's identity.

"Doug? That you, boy?" John jerked on the reins and the horse pulled up.

The child's eyes grew as big as a river lizard's. The smile coming from John was real.

"I swear, boy, I thought you was Cleveland, you is so big." John jumped down from the buggy, holding the reins in his hand.

"Papa? Papa is that really you?" His son's voice was still that of a child, and he was glad.

The troubles of a man would come on his boy soon enough. John wanted to make these last years of his Doug's childhood filled with some of the fun and ease the lad had missed.

"Well, what other man come ridin' up to yo' barn? 'Cose it's yo' papa." John swooped the boy into his arms, lifting his feet off the ground.

The child flung his arms around his father's neck.

"I knew you'd come back. I knew it mo' than anything." He squeezed his skinny arms all the tighter.

"Whoa, boy. You 'bout to choke me to death." John set the boy back on his feet and held him by his shoulders. "Whoee. You sho' have growed up. I hope all those new clothes I got fo' you will still fit." He jerked a head toward the box as he winked at his son.

Doug started toward the buggy, but with John's hands on his shoulders, his feet marked time in place. John's laugh boomed out in the afternoon sun.

"Papa, is you rich now?" Doug stopped his fruitless running and looked John up and down.

"Yo' papa's always been a rich man. I got you, yo' brothers, yo' little sister, Lottie, and yo' momma. It's just that now, I can show you all how rich I am."

The stunned look on Doug's face kept bringing on the pleasure. John couldn't stop the laughter, and it felt good. He hugged the boy again.

"Say, why ain't you in the fields? Ain't this the first day of plantin'?" He gave the boy a crooked half smile.

"How'd you know that? You must be a magic man and a rich man." The surprise on his son's face slowly gave way to a broadening smile.

John closed his eyes for an instant. All those Nashville months melted away in just this instant. He looked at his boy again.

"Is Cleveland in the fields with yo' momma?" Through all the joy he was feeling, the slim sound of silence broke through.

More curious than anything else, John stood up and walked around the buggy. With a hand over his eyes to block out the mid-afternoon sun, he scanned the fields near the path. Where were Annalaura and the other children? At least some of them should have been within seeing distance.

"Cleveland's plantin' tobacca with the new man and Momma's upstairs. She sent me down to bring up the pork chops."

For the first time, John saw the covered platter lying on the ground. Doug must have set it there when he first spotted the buggy.

"Pork chops? Where'd you all get pork chops in May?" That cracker McNaughton was being awful generous to the new hired man.

To get pork chops at the beginning of planting would have taken an advance so big that most tenants could never pay it back. Too excited to notice Doug's attempts at a stuttered answer, John grabbed the box from the buggy and headed into the barn.

Doug trotted along behind him. John barely noticed that there were now three cows in the stalls, and the oinking of the four sows in their pen just on the other side of the wall completely escaped his conscious mind in his haste to climb the ladder to the loft and Annalaura. He nearly stepped on one of the two dozen chickens in his way before he reached the bottom rung of the ladder.

"Papa's come home. Momma, Papa's back." Doug's high-pitched voice reached the landing before John could quietly slide the gift box on the floor.

He turned and tried to shush his son, just as the boy started his climb.

"Shh. Papa wants it to be a surprise." He was too late.

The oval-shaped face of a pigtailed girl of about six looked down at him.

First, the child scrunched her eyes in confusion then got down on one knee to peer at the face just coming into view. John put a finger to his lips to silence Lottie. But, like her brother, she was having none of it.

"Momma, Momma. It's Papa. I know it is. It's my papa." She jumped straight to her feet like a Jumpin' Jill.

John reached out a hand to stop his daughter from tumbling down the landing opening right on top of him. As he cleared the last rung, he picked her up as she wrapped both arms and both legs around him. The little girl covered his neck and face with kisses. He twirled Lottie around the room in a dance of triumph, bumping into a wall. He paid no notice as his eyes caught their first fleeting glimpse of Annalaura, but a full view of her was blocked by Henry hurtling at him faster than one of those Kentucky racehorses.

"Lottie, is he really my papa?" Little Henry pulled at the black high-topped shoe on his sister's foot.

The thought that the shoe was new and well fitted burst on his brain and fled when he saw Doug dive into the gift box.

"Papa brought us some presents," Doug announced as Lottie squirmed out of her father's arms and nearly ran over Henry getting to the box.

"Hold on, now. There's presents for everybody. Lottie, you and Henry, let Doug pass 'em out." He reached for his new breast pocket handkerchief when his eyes finally lit on the woman sitting at the table.

The kerosene lamp highlighted the little spray of sun coming from the window over Cleveland's alcove and played across her face. He was almost sure it was his Annalaura, but that look of surprise, mixed with a goodly amount of fright, confused him. He took a step toward her but stopped when he saw her face go pale underneath her brown skin.

"Annalaura, I knows I been gone awhile, but I can explain." Of all the faces he had pictured on his wife when she greeted his return, he hadn't seen this one.

He knew she'd be spitting mad, but this wasn't anger that he saw. He took another step toward her and thought she was going to fall right off that chair onto the floor. Chair? He remembered that when he left, the family only possessed two crates for sitting and no chairs. He waited for her to speak, and when her mouth looked like it was frozen into a face that had seen a lynching, John's heart picked up a pace. If it wasn't anger that he was seeing on Annalaura's face, then she must have passed anger and gone straight to don't care. He couldn't bear the thought. He unfastened the bottom button of his new suit and pulled out his shirt.

"Now, Annalaura, ain't no reason fo' you to take on so. I been workin' fo' us." John slipped a hand under the waistband of his pants and fumbled with his money belt. He undid the clasp and pulled out bills, piling them on the table. Fives, tens, and even twenties mounted. He kept his eyes on his wife who seemed to be barely breathing.

"Annalaura, darlin', I got us over a thousand dollars here." He swept his hand over the pile, knocking a few of the bills in her direction. "Sugar, we gonna get us our own place. Our own farm. Girl, stop yo' frownin' and let's get to dancin'." He stepped around the table and pulled Annalaura from her chair.

His arm moved to encircle her waist.

"Mmm" strangled out of his voice box. "A... An..." His open hand lay inches from where her tight belly used to be. He stared at that big bulging thing that poked out in front of Annalaura.

The look of the living space rushed at him. He spotted the pot-bellied stove that hadn't been there when he left. He saw the tin of candied yams warming on the grate. The sweet scent of melting butter and sugar that should have filled his nostrils with pleasure brought only the taste of bile with it.

He tried to swallow but choked on a mouthful of nothingness. His arms moved without him willing them. He pushed Annalaura away.

"That." The word exploded out of him. He pointed at her stomach. She stood stock-still.

"Annalaura?" The word squeaked out of him. Tears puddled in his wife's eyes. "Annalaura...you ain't...that...ba...you cain't..." He blinked his eyes to make the sight disappear. "What's in yo' belly, woman?" The words burst out of his dry throat.

He took in a deep breath, ready to hear that his wife had contracted some dreaded woman's disease. Something that men never learned about until their own women caught it. Something so terrible that it made the belly swell up like the woman had been poisoned. John grabbed the back of a second chair. His mind registered that the chair hadn't been there when he left. He swallowed, steeling himself, waiting for Annalaura to tell him that she was about to die of some awful thing that had no medicine cure.

"What's in there?"

She stood silent.

"What's inside you?" His voice clanged in his ears.

Annalaura's eyes turned in his direction.

"I want to know what's in yo' belly." He stepped toward her. "Tell me now, Annalaura. Damn it to hell, tell me now."

John didn't feel his arm move out from his shoulder, but he heard, rather than felt, the smacking sound the back of his hand made across Annalaura's cheek. He saw, rather than felt, her head snap back, and to the side as she stumbled against the wall that hadn't been there when he left. He sensed, rather than heard or felt, the commotion from his children behind him. He walked closer to his wife, the left side of her body punched into the wall, sliding down the whitewashed surface. Her hands hung limp in the folds of her dress. Her mouth stayed silent. Grabbing her shoulders, he roughly pulled her back to her feet.

"Annalaura, you ain't done this to me. Tell me in the name of God Almighty that you ain't done this to me." He didn't know how long he shook her, but the sound of her head banging against the wall finally entered his ears along with cries from either Lottie or Henry. He couldn't tell which. He spun his wife around and closer to the kerosene lamp on the table. His hand ran roughly over her stomach. The hardness greeting his touch caused him to throw her down on the table.

"Papa, no." He thought it was Lottie's scream this time, but it could just as well have been Doug. He pushed whoever it was away before he pulled Annalaura off the table and spun her around again. This time his arm flailed out at the other side of her face. The taste of someone else's blood felt strange flying onto his own lips.

"You ain't betrayed me, Annalaura. I know you ain't." He doubled his fist and smashed her in the right eye.

She began to slide down the wall again.

"You is the best woman in all this world." Pulling her to her feet again, his fist pounded at her left breast.

"You is the sweetest, smartest woman on all God's earth." There was moaning in his ears, but it wasn't coming from Annalaura. He had the strange impression that it was his own pain leaping out of some deep place.

"I would give this world fo' you. I done everythin' fo' you." This time his fist found her nose, then her chest, her ear, her eye.

Each time she slumped to the floor with her arms limp at her side. He pulled her to her feet. Still she said nothing.

"You gonna tell me, woman. You gonna tell me the bastard's name who give you this." He punched her in the stomach, the forehead, and with doubled fist, aimed back at her stomach.

Annalaura bent over and dropped to the floor, drawing her knees as close to her chest as she could.

"Kill me. Kill me, now. Just don't let yo' children see." She lay there with her eyes closed, ready.

"You damn right I'm gonna kill you, but I'm gonna kill that nigger first. I'm gonna kill him and make you watch." He reached down to drag her to her feet again when he felt something sharp jab him straight in the backside.

"Let her be. You ain't got no right to hit my momma. It was all 'cause of you." The voice crackled between boyhood and manhood.

John dropped Annalaura back to the floor and turned to look into the face of a wild-eyed twelve-year-old. The trembling boy held the pitchfork in his hands, and his eyes held the determined look of a man.

"Cleveland? Cleveland, is that you, boy?" He stared at his eldest, torn between joy and the most misery he had ever known. "Yo' momma, she..." Without taking his eyes off the boy, he gestured behind him to the fallen Annalaura.

"She did it fo' you. She did it fo' you." The boy's voice shook.

In his fear for his son, John searched out his other children. Over in the sleeping alcove, Doug held both Henry and Lottie close. Henry had his head buried under the bedcovers as he turned his back on the scene. Lottie and Doug sat frozen in terror. John knew it couldn't be helped. Someday he would explain their mother's betrayal, some-day…Cleveland's words started to notch their way into his brain.

"Did it fo' me? What did she do fo' me, son?" He gained control over his voice.

"You ain't no good. You ain't no damn good." Cleveland jabbed the pitchfork in the air. "You left us all here to starve and freeze. You took all the money and all the food. What was she supposed to do?"

"Shut up, Cleveland. On your pa's life, shut up." Annalaura swayed despite her grip on the back of the chair.

John, startled, looked down at his now open palms. He lifted his head. Confusion roiled in his belly. The cupboards were full, his children wore new shoes and fresh clothes. He turned back to Annalaura.

"What did yo' momma do fo' me?" He glanced down at the floor where he stood. When he left, it had been full of knotholes. Now the boards were closely spaced, knot free, and good-sized rugs covered the sleeping places. How did Annalaura come by them? No Negro in Lawnover had money for such fancy this time of year, nor at any other time. No Negro…

"Cleveland, what did yo' momma do fo' me?" Fire and ice fought inside him.

"Cleveland." Annalaura croaked out the command.

John stared at the sleeping quarters. Where a thin sheet of cloth had been their only privacy, now stood a full wall and proper doorway.

"We was hungry," Cleveland called out. "We ain't had nothin' to eat but some dandelions in a whole pot of water. He gave her money and food and clothes fo' us. It was fo' us." The boy lowered the pitchfork, but John could see the firm grip his son still kept on the handle.

"No, son." Annalaura's knees buckled.

"If she hadn't let him stay, he was gonna throw us off the place." Cleveland's words came in jerks between sobs.

Out of the corner of his eye, John watched Lottie pull a porcelain-headed doll with light brown hair off Cleveland's bed and hug it to her little chest. Toys, food and plenty of it, new clothes. Even the dishes were new. He looked over at Annalaura who staggered to her feet. Her eyes stared frantic warnings at Cleveland.

"Boy…" John menaced.

"I'll tell you if you promise to never touch my momma again." Cleveland raised the pitchfork chest high. "If you don't, I'll kill you myself."

He turned to the woman he had always thought too good to be his wife. The swelling had almost closed both eyes shut. Blood ran from nose to mouth. A button was ripped off the bodice of her dress. A cut sliced across her forehead.

"You ain't had no right to go with no other man no matter what I done." The words croaked out.

Cleveland aimed the pitchfork directly at his back.

"I ain't meant to wrong you by leavin'. I did it fo' business, but if you think it was fo' somethin' else, then I reckon I'm sorry you didn't put mo' stock in me than that." Bile filled his throat. "But, ain't nothin' I could have done was so bad that you had to do this." Rage stoked his mind. "You ain't had no right to go with another man. You is my wife." He started toward her. The pitchfork scraped his side.

"She didn't go with another man." Cleveland's voice rose to the rafters. "Mr. Alex went with her."

"Say what?" John felt frozen in mid-step.

"She ain't had no say-so. Even I know that." Cleveland circled the pitchfork next to John's heart.

What difference did it make if his own son ran him through with a farm tool? His boy's words had already done that damage and more. "It's all right, boy. I ain't gonna touch…touch yo' momma if you is tellin' me that the owner of this here farm…Mc…he the one who…McN…bought all these things for you…Mm…and yo' momma. McNaughton." The name stuck in his throat.

"I am telling you that." Cleveland lowered the pitchfork.

Most of the air swept out of John's lungs. He turned to Annalaura and worked his mouth. Pictures flooded his head. Bad pictures. Awful pictures. His wife naked. White man...nak...

"I knows I wronged you, John." Annalaura stood behind the new kitchen chair, both hands on its back. "I knows I ain't never been the wife I should have been to you. I was just too 'shamed to tell you." The words came out of a mouth full of blood.

The roaring in John's ears made listening hard.

"Cleveland's tellin' you the truth. Mr. Alex did give us all this stuff, and he has been sniffin' around here a lot, and I knew what he wanted." She rubbed at her stomach. "You is right to beat me. He said he would bring food if I let him have breakfast here some mornings befo' I went to the fields."

"And you ain't known better than to trust a white man's word?" John nodded his head in disbelief.

"I know what you say is true. I was weak. I couldn't think how else to feed my children."

"And you let him climb into yo' bed...my bed...and give you a baby 'cause you was weak and couldn't think?" The rage roared back.

"No," Annalaura screamed as her figure swayed in front of his face. "What Cleveland saw was true. Mr. Alex was here in the mornings for breakfast, but this ain't his baby."

Out of the corner of his eye, John watched the surprise in Cleveland's face.

"This here ain't his baby." Pleading edged out of Annalaura's voice.

John inched closer. "A white man brings you food and gives you presents and you gonna tell me it ain't his baby? Who the hell baby is it then?"

His fists doubled up just as the pitchfork jabbed him in the small of the back.

"He was gonna throw us off the place if we didn't bring in the tobacco. You was gone fo' months. I thought you was never comin' back. Me, Cleveland, Doug, and Lottie worked so hard, but it wasn't enough. Even little Henry carried water to the fields. We just couldn't do it by ourselves, John." Her garbled voice rose to a shriek.

John nodded his no.

"Mr. Alex could see that. White men don't like to lose no money.

He had to hire him an extra man just fo' the harvest." She ran out of air and had to stop to gulp in more.

"And?" Agitation rumbled in his chest. John turned to watch Cleveland.

The pitchfork lowered a fraction.

"I thought you was never comin' back. That I was too plain for a man like you." Her voice roamed over the loft. "That hired man make me feel like somethin' special. He talked real nice to me. Said sweet things to me." Annalaura struggled to grab in enough air to get out the rest of her words.

A pain thumped across John's chest.

"He wasn't good like you…you know…in the night." She gained control of her voice. "But I was lonely. He made me feel like a woman again." The tears started to fall.

Something wasn't right. Had his wife betrayed him with two men? Whose baby was this?

"The hired man? You laid with the hired man?" He advanced on her again. "If you is lying to me, Annalaura, I'll kill you." The pitchfork dug in deep enough to draw blood. John didn't care. "Where the hell is this Negro? What's his name? I'll beat the truth out of him, myself." He wanted to retrieve his valise from the buggy. His pistol lay inside.

The hard-to-understand words came out of Annalaura's mouth again.

"He ain't from around here. Kentucky. Gone back. Ask anybody." Her lips were now so swollen that he could barely hear her at all.

What was the truth? Had his wife slept, on purpose, with a colored hired man? Or, had she been forced into bed by a white man? She was lying, but about which one? He wanted to kill her and he wanted to take her in his arms all at the same time. He wanted to tell her how much he loved her. He turned to Cleveland.

"Son, I ain't gonna hit yo' momma no mo'. But I ain't lettin' her keep all this trash either. You can hold that pitchfork on me all you like, but I'm throwing this junk outta here." John grabbed the kitchen chair from Annalaura's hands, walked to the landing, and tossed it down the ladder.

"Doug, you and Lottie help me get rid of this mess while Cleveland keeps hold of that pitchfork." As he swept up the new dishes, he turned to his wife.

"I ain't sure who you laid with, but I do know it won't be me 'til I kill whoever he be. I ain't stayin' here tonight."

# CHAPTER TWENTY

"Why did Papa throw my dolly away?" Lottie sat on the floor in front of Cleveland's alcove, her chin on Annalaura's shoulder.

Henry had his head in his mother's lap, his fingers rubbing at a spot of blood on her dress.

"Papa bought you a better doll, a prettier doll." It took all of Annalaura's fast-ebbing strength to calm her children's fears as she waited for John to thunder back into the house.

Her husband and Doug had wrestled almost everything Alex had given her down the ladder and out of the barn. The commotion had set the cows to mooing. John had even tried to dislodge the potbellied stove but, thankfully, even in his frantic state, he had realized that he couldn't put his hands on a red-hot stove. All of the clothes, pots, pans, toys, and especially the bed linens had been bundled up and dropped straight down the ladder opening. Annalaura reckoned that a better part of an hour had passed since John began sweeping the loft clean.

"But, why can't I keep both dolls?" Lottie, the tear tracks barely dry on her face, kicked at the doll John had pulled from his valise, knocking off one of the button eyes in the process.

Lottie had screamed in confusion and fright when her father snatched the other doll from her hands.

"Why is Papa mad at us, Momma?" Henry raised up from her lap and let his hand dab lightly at the swelling place on her cheek. "Did we do somethin' very, very bad?" Fright leapt out of the boy's eyes.

Annalaura laid her hand against his skinny wrist and let her youngest's open palm rest on her bruised cheek.

"Hush, now, both of you. Papa's not mad at any of you all." She slipped an arm around Lottie and pulled her tighter, though Annalaura's ribs pained her so much she could barely take in a breath.

"Papa's mad at Cleveland. I know he is 'cause Cleve's 'bout to stick him with that fork." Lottie started to whimper.

"No such thing. It's just that Papa's been gone so long, and he left me to take care of everthin' here and I didn't do a good 'nough job of it. It's me Papa's mad at, not you all." She hugged both children, but this time she couldn't stop the little cry of pain that made its way out of her mouth.

"But you took care of us good, Momma. Why ain't Papa happy?" Lottie ran her hand over Annalaura's torn dress.

"I done the best I could, but it's not what Papa wanted. He'll be mad fo' a little while, then he'll be back laughin' and playin' with you children befo' you know it." Annalaura didn't believe a word she had just uttered. Maybe these two young ones would take her story for truth, but Doug and Cleveland would be ten times harder to convince.

"He gone, Momma. Say he won't be back 'til he ki…" Doug, just clearing the last rung of the ladder, looked toward the little group sitting on the floor. "He say he won't be back 'til after he k-i-l-l the man."

Annalaura closed her eyes and wondered why she wanted to smile when Armageddon was falling all over her head. Thanks to Alex, her second son had finished a whole year of school, and the colored teacher told her that Doug could read, write, do his numbers, and spell better than most third graders.

"How you know he gone?" She let her sore face brush the top of Lottie's pigtails.

"He made me pile everthin' down by the lane and he took off in his buggy. He say he might k-i-l-l him two men." Doug dropped to the floor in front of Annalaura. He pushed the heads of Lottie and Henry deeper into their mother's body. "He took his pistol with him."

Annalaura had shown tears to John, but that was just to buy herself time to work out a plan that would keep her hotheaded husband alive. She tried to move the little ones away, but fell back against the alcove.

"What's the matter, Momma?" Henry jumped off her lap. "I won't hurt you. I won't hurt you, never." He threw his arms around her neck sending shoots of pain through half her body.

She kissed the tip of his ear. "Help me get to my feet. We got to go."

With Doug pulling and Lottie and Henry pushing, she struggled upright.

"Get yo' coats." Then, she remembered.

John had taken all their clothes, except what they had on their backs. "Never mind."

"Momma, where is we goin'?" Doug went down first and waited for Lottie, then Henry, to follow.

"Aunt Becky's." Annalaura started down. With each step jarring something loose in her belly.

&#8734;

The sun told her the time was about five o'clock, and the white farmers at the planting dinner should be heading to the Lawnover tavern for more celebrating. With the misery of her swelling eye shooting pain down the side of her face, Annalaura focused on the day of the week. Today was one of the times Alex had agreed to stay away from the barn, but he had made a poor job of honoring his promise to visit no more than one night a week. Most weeks, he came three, and sometimes four, evenings. Some nights, he just laid with her in his arms, whispering sweet things, and nothing more. Those were the times when she most had to remind herself just who Alex was.

Her feet weren't carrying her as fast as she needed. Now, everything on her hurt, and it had taken her almost five minutes to walk the little distance down the path to the lane. Standing over the pile of the family's former belongings, she spotted Cleveland, the pitchfork still in his hands. She saw little Lottie crane her neck toward the pile, her eyes searching for her lost doll. When Annalaura finally reached her eldest son, she slipped an arm around his shoulders and kissed his forehead.

"I knows you is too big to kiss, but I thanks you fo' what you done today." She whispered. "But I don't never want to see, nor hear tell, of you raisin' a hand against yo' father again. This fuss is between him and me. It ain't got nothin' to do with you." She let him go but kept a hand on his shoulder.

"I ain't never gonna let any man beat you like that ever again." Cleveland wagged his head at her.

"Cleveland, this ain't yo' worry, but I do have to ask a big favor of you." She squeezed the back of his shoulder. "I wants you to tell these words to Mr. Alex 'xactly like I tells them to you." The top of her boy's head still only came to the tip of her nose. To Annalaura, he was still her baby.

"Yes'm." His voice sounded deeper.

"Now, I'm thinkin' Mr. Alex shouldn't be botherin' us tonight, but if he do come, this is what I wants you to tell him."

Cleveland dug the pitchfork into the dirt next to the pile of clothes, chairs, toys, and even books. He kept both hands on the handle and leaned over it.

"Tell him that yo' papa, John Welles, is home. Tell him that John has a pistol with him." She stopped to make sure Cleveland understood the words.

"I can tell him that, Momma." Cleveland put a foot on the fork and pushed it deeper into the ground.

"Tell him the gun is fo' the hired hand, Isaiah Harris." Pain in her ribs rocked Annalaura. "Say to Mr. Alex, John is lookin' fo' the father. Isaiah." She waited for that glimpse of agreement in her son's eyes. "Tell it back to me, son."

"Papa's home, and he got a gun and he's goin' after Mr. Harris. I knows what to say, Momma." He looked down at the fork and the little piles of dirt it had scuffed up.

"That's good, son, but there is mo'. Make sure Mr. Alex knows that it ain't in yo' papa's head to do no harm to nobody else. He just wants to talk to Isaiah Harris. You understand that, Cleveland?"

"No, Momma. I don't understand. Papa said he might kill him two men. Ain't one of 'em Mr. Alex?" Cleveland stared straight at her.

She cupped her boy's face in her hands and pressed hard into his cheeks.

"Cleveland, you is twelve, and you is old enough to understand that no colored man can never, ever even think harm comin' to a white man.

If the words come out of a black man's mouth, even if he don't really mean them, terrible, terrible things could happen. Things worse than cuttin' off the head of John the Baptist."

"Almost as bad as Jesus dyin' on the cross?" Cleveland's eyes grew large.

"Just 'bout as bad. What you think you heard yo' papa say ain't what he really means. No need to tell that to Mr. Alex, now is there?"

"No'm." Cleveland grimaced, and Annalaura loosened her grip.

"Make sure Mr. Alex knows that yo' papa only wants to speak to the hired man. Just the hired man. And, Cleveland, one mo' thing. Tell Mr. Alex it would be a kindness to me if he wouldn't tell nobody else that John is back with a pistol. Tell him I'd be pleased to have him visit us in the fields next week after my husband has come back to himself but not a day befo'." She tightened her grip on his cheeks again. "Cleveland, can you 'member all that? Mr. Alex ain't to come here fo' breakfast never no mo'."

Her son's hand reached for her wrist and broke her hold.

"Momma, I remembers every word."

# CHAPTER TWENTY-ONE

The walk up the lane had taken the better part of an hour because of Annalaura. Her good sense told her that she had to get to Aunt Becky before John came back with the truth. She didn't need to lift her bruised face toward the sky to tell her that the sun was about to set on this day in no more than three hours. But the faster she walked on the wagon-rutted dirt road, the more air she had to take in, and with each breath, the more her ribs hurt. Doug had done a good job of keeping the young ones from overly fretting each time she stopped to steady herself at road edge. When they reached Thornton land, and Aunt Becky's cabin, Doug called out to the old woman, but all three children knocked and kicked at the door. Becky, with her all-seeing eyes, took one look at her niece and buried the scowl that crossed her brow.

"I knows it's too early fo' supper. Not but six o'clock, but I bets you chil'ren would like some peach preserves on some of Aunt Becky's biscuits, now wouldn't you?" Rebecca's voice did not match her face. "You all come set a spell."

Auntie slathered preserves on bread, grabbed two shawls to warm the children, and shoved all three outside. Annalaura eased herself onto the wrought-iron bed, grateful to Rebecca.

"Gal, let me take a look at them cuts." Becky rushed over to Annalaura, her arms full of jars and a tin box. She set them all down next to the pitcher of cool water. "What he hit you with?" Rebecca turned up the wick of the kerosene lamp set on a crate by the bed.

"Umm." Annalaura grimaced.

Becky stood over her niece. The woman's hands lifted Annalaura's chin. Rebecca took a quick look at the battered face, then ran her hands from neck to waist. "Don't look like mo' than his fists, but he done punched you in the belly."

"Aahh." Annalaura grimaced as another pain went through her middle.

With hands flying like they were thirty years younger, Becky opened jars and began mixing ointments, salves, and foul-smelling liquids on the upturned lid of one Mason jar. With her quick fingers, she dabbed Annalaura's cuts. Auntie opened another jar and poured a small amount of honey-thick unguent into the palm of one hand. She rubbed her hands together and patted the lotion all over Annalaura's bruised and battered places.

"Don't look like he touched the chil'ren, praise the Lawd. Bad business when a white man beat on you. If he gets mad 'nough, he takes it out on the whole family. Gal, did you tell him no when you shoulda said yes?" Becky's hands skimmed over Annalaura's lower body. "Ain't nothin' broke." Her inspection done, she sat on the bed beside her niece.

Whatever her aunt had doused over her started its work. Annalaura let her heavy eyelids close. Becky's fuzzy words swept in and out of her ears. Annalaura struggled to ease herself up in the bed. This was not the time to give in to sleep. She pushed away the damp cloth Becky tapped over her eyes.

"White man?" Her aunt's words registered in her head. "No. Alex ain't never hit me. John ..." She fell back onto Becky's feather pillow.

"Say what? If it wasn't McNaughton, who was...?"

"God help me, Aunt Becky. John, he's back." She could feel her tongue swell in her mouth.

"What you sayin'? John Welles is back, and he done found you like this?" Becky pressed open the tin box and took out a pinch of dried leaves. She dipped them in a fresh-poured cup of water and lifted the scruffy enameled container to Annalaura's lips. The elixir poured over Annalaura's chin, and down the front of her dress. She lifted a hand to right the cup.

"I ain't told him who it was." Annalaura managed a sip just as another pain gripped her midsection.

"John's back, Lawd Almighty. It's troubles. Troubles." Becky closed her eyes and began to rock on the bed.

In a voice so tiny it could barely be heard, the old Cherokee began to mutter words that Annalaura knew weren't English. Annalaura propped herself on the pillow and raised her voice as loud as she could muster.

"John. He don't know 'bout the father."

"Hard to keep a man from knowin' who the daddy. When a man think it ain't him, he gonna worry everybody 'til he finds out who. Old Ben Thornton, he knew 'bout Johnny." Becky's eyes refocused on Annalaura's face. She reached over again and wet the cloth in the tea mixture and laid it over her niece's eyes.

"The hired man. I told John it was the hired man." Annalaura felt Becky press gently down on the wet cloth, and its coolness soothed her.

"Hmm. Hired man, you say? That's good 'nough fo' a beatin' but not a killin'." Becky kept the cloth over Annalaura's face. "How long 'til yo' John find this hired man, and the truth?" She pulled the cloth away.

"I pray to God he don't never find out. I hear Isaiah gone home to Kentucky." Another pain wracked her, and she rolled to her side and drew up her knees.

"You'd better be hopin' that baby ain't comin'. That them pains is just from the blow he put on you." Becky handed the blue enameled cup to her niece as she moved toward the kitchen safe.

Annalaura heard her aunt rummaging through the drawers and shelves. Becky returned to stand over the bed.

"I'm goin' to Hettie's. Takin' the children with me," Becky announced.

"The children?" Annalaura's heart thumped through her sore ribs. "You can't take them to Hettie. Mr. Ben Roy will ask what they doin' there. You knows Mr. Ben Roy's a beatin' man. He'll make Hettie or the children tell him. If he finds out John's back with a pistol..." Annalaura grabbed the cloth and laid it tight over her forehead.

"Ben Roy?" Becky let a little smile creep across her face when Annalaura slid the cloth up her forehead to peek at her aunt. "Ain't no need to be 'fraid of Ben Roy. I still got the paper hid real good."

A new shot of fear filled Annalaura. "Aunt Becky. I can't have my children with Hettie if Mr. Ben Roy is at her cabin. It won't take nothin' but a minute fo' him to find out 'bout…'bout…"

Rebecca's eyes began their backward shift to that other place where her mind sometimes dwelled.

"I done tole you, don't worry none 'bout Ben Roy Thornton. I gots the paper hid real good."

"What paper, Auntie?" Annalaura raised herself on the feather pillow, resting her back against its softness. She reached out a hand to pull Becky down to her. If she could lay her hands against skin, look the old woman straight in the eye, maybe she could bring her back before her auntie's lost mind put her children in harm's way.

"I'm takin' yo' chil'ren to Hettie's, then I'm gonna look fo' some of them herbs what will keep that baby from comin'. Baby ain't ripe yet." Becky jerked her arm away.

"It ain't the baby." Annalaura regretted the scream that forced itself through her broken lips. "I know when a baby's coming. I done had fo' of them. This don't feel nothin' like that. These pains will go away directly." A spasm rocked Annalaura. "Doug can watch the little ones while you go fo' the medicines. No need to bother with Hettie." Another pain jolted her body.

"Might be you can't stay here. Might be best fo' you to get gone." Becky's eyes were halfway to that other place. "Ben Roy's buggy. I'll speak to him." Becky turned toward the center of the room while Annalaura frantically reached out to grab her arm.

"No, Auntie. Don't go botherin' Mr. Ben Roy." The effort had lurched her to the edge of the bed, and a cramp in her stomach caught and held her breath.

"You don't know nothin' 'bout it, do you, girl?" Becky moved back to the safe and dropped to her knees. Bending down, she pulled out a lower drawer and swept an arm inside.

Fighting the spasm that didn't want to let her go, Annalaura heard Becky tapping on something for long seconds. She turned her head just enough to catch Rebecca's crooked grin as her aunt slammed the drawer shut and stood up. Her hands empty, Rebecca walked back to the bed.

"I ain't never tole you straight out, but I reckon it's time you finds out." Becky sat down beside her niece. "Ben Roy do what I say 'cause I got the paper." Becky looked toward the safe.

"The herbs, Aunt Becky. Go get 'em." Annalaura didn't know which hurt more, the pains in her body growing worse every five minutes, or her aunt, lost in that other world. "I got to get my head clear so I can keep John from this bad trouble I done made fo' him. Aunt Becky, I needs you to help me figure out what's best."

"I is helpin' you, girl. I'm takin' Ben Roy's buggy."

Annalaura snapped her hand around Becky's skinny arm. The shadows lurking in her mind of Alex shot dead and John Welles hanging from a tree, all because she hadn't tried hard enough to outmaneuver Alex, flooded into her head.

"Old Ben give 'most five hundred dollars for that horse." Becky's eyes stared at a far corner of the bed.

"Umm." With Rebecca sliding to that other place, Annalaura had to conserve what little strength she had to think of a way to keep John and Alex apart. "Aunt Becky, John gonna find out real soon 'bout the hired man." Why was the old woman rambling about events long past when fresh disaster faced them all in a day's time?

"Five hundred dollars, he did." Becky rocked on the bed.

"I need help. I got to find me a way to get that gun away from John befo' he . . . befo' he kills Al . . . I can't allow that." Annalaura pinched down hard on her aunt's arm. "But I can't do nothin' 'til these hurts leave me. Please get me some of yo' herbs?"

"Help you?" Becky patted Annalaura's arm. "That's what I'm tellin' you, gal. Ben Roy gonna give us the buggy. I'm gonna find someplace to put you 'til the baby come. Someplace where yo' John can't find you. Alexander McNaughton neither." Becky stood and walked to the door, where Lottie, Henry, and Doug snuggled under her warm wraps.

Annalaura eased up in bed and struggled to dangle her feet over the edge of the mattress. She couldn't let Becky take her children.

"Not Mr. Ben Roy. Please don't go to Hettie. Mr. Ben Roy could be there." The pain in her ribs wrenched the breath out of her.

Almost at the door, Rebecca turned back toward her.

"Ain't you heard nothin' I said? Ben Roy ain't gonna tell nobody a word, and he ain't gonna start no white man's ruckus either 'cause he knows what I know." With quick steps Becky crossed the floor and sat down on the bed beside her niece. She slipped an arm around Annalaura's shoulders and leaned her niece's head against her chest. Her aunt's comforting arms helped ease some of the pain.

"I'm takin' yo' chil'ren to Hettie. I'm gonna take Ben Roy's buggy to find somebody to take you in fo' one or two weeks 'til yo' baby come. Where can't neither John nor Mr. Alexander get to you. Baby stay with them awhile. If the baby's color come in dark, maybe John might let you bring it back. Now, I'm gonna pick you some herbs to keep that baby in you 'til its time. Ben Roy ain't gonna bother none of yo' chil'ren, 'cause he know I got the paper."

# CHAPTER TWENTY-TWO

The sun readied to shut down for the day, though Alex had tried by hours to beat its demise, and get back to the mid-forty. After dropping the new hired man off there right before sunup, he had spent the time before the planting and prayer dinner setting up the rest of his acres. He hadn't even gotten to the Thornton place until close to three o'clock. With Reverend Hawkins droning his prayer a good thirty minutes, the dinner hadn't been over 'til way past six. After asking Tillie and Wiley George to check on Eula, who had strangely gone home before he arrived, Alex finally made his excuses and left to pick up the hired man and bring him back to the main barn. Since Eula had taken the buckboard, he borrowed Ben Roy's high-seated wagon and Fedora's slow-stepping mare.

Alex watched the day's shadows grow longer as he whipped at the horse. If he could get to the mid-forty twenty minutes before nightfall, he might find Laura before the hired man came in from sowing seed. He knew this was not one of those nights she had agreed to his staying, but a quick kiss and a touch could do no harm to the baby. He gave the reins two rapid flicks, but Fedora's horse, unaccustomed to a quick pace, decided to slow down even more. Wishing for the gray, Alex finally marshaled the old animal down the lane. He trained his eyes on every foot of the familiar road knowing that when he cleared the stand of sycamores a hundred yards distant, the path to the mid-forty would be no more than a minute's ride away. He looked up at a sky filled with deepening reds, oranges, and yellows. Alex reckoned he still had fifteen minutes of daylight before cooling evening arrived. He hoped the hired man was as good a worker as he pretended, and wouldn't leave the new-plowed fields until after the sun had settled in for the night.

In answer to Alex's constant snap at her flank, the horse finally got the message and quickened her pace. With the sycamores behind him, he spotted the path and the figure of a standing child, dwarfed by a large pile of what looked like household furnishings. Sticking out of the heap were the four wooden legs of a chair turned bottom-side up. As Alex neared the path, he pulled on the reins so hard the horse nearly reared up on her hind legs. Young Cleveland took a half step forward to greet him.

"What's this?" Puzzled, Alex pointed to the pots, shoes, coats, skillets, quilts, dresses, cotton sheets, dolls, and dishes.

Half buried beneath the blanket that just two days ago had served as the cover separating the open alcove doorway from the rest of the living quarters was Laura's blue serge coat. The new hired man, sitting on the ground at the edge of the pile, stood up, removed his sweat-stained straw hat, and walked a few steps back into the fields. He could feel the man's eyes sidle in his direction. Alex grunted his dissatisfaction at the tenant's too-early quitting time. Cleveland looked at the family's belongings and then turned back, his eyes staring down at the wheels on the buckboard. The boy shrugged his shoulders as Alex followed the tilt of the lad's head.

"Momma says to tell you that my papa done come home."

Alex shook his head trying to clear his ears. The boy stood somewhere between pleased, confused, and frightened.

"Your papa? John Welles is home?" The man had been gone almost a year without a word to anyone. What the hell was the nigger doing back in Lawnover? Alex pulled out his big square handkerchief and swiped it across his forehead.

Cleveland nodded his head. Welles back in town, and for what? Alex looked at the tumble of clothes and furniture. Whatever brought the nigger back to town couldn't hold a man like him for long. Unless…Alex glanced over at the hired man who stood flat-footed, slowly twisting his hat in his hands. Welles was no ordinary hired hand. He would never stand there holding a cap with his neck bowed like he meant it. John

Welles would always tip that head just short of respectful or let those eyes roll up a little bit too high. An uppity nigger like John Welles was nothing but trouble. Alex took another look at Cleveland and sat up in the buckboard. The boy dragged the toe of his shoe back and forth in the dirt.

"Welles is home?" The question wrenched Alex's body just as much as the first time he asked.

"Yes, suh, he done come home. Momma say to tell you that he lookin' for Mr. Harris."

"Harris? You talkin' 'bout the hired man that helped bring in last fall's harvest?" The very man Laura had claimed as father of their baby?

Alex peered at Cleveland. Laura must have told John Welles the same story she tried to peddle to him about who'd fathered her baby. He jumped down from the buckboard and walked over to Cleveland, laying a hand on the boy's shoulder. Out of the corner of his eye, he spotted the half-surprised, half-scared look on the face of the hired man.

"What else did your momma tell you to say, and why ain't she here tellin' me all this herself?"

Cleveland raised his eyes to Alex's shoulder. "She say that I is to tell you that papa have a gun, but it's just to take with him when he go to talk to Mr. Harris." The boy recited the words his mother most likely put in his mouth.

Alex kept his hand on the boy.

"Momma say the gun ain't fo' hurtin' nobody. Not even Mr. Harris. She say Papa just a little upset with that hired hand."

"Gun? You say yo' papa's runnin' the countryside with a pistol?" Alex lifted his eyes to the barn window. "Is your momma in the barn?" He took a step up the path while the hired man walked even deeper into the darkening rows of new-sown seed.

"No, suh. She ain't in the barn." Cleveland's voice sounded alarm. "She took Doug and the little ones over to Aunt Becky's."

"Where's your papa? Did he go with Laura?" Alex dismissed his misspeak. Calling Annalaura by the special name he'd given her could make no difference to any of them if John Welles really was back, and with a pistol.

Cleveland stood silent. Alex walked toward the barn, rushing thoughts flooding his head. Welles was going after the man Laura claimed to be the father of her baby. A nigger, a pistol, a wife big with another man's baby were a dangerous combination. Alex stopped and stared at the window in the loft—that very same window where the moon had played its silvery light across Laura's naked body when the two made love. A dozen horrors ran through his mind. What had Welles done to Laura?

He struggled to keep his face and body still as he turned back to Cleveland. One thing was certain. John Welles could not stay long in Lawnover.

"Is your papa with your mother?" He whirled around to Cleveland.

"No suh, he ain't." The boy finally answered. "Don't rightly know where he is. Momma say Mr. Harris might be in Kentucky."

If the boy was telling the truth, Alex thanked God. Laura was safe for now. He gestured to the hired man to climb into the wagon. The beginnings of a plan laid itself out in his mind. John Welles couldn't be allowed to stay in Lawnover. He couldn't be allowed to lay claim to Laura.

"Momma say I was to ask you one mo' thing." The boy's voice was a whisper as the hired man sat in the back of the buckboard, his knees hunched up to his chin, his face turned toward the chirp of a night bird.

"Yeah?"

"She say it would please her most kindly, suh, if you come visit us in the fields next week, but not a day befo'. She say, it might be good if you don't visit fo' breakfast no time soon, neither." Cleveland sucked in a lip. "One last thing, Mr. Alex, suh. Momma say please kindly take these things away."

The Lawnover store hadn't changed much in the thirty years Alex had gone in and out of its doors, other than that Bobby E. Lee Thompson ran it now since his daddy, Andrew Jackson Thompson, died some ten years back. The building still had that big main room that sold everything from galvanized nails to barrels of brined pigs' feet to baling wire to calico cloth for the women. The two windows on opposite sides of its walls, along with the kerosene lamp that burned even in the daytime, gave the place enough light so a woman could tell blue thread from black. Bobby Lee and his wife worked the store pretty much through all the daylight hours six days a week, twelve months a year. During the busy times, like right after the money came in for harvest and when supplies were needed to set up the hired hands for planting, Bobby Lee would take on extra help from some of the poor whites down by the tracks.

Bobby Lee only allowed colored to come in two at a time and that was only a man and a woman, or two women together, but never two colored men at once. No matter how many niggers were lined up outside, and sometimes there might be a dozen or more, or how many were inside, Alex knew they had to step aside when he walked up to the counter. Even if the colored woman had her calico yard goods already cut for her and her coins on the countertop ready to push to Bobby Lee's wife, careful not to let black hands touch white, she would have to stand aside 'til he got his business done.

Between his and the Thornton farms, the Lawnover store was the only buying place for five miles around, for either colored or white. Every farmer in the area had to stop by Bobby Lee's at least once a month for supplies. It hadn't been two weeks since he brought Eula by to pick up some lard and some sugar while he went to the blacksmith shop to repair the rein for the gray.

Tonight, it was after dark and the main business at the Lawnover store was over. The front door was closed and latched. Once he dropped off the hired man at the McNaughton barn, Alex switched over to Eula's horse and rig. Coaxing the buckboard and the wheezing horse around back of Bobby Lee's, he led the animal to the water trough. Inside the store he heard the guffaws.

On just about every Friday and Saturday night when it wasn't harvest or planting time, Bobby Lee opened up his back room for poker games. Most game nights, five or six of the farmers would gather. But the first day of planting was special. After all that planting and praying, the men needed to end the evening with a little fun. Bobby Lee might have twelve or fifteen farmers in the store tonight. If he did, he would put the overflow into the store's main room. Only the regulars like Ben Roy got the use of the back room.

While Eula's horse was taking in water, Alex walked across the way from the hitching post to the log-hewn building. The light from the lamps shone through all three windows, even though curtains covered only the one in the back gaming room.

It took Alex a few seconds after he opened the back door to spot Ben Roy through the brown cloud of cigar, cigarette, and pipe smoke. The kerosene lamp sitting on a highboy right over Ben Roy's left shoulder shone directly at the back door and at any newcomer. Those already in the room knew who had entered long before the late arrival could make out the outlines of the pork and pickle barrels lining the walls.

"Alex, 'bout time you got yo' ass over here. Come on and set a spell." Ben Roy had left the planting party at his own home no more than thirty minutes before Alex, yet he was already halfway into a tall Mason jar of Tennessee whiskey. His eyes finally adjusting to the glare, Alex made out the uncovered planked wood table set in the center of the room surrounded by boxes and barrels piled shoulder-high to the timbered ceiling. He made his way around a big barrel of sweet-smelling sorghum molasses and another of pickling hog heads. Ben Roy, Wiley George, two other Thornton kin, a farmer from down county, and Bobby Lee were already sitting around the table, with Ben Roy fumbling with a deck of cards. Bobby Lee grabbed a Mason jar of whiskey as he scuffed his chair backward to stand. Clapping Alex on the back, he almost pushed him into the vacated chair.

"Here's hopin' your new hired nigger sticks it out this time." Bobby Lee reached for the first in a line of empty Mason jars standing next to the kerosene lamp behind Ben Roy's head. Bobby Lee handed the container to Alex. "If that nigger don't work out, I got another one been 'round here beggin' fo' a place to farm." Bobby Lee took a swig of his own homegrown mash liquor and walked through the door into the front of the store where, judging from the cursing, spitting, and laughing, the second poker game was heating up.

"We're playin' five-card stud, no deuces wild." Ben Roy tapped the table. A mound of greenbacks lay in the center.

Alex frowned. Big money like that usually was gambled only after harvest. Planting time most often meant quarters and half dollars, but any game with Ben Roy usually called for betting more than most farmers wanted. Winning or losing at gambling was not what had brought Alex to Lawnover this night, but he had to bide his time. He pulled out a crumpled dollar bill from his overalls pocket and tossed it on top of the money pile.

"Hey, Bobby Lee. Send Hettie in here." Ben Roy shuffled the cards as his shout went through the closed door into the main room.

"I sho' hope to hell you give me some better cards than you did last time." Wiley George looked like he was enjoying the aftermath of his second Thornton family planting party just a little too much.

Alex saw the other Thornton kin shoot the boy a warning look.

"It ain't the cards, you dumb bastard. You hold your liquor worse than a ho in a white trash bawdy house." Ben Roy chomped down on his cigar as he deliberately blew a large puff of smoke into his son-in-law's face.

Wiley George turned his head and took another gulp from his Mason jar. Ben Roy slapped a card face down in front of Alex.

"Lay offen him, Ben Roy. You know Tillie and that new baby got him goin'." The Thornton middle son, who thought he was the better man with his fists, challenged Ben Roy.

Another card slipped down in front of Alex, its blue-patterned back staring up at him. He let his eyes travel around the table checking on how much liquor had been consumed. This was not the night he wanted the Thornton boys engaged in one of their usual brawls. He needed them at least part of the way sober to hear what he had to say.

"I reckon I know what's got Wiley George goin' as good as you do." Ben Roy grunted at his younger brother just as Hettie opened the door carrying a big jug in one hand and a platter of leftover fried chicken in the other.

"Set that chicken plate down and pour us some more whiskey, gal." Ben Roy barely looked at his woman as he laid a third card in front of Alex.

Hettie bobbed her head as she circled the table behind the men, refilling each glass jar. As she poured Bobby Lee's home brew, Alex sneaked quick glances at the woman every man at that table knew Ben Roy had bedded for six years. Her brown skin may have been a shade fairer than Laura's, and her hair might have been a degree less crinkly. It was hard to tell about the shape of her body since she still carried much of the weight of her last pregnancy. But, remembering her from before, Alex could tell Hettie's body had never been a match for his Laura's curves and firmness.

He admitted that Ben Roy's Hettie once had a bit of prettiness about her, but that had faded over the years. He looked up toward her eyes. Sensing his glance, she looked back at him, paled, became flustered, remembered her place, and hurriedly backed away. Ben Roy hadn't noticed any of his woman's antics as he laid the fourth card before Alex.

"Wiley George, you think you gonna need some help readin' them cards?" Ben Roy snickered at his son-in-law as he placed the fifth and last card in front of Alex.

"Don't need no help readin' aces and kings." Wiley George clipped off the answering challenge.

Everybody knew that Ben Roy hadn't exactly approved of Tillie's marriage to the Jamison boy, just as he'd never fully accepted Alex into the Thornton ranks. Both the Jamisons and the McNaughtons farmed acres too few and too poor to be in league with the Thorntons. Alex recognized the taunts Ben Roy tossed at Wiley as the same ones he'd fielded himself more times than he cared to remember. Alex picked up his poker cards. Besides a pair of nines, he held only an ace, queen, and jack, all of mixed suits. Gambling had never meant anything to him other than throwing away good money. Sometimes, he thought that was the reason Ben Roy insisted that he sit in on the poker games with him. It was just another opportunity to flaunt the Thornton money in front of a struggling in-law.

"If you readin' aces and kings, Wiley George, sweeten the pot." The Thornton cousin sat with his lids half closed.

"Big talk is the onliest thing Wiley George got of any size on him. Everything else is puny." Ben Roy guffawed at his own slight of his daughter's husband. He pulled out another dollar bill from his work shirt pocket and pitched it onto the pile of money in the center of the table.

Hettie, the jug now almost empty, turned to leave. Without looking in her direction, Ben Roy jerked hard on a handful of skirt. As she stumbled toward him, he lifted the back of the garment and ran his hand up the inside of her legs and jammed it hard into her crotch. Hettie flinched. A little grunt escape her mouth as Ben Roy rubbed his hand back and forth between her thighs. He let the skirt fall back in place as he delivered a smack across her butt hard enough to push the unhappy woman against the edge of the round table.

"Now, git," he commanded as he looked around the table for player hits.

The younger Thornton brother signaled for two card replacements while Alex stared at his own. Despite his poor poker-playing skills, he had to play it smart tonight. When Ben Roy trained his eyes on him, Alex discarded the three face cards and kept his pair of nines.

"You either got a hell of a hand or you need help readin' them cards like Wiley George." Ben Roy stubbed out his cigar and tossed it in the spittoon.

Alex picked up a jack, a seven, and a third nine. All Thornton eyes were trained on him. With his trio of nines he doubted he could beat a Thornton hand, but they were waiting for him to add to the pot. He reached back into his overalls pocket and pulled out a silver dollar. The thud it made as it crashed into the mound of bills was the only sound coming from him. If he'd learned one thing in twenty years of poker playing against Ben Roy, it was to keep a straight face and do almost no talking, no matter how hard his brother-in-law pushed.

"Bobby Lee ain't skimpy with his liquor. Take yo'self a drink, Alex." Ben Roy dealt himself one card before his eyes scanned the table for new hits.

Funny, Alex reminded himself, how Ben Roy's flat, brown eyes looked just like Eula's. Looking into them was like looking into the bottom of a dirty coal pail.

As if by invitation, the belly-bulging down county farmer picked up his Mason jar and drained it empty. All in a pile, the farmer spread out his cards—two pairs—one of eights, the other of fours. The man clamped both hands around the empty Mason jar. Wiley George threw his cards on the table in disgust. A little grin started to play across the farmer's face as he eased up on his grip of the Mason jar and leaned toward the mound of bills.

"Well…" The broad-shouldered Thornton cousin drew out the word as though it had four syllables as he laid his cards on the table. Three sevens showed their faces.

A grimace flickered across Ben Roy's lips but disappeared. He turned down his cards. The younger Thornton brother reached for his own Mason jar. The down county farmer looked as though his prized sow had just been shot. Once again, all eyes at the table waited for Alex. He checked his cards again—a seven, a jack, and three nines still remained in his hand. Slowly, he spread them on the table.

"I'll be damned," Ben Roy exploded, "if brother-in-law ain't won a round." He pushed the pile of money toward Alex while the older cousin scowled. "Hettie, gal, get yo' ass in here." Ben Roy pulled at the string dangling from his pocket and lifted out the packet of chewing tobacco. He pinched off a wad and popped it into his mouth.

With as much casualness as he could muster, Alex slid the entire pile of money back to the center of the table. He needed far more than money this night. "Let it ride. Deal, Ben Roy."

As the playing cards circulated to the left, Alex studied each man trying to judge the best time to bring up the subject. Liquor was having its way, with Wiley George leading the pack. After losing the last hand, the down county farmer looked even more glum and desperate as he tapped at his empty glass. The Thornton kin still held sober enough heads. Ben Roy was the key to getting rid of Welles. The fifth card landed just beside Alex's hand. He picked up all five. An ace, two queens, and two tens greeted him. He knew he already held a better than decent hand as he picked up his untouched Mason jar and brought it to his lips. He saw Ben Roy give him just the slightest nod of approval as the dealer scanned the table for hits.

"Damn it to hell, Ben Roy. Can every goddam card you deal me be this bad?" Wiley George threw down four of his five cards.

"Watch yo' mouth." Ben Roy's younger brother looked out from under his sandy brows to warn his niece's husband.

"He don't mean nothin' by it. Wiley George is just havin' a bad run of the cards. Can happen to anybody." While the wide-shouldered cousin tried his hand at peacemaking, Alex went over his words again in his head.

"Better not have meant nothin' by it. Wiley George, just what are you good for? You can't play cards, you can't hold your liquor, and you can't take decent care of my daughter. Just what the hell can you do?" The sound of Ben Roy's rising voice brought Bobby Lee through the door and into the back room.

"You boys need somethin' else in here?" Bobby Lee walked over to the platter of chicken sitting on top of a barrel of crackers.

Now was Alex's chance.

"Bobby Lee, you say you got another nigger lookin' for hired-out work?" Alex glanced up at the proprietor but kept focus on Ben Roy. Alex swept the table with his cards once.

Ben Roy laid another card face down in front of his brother-in-law.

"I thought you was gonna be all right with the nigger you just got." Bobby Lee came over with the chicken. The down county man took himself a thigh but the three Thorntons waved the proprietor off.

"That's just it. I don't need no more hired men. Got too many as it is. That no good nigger John Welles come back today." He said the name slow, but he held his voice steady like it didn't really mean everything to him.

Alex let his eyes slide to his left to catch a secret glimpse of Ben Roy who sat staring at his cards. Had his brother-in-law heard? "Place don't need two hired men. 'Sides, that Welles is a no good nigger." Alex picked up his fifth card and slipped the third queen in next to the other two. He had a full house. He knew not to make a move, not to make a big show of adding more money, nor to let a muscle twitch in his face. In this game of poker, when a man held a winning hand, he was the only one in the room who needed to know until the very last second. But the hand Alex most wanted to win wasn't in any poker game. Nothing but besting John Welles would do, and Laura was the pot.

"What's no good 'bout this nigger?" The down county man asked as he scowled at his own cards. He swept the table twice, and Ben Roy delivered him two fresh replacements.

"Left me right after plantin' last year and ain't showed his black ass back in these parts 'til today." Alex slowly reached back into his overalls pocket and pulled out a second silver dollar. He slid it casually toward the pile of cash.

Ben Roy caught the action.

"How'd you come to know he was back in Lawnover?" Ben Roy let the words slip out of his mouth.

Alex watched him discard a card, take a replacement, and pull out two greenbacks from his shirt pocket. Slowly, he shifted his eyes to Alex as he waved Bobby Lee and his chicken platter away. Bobby carried it into the main room. Alex's mind was a whirlwind of concoctions.

"Picked up my new hired man on the mid-forty tonight. Welles's boy said his pa had been back around the place." Alex held on to his cards, careful to avoid Ben Roy's eyes.

"The nigger's back on the mid-forty? I thought for sure he was gone." Wiley George spread all five of his cards face down on the table as he finished off the contents of his Mason jar.

Alex didn't answer.

"You seein'?" Ben Roy jabbed a finger in the direction of his cousin, who laid out three tens. The self-appointed Thornton patriarch turned to his younger brother, who laid his own cards face down on the table. Ben Roy lifted his fisted cards to his chest and turned back to Alex.

"That Welles nigger come back to look in on his wife and kids is all. He's got a mad on 'cause his wife's got a full belly." Ben Roy shrugged his shoulders. He paid more attention to the down county farmer when the man laid his cards face down on the table than he did to Alex.

Though the May evening was mild, Alex watched the sweat pop out across the hapless farmer's forehead. Other than making for a tight summer, what did the man have to fret over? It was nothing compared with the danger to Laura.

"Once he smacks her around a time or two, gets it out of his system, he'll haul his ass out of Lawnover." Ben Roy inclined his head toward Alex. "Take Bobby Lee's new man to help you out." Ben Roy made no move to show his cards.

The thought of Welles laying a hand on Laura brought up the bile in

Alex's throat. "That ain't likely. Not on his own. John Welles is an uppity nigger. His boy says he carryin' 'round a pistol." Alex let the words ease out of his mouth.

Hettie reentered the room carrying the jug. Judging by how tightly she gripped it, Alex guessed it to be full. Ben Roy's eyes were still on his own cards.

"Pistol? A nigger don't need to be carryin' 'round no pistol." Wiley George held out his Mason jar to Hettie for a refill.

"Hettie." Ben Roy's voice bounced off the bundles, barrels, and crates in the small back room as the woman stiffened her body. "Don't pour that fool another drop. I ain't carryin' no drunk home to my daughter." He turned back to Alex. "Niggers carry pistols even when they go to church. Long as he ain't carryin' it 'gainst no white man, who the hell cares?"

"The boy says Welles is after the father of his woman's baby." Alex locked eyes with Ben Roy, who had taken a quick glance away from the cards in his hand. "A nigger takin' a gun to church is one thing, but a crazy nigger with a loaded gun runnin' 'round the countryside after dark is another. It just might go off against anybody." Alex held Ben Roy's gaze.

"Maybe Welles slipped back in town last fall for a night with his woman. Who's to say who the baby's daddy is." Ben Roy still held his cards clamped to his chest.

"That nigger needs that gun taken away." The down county farmer held up his glass for a refill.

Ben Roy rolled his eyes, and the farmer set his Mason jar on the table only a quarter filled.

"You gonna be the one to take a loaded gun from a crazy nigger?" Ben Roy turned to spit out the wad of tobacco he'd been chewing, spewing brown specks across the table. "Alex, let's see yo' cards."

Alex's throat had gone dry. He couldn't speak even if he had thought it a good idea. Instead, he laid out his three queens and two tens. He heard Ben Roy's big intake of breath just as Hettie approached him with the jug in her hand. Alex glanced down at his own almost full Mason jar and nodded her off. Ben Roy threw his cards to the table and pushed the pot toward Alex.

"You been a lucky bastard tonight." Ben Roy leveled his eyes at him. "Every man here wants to have that kind of luck this plantin' season. Put the seed in the ground, hope the rains come and go when they 'sposed to, fight the bugs so they don't eat yo' leaves down to nothin', and pray that the niggers will stay in the fields. That's the kind of luck every white man wants this time of year. Unless it's somethin' powerful bad, ain't no need to rile 'niggers up with trouble right at the start of the season. A nigger will sit down quicker than he'll work. Don't give 'em no excuses."

"A nigger runnin' around the county with a loaded pistol, crazy enough to shoot anything and anybody, white or black, coming across his path...that's bad business." Alex looked at the men at the table.

Hettie, on her way to Ben Roy, slipped on some of the tobacco spittle that had escaped the spittoon. Catching herself, she splashed the liquor mash from the jug across the poker table and onto the deck of cards Ben Roy was shuffling. Almost before Alex could blink, Ben Roy's arm shot out and backhanded Hettie hard across her new-mother's nursing breasts. Still clutching the spilling jug, the woman doubled over in pain, one hand at the bodice of her dress.

"Leave the damn jug and get the hell out of here." Ben Roy wiped the cards clean on his shirt as Hettie scurried into the main room.

"I reckon that's it for me for tonight." The down county farmer scraped back in his chair. He turned to Alex and almost whispered to him. "If you need me to help out with that uppity nigger after harvest, you come on back to me."

"Me, too." The Thornton cousin gulped down the contents of his Mason jar. "That nigger's just gonna run around a lot tonight. Get liquored up and find him some other nigger to shoot in the ass. He'll light out of here by mornin'."

"Two or three men tonight to catch this nigger and run him out of Lawnover is all I need. He ain't nothin' but trouble for all of us." Alex swept his eyes between the Thornton kin and Ben Roy.

Pushing the deck of cards aside, his brother-in-law leaned toward Alex, his elbow on the table, his hand half shielding his mouth.

"Hettie's got a tan-skinned cousin come to visit. Not but thirteen, but

I reckon she's trainable. You can help yo'self to her." Ben Roy may as well have been speaking Geechee to Alex's ears.

Why in hell would his brother-in-law think he wanted a thirteen-year-old child? All Alex could do was nod his head in confusion.

"I don't mean nothin' serious like a lynch..." Alex realized his error almost before he spotted Ben Roy's eyes narrow to half the size of a lead point. He shook his head and hurried off the memory. "Just need a few of us to make sure Welles knows he ain't welcome back here in Lawnover after what he done." Alex watched Ben Roy shift a new wad of chewin' tobacco from one cheek to the other. He said nothing, as Wiley George, his head bowed down toward the bare table, glared at his father-in-law under his brows.

"You talkin' 'bout bad business." Wiley studied the tabletop like he was counting each and every stain ever put on it by a poker player. "Seems like to me, it's bad business if a nigger don't stay put when it's plantin' and growin' season. Ain't right for niggers to shuffle off whenever they take a notion. Only a white man gets to say when a nigger goes or stays." He finally lifted his eyes to his father-in-law. "If you was to ask me, I'd say that nigger needs to learn him a good lesson that'll stick with the rest of 'em too." Before Wiley George finished, the down county farmer pushed back from the table and rose to his feet.

"I just put me in eighty acres with one pretty good hired hand. But he got six children, five of 'em boys and all of 'em of good size." The farmer patted at his now flat pockets as he turned toward Alex. "With all them children, I don't reckon my nigger gonna run off, lessen there's trouble. I can't afford to have me no trouble." He held out a hand to shake Alex's. "If that nigger of your'n still needs a good learnin', call on me after harvest."

"Alex, you ain't got nothin' to worry about. My money's on that nigger gettin' out of town once he's done a little drinkin', some fightin', and a little harmless shootin' with that pistol. He ain't gonna turn that thing on nobody." The cousin flexed his shoulders as he stood, stretched, and reached for his empty Mason jar. His own jar refilled, the man slipped into the front room. Ben Roy turned toward his younger brother.

"Take this fool on home to Tillie." He jerked his head toward Wiley George. "God gave a goose a double more portion of good sense than he gave you. You don't stir niggers up when you need 'em. Fifteen farmers in this part of Montgomery County depend on them hired hands to bring in the crop, and you tell me that now is the time to teach them a lesson?" He turned full on his son-in-law. "When you rile up niggers, it better be over somethin' good and worth the trouble 'cause there could be hell to pay for more years than you wanna count." Ben Roy slumped back into his chair.

The younger Thornton grabbed Wiley by his collar and pulled him to his feet. The new father wobbled. Little brother Thornton put an arm around Wiley's waist and half dragged his niece's husband out of the room. Ben Roy reached for the last of his whiskey. He drained the jar, started to call out Hettie's name, changed his mind, and turned to Alex in the now quiet room.

"Funny 'bout the Welles woman and her baby. Been a lot of years between babies." Ben Roy leaned his elbows on the table. "You sure 'bout that baby? Could some other nigger have come in there?" Ben Roy kept his voice low.

"I'm damn sure 'bout the baby." Alex's voice boomed.

"If you're sure 'bout the daddy . . . high-yella babies born in these parts stay put for a lot of years." Ben Roy let loose with a stream of tobacco juice. "Anybody can take a look at the child from time to time. 'Course, could be troublesome if the momma moves on to some other man's acres." Eula's brother nodded his head toward the door leading to Bobby Lee's main room. "If a man took a mind, he could always come here to town and take himself a good look. The momma's bound to bring the child by at least once a month." Ben Roy squeezed the deck of cards tight between his hands. "Baby or no, if that nigger goes back to his woman, works like he ought, and don't cause no trouble, ain't nobody in Lawnover gonna drive him out of town. That's the way of it."

The building went silent in Alex's ears. The clink of jars and the cussing from the front room stopped. The smell of sorghum turned sickly sweet. The brined pickled feet let off the stench of ten-day-old pig slop. Alex worked his mouth to speak. Ben Roy raised a hand that the kerosene lamp silhouetted against the table into long snakelike shadows.

"You think about Hettie's tan-skinned cousin." Ben Roy laid a hand on Alex's arm. "We can't have no trouble right now. Troubles don't go 'way quick. They have a way of lingerin' long and makin' you wish things was different…wishin' you hadn't done what you…" Ben Roy's eyes drifted off an instant. "Get on back to my sister, and do what you ought, 'cause that other one, that Annalaura, she ain't available no more. Her husband's done come home. As long as he's around, she's John Welles's woman. Ain't nobody 'round here gonna run him outta town over a baby that belongs to another man. You got to let her go, Alex. It's over."

# CHAPTER TWENTY-THREE

The clock struck eleven. Eula felt each strike bounce around her skull like apples jostling in a bushel barrel after the fall pick. Funny the things that went through her head since she'd taken to the bed nine hours ago. Not that she slept even one minute in all that time. Silly little things played in her head as she lay there staring at her white-painted ceiling and daisy-printed wallpaper, like the first time the grandfather clock had appeared in her parlor. It hadn't been new, of course. She and Alex couldn't afford new, especially in those days. Alex had bought it at a Clarksville auction just two weeks after her thirtieth birthday. She always liked to make believe it was a present from her loving husband, though Alex never said one way or the other.

The clock's time had always been right on the strike. Eleven o'clock, and she hadn't spoken more than a dozen words to her husband in days. As good as she'd been at keeping the pain away these long hours, sometimes those other thoughts crept into her head. Sister-in-law Belle, missing preserves, the money box lighter by five hundred dollars, all insisted on tumbling around in her brain, forcing her to find another way to push them back into that wonderful place of forgetfulness.

Eula rolled to her side and remembered something about her feet. She had managed to unbutton her shoes and drop them to the floor when the clock had struck six. But she still hadn't found a way to remove any of her other clothing. The house was dark with the moon only a quarter full. Since it was just mid-May, Eula guessed it was probably still chilly enough this close to midnight to warrant slipping under the bed-covers. She would have done just that if her body and head could tell the difference between hot and cold, comfort and misery. Everything about her had gone numb. Other than the hourly sounding of the grandfather clock, her ears had mercifully screened out the sound of Belle Thornton's hateful voice.

Lying on her right side, Eula could see her half-shut bedroom door. She wanted no part of staring at a closed door, counting the hours until it would open. She rolled over to her left and lay her forearm across her eyes. Now she would neither hear, nor see, nor feel Alex, but the smell of him lingered on the feather pillow next to her head. She lowered her hand to cover her nose and mouth and went into herself so deep that the sound of the back porch and kitchen doors opening and boot steps strid-ing across her pinewood floors made no more sense to her than the little breeze tickling her windowpane.

She finally caught snatches of sound that might have come from a human voice, but she had the feeling that the noises had been repeated over and over before they woke up her ears. It was the unsure touch on her shoulder, rather than the harsh light from the kerosene lamp, that finally forced her mind to come partway back from that quiet, safe place.

"Eula?" The voice was uncertain, concerned, and befuddled.

She kept her eyes shut. She had no desire to answer that strained voice.

"Eula Mae, you sick or somethin'?" The voice sounded tired and downcast.

The hand on her shoulder became more certain, more familiar. Though she didn't want it to happen, her ears slowly returned to their job. Outside, somewhere, that little breeze kicked up and rustled the new leaves on the trees. A weight dipping down the small slice of mat-tress behind her back told her that the owner of the voice had leaned on the bed.

"Can I get you somethin'?" The sound of the voice shot through her.

Eula rolled on her back as though a good gust of wind had pushed her over. Her eyes flickered open.

"What?" She stared at her husband.

Alex stepped back from the bed and lowered the lamp to her face. He scanned the light down to her feet. The glare blinded her and kept her from reading his eyes to see what could have possibly prompted him to offer help for the first time since their baby di . . . She pushed that thought back into its rightful hiding place.

"How long you been layin' here? It's 'bout midnight, and you still got your clothes on." The sound of concern in her husband's voice woke up her worry over him, and she hated the bother.

Eula raised herself against the headboard, a pillow at her back. She reached to move the light aside. As Alex set the lamp on the bureau next to their bed, she caught a glimpse of his face. His mouth drawn down in weariness, his cheeks sunken in, he looked as though he had lost the battle with a swarm of tobacco locust.

"I'm a little tired, is all. But, you . . . uh . . . didn't the plantin' go well?"

She readied herself for his quick burst of annoyance and then his silence. It was not as if anything she ever did could warrant true anger, or any other real feelings from the man. Whatever outbursts he had were over in less than a minute.

"Plantin' went just fine. Could do as well as last year. You sure you all right?" He walked around to his side of the bed, unbuttoning the straps of his overalls as he moved.

Eula mustered a shake of her head. Thoughts of missing money and missing nights rammed back into her mind. She watched Alex as he sat on the edge of the bed and removed his boots. The light from the kerosene lamp cast his face in half shadows. She hadn't noticed before that his eyes looked quite so haggard.

"You feelin' a little warm?" Alex placed the back of his hand against her forehead.

"I'm mmm." If she dared, her hand would have been the one to reach up and check Alex for fever. Her husband was not in the habit of checking on her health.

"Let's get you under the covers." Dropping his overalls, he scooted onto the bed.

Alex reached behind her neck to unbutton the apron she had put on fresh this morning. Before she could reach out a hand to stop him, he pulled the garment away, the untied strings dangling, as he dropped the muslin to the floor. What was he doing? Alex tugged her shirtwaist free from her skirt. When his hands reached for the buttons, Eula slapped her hand over his before she could get her mind clear.

"No. I mean...I'm not sick. I'm just tired. I can do it myself." Her hands fumbled with the buttons before she remembered that she had never taken off her clothes with him looking straight at her.

If he did happen to walk in when she was changing, she just stepped behind the chifforobe door, turned her back, stepped out of her petticoat, and pulled her nightdress over her head, all before he could notice. Eula's shaking hand stalled at the button over the center of her chest. Alex finished the job. She held her breath as he pushed the shirtwaist back from her shoulders. In her confusion, she moved to stop him before she remembered. She only wore an everyday chemise. Was Alex checking to see if she was properly dressed?

"My corset, too tight. I need..."

"I reckon I haven't always given you what you need. If you need a new corset, I'll make a way..." He slipped her skirt over her feet as he looked at her face.

What she needed? Make a way? Did Alex have some money worry that a good wife should have seen and sorted out?

"You already give me everything I need." She stumbled out the words as the dwindling numbers in her journal and the missing supplies in her storeroom flashed into her head. Did the missing food and five hundred dollars go to pay some debt he hadn't bothered her with? Had Ben Roy gambled her husband out of some money?

Alex slipped his hands to the bottom of her chemise and began sliding it up her chest.

Eula laid a hand at her bodice and scrunched up the loose fabric, her fingers squeezed so tight, they hurt. Alex hadn't seen her naked in full kerosene light in years. He pulled her back when she reached out an arm to turn down the wick. Had she paid more mind than she ought to her mean-spirited sister-in-law? Was the real answer gambling?

"No, I ain't given you everythin' you need. I ain't never told you how much it means to me how well you run this place."

The quick intake of her own breath sent a charge straight to Eula's chest. She let her tight grip on the chemise loosen. Her eyes closed as Alex slipped the garment over her head. Her hands dropped to her side, not bothering to cover the breasts she always tried to hide from him because they hung flat like empty socks laid over darning balls. How could she have let herself pay mind to the silly prattlings of her brother's wife? The real problem lay with Ben Roy all along.

Alex ran his hands the short distance from her belly to her breasts. A breath came in and she held it as his hands worked her drooping bosom up her chest to where it used to blossom.

The heat coming from his body as he leaned in and parted her lips with his mouth set up a sweat all over her own body. Something foreign found its way into her mouth. Was it his tongue? She wasn't sure if he was searching for hers or not, but she let him find it. He pressed her bunched up breasts against his rough cotton work shirt as his tongue encircled hers. Before she could will it back, her tongue pressed into his.

She let her arms slowly reach around his back. Alex kissed and released her lips. She flickered open her eyes just as his flushed face drifted to the windowpane.

"I'm glad it was you." Alex's voice carried the husky sound of a man struggling for breath. "You and not Bessie." He moved to take off his shirt and summer drawers.

Eula chanced a look at his manhood. It had made a full rise.

"What?" She managed to get out the word. She saw her reflection in Alex's eyes. Did he really want her like a man hankers after a woman? The calendar on the pantry door conjured up in her head. Tonight was Tuesday.

"I could of been like Wiley George. Worse." Alex leaned toward her again and kissed her on the cheek.

"You're the best wife I could of found. I'm not a man born with good luck. Without you, I reckon I would of lost this farm." Alex's hand squeezed her breast.

She sucked in her upper lip. Alex was saying things to her that he'd never said before. Though she'd never known her husband to be a drinking man, did his words come from the liquor served at Bobby Lee's? Was that how Ben Roy did it...got her husband drunk?

"No such thing. I just always want to be a help to you because I..."

Eula had loved Alex for sure since right after the baby, but she had never said the words. And Alex had never said them to her. Was he trying to say them now? His eyes caught hers again.

"It was a blessin' that Bessie was taken when I finally came callin'. She was pretty, but pretty goes away and a man's left with nothin'." He slid his hands around her hips to the place where her thick thighs came together.

Alex let his fingers stroke midway between knee and her triangle. Eula grabbed another gulp of air. With the pleasure feeling spreading between her legs, she looked at Alex as his eyes fixed on the windowpane.

"Every man in Lawnover was wantin' Bessie. I don't fault you for that." She let out a sigh as her legs spread open just enough to still remain a lady. How silly she'd been.

Alex's hands moved between her thighs. "It wasn't much I done to get you, but I'm thankful you're my wife." His fingers played in her triangle.

Eula bent her knees. The muscles in her thighs jerked. She fought to press her hips into the mattress to stop her legs from throwing themselves wide open. She battled the urge to lift her arms to his shoulders and pull him down on top of her. Scattered bits of pictures jumped into her mind, and right out, again. Liquor, money troubles, swollen-bellied colored women, the lying lips of sister-in-law Belle, Alex's love for her finally finding words, spun around in her head.

Eula arched her body into Alex, allowing her legs to gap a little more than was proper for a lady. She sniffed to catch the smell of Bobby Lee's home brew. She wanted to swim in Alex's embrace, answer his exploring fingers, sweep his ears with words of her love, but if he was a little bit drunk, he would hate her in the morning for behaving like a roadhouse tart.

"Alex, I try to be the wife you deserve because I know I haven't served you as well as I ought." She grunted as Alex's finger slid all the way inside her triangle. He had never touched her like that in all their years together.

"Mmm" was all she heard from him.

"A good wife…ooh…a good woman…my Lord…would give you babies…oh, my Jesus…I…"

"No. No talk 'bout babies tonight. No talk 'bout babies, ever." His voice sounded gruff as he pushed his manhood inside her.

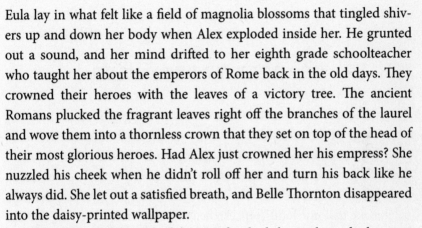

Eula lay in what felt like a field of magnolia blossoms that tingled shivers up and down her body when Alex exploded inside her. He grunted out a sound, and her mind drifted to her eighth grade schoolteacher who taught her about the emperors of Rome back in the old days. They crowned their heroes with the leaves of a victory tree. The ancient Romans plucked the fragrant leaves right off the branches of the laurel and wove them into a thornless crown that they set on top of the head of their most glorious heroes. Had Alex just crowned her his empress? She nuzzled his cheek when he didn't roll off her and turn his back like he always did. She let out a satisfied breath, and Belle Thornton disappeared into the daisy-printed wallpaper.

Though her husband hadn't quite finished the sixth grade, he must have heard parts of that lesson on the laurel wreath because the slightly garbled word bursting from his mouth was loud, drawn out, and laid sweet upon Eula's ears.

"L-a-u-r-a."

# CHAPTER TWENTY-FOUR

Something was wrong. Something was very wrong. Annalaura had long since let go of the idea that the pains twisting her body came from the blow John had struck at her belly. Though none of her other four children had given her this much trouble getting born—nor had they come early—they had all belonged to John Welles. How could the father of a baby make that much difference? The latest big cramp eased up, and she let herself take a breath. She grabbed at one of the three cotton blankets Aunt Becky kept tucked across the bed, summer and winter. All three were drenched in her sweat, her broken bag of waters, and the plug of blood from her womb. This baby was definitely on its way, and Aunt Becky hadn't come back to the cabin.

Hours had passed since Rebecca took the three youngest over to Hettie's. The ever-closer pains faded most of Annalaura's sense of time, but the low-burning embers in the fireplace told her daybreak couldn't be more than four hours away. Even the kerosene in the lamp burned low.

Right after Becky had gone with the children, her water broke and it was all she could manage to struggle out of her shirtwaist and skirt. Annalaura had tried to raise herself to refill the lamp and stoke the fire only to be wracked by a pain so strong that even setting one foot on the floor caused her more misery. Was she suffering God's punishment for her sin against her husband?

Another cramp wrapped her in its grip. Where was Becky, or even Cleveland? She had stationed her boy at the gate to warn Alex away. After that, Cleveland had probably gone back to the loft to await her return, not knowing how much his mother needed him. She prayed that her boy would worry over her and come to Becky's cabin. As she held her breath against what felt like a thousand tobacco spearing sticks jabbing at her insides all at once, she remembered that her twelve-year-old would never venture out after dark. John had trained him too well. Colored weren't safe wandering the lanes after nightfall.

The spearing sticks did their worst and started to ease up. If she had a timepiece, she would know for sure that the pains knotting her insides into a hot ball of fire were coming every five minutes and holding tight for well over a minute. She frowned, drew in a long breath, and knew her body was tiring much too soon.

It wasn't that dying would be so bad. Being dead had to be better than the double portion of misery dragging her with it right now. But who would see after her children? John would take Cleveland and Doug— they were of some size to be of help. But Lottie was a girl, and Henry, little more than a baby.

Before she could worry after her two youngest, a poker, hot from a roaring fire with butchering knives stuck all around it, let itself loose in her belly. Annalaura rocked on the bed in time to her own yells in a fight with the red-hot poker. In what felt like an hour, she sensed the poker start its cooldown. Her forehead beaded in sweat, her breath came in short bursts. She lay gasping on her aunt's bed, the covers tumbled tight around her. A hard tap at the door brought the first burst of hope.

"Push it open, Cleveland. Hurry." She fell back against the flat feather pillow. "Yo' momma needs you bad."

The door slammed against the cabin wall, but her head felt too heavy to lift.

"My God, Laura, is it the baby?" The sound of Alex's boots bounded across the room. Her world suddenly turned to a blur of dying fireplaces, dying lamplight, and dying men.

"No. Gawd. Ughh. Mmm. You can't be here. You got to leave. Cleveland…"

"It's the baby." Alex's hands tumbled over themselves in his hurry to untangle the blankets that wrapped her body.

He bent closer, looked into her face, and frowned. Annalaura took in short, quick breaths. She had to warn him out of the cabin. He stopped fumbling with the covers and moved to retrieve the low-burning lamp.

"Shit. Where's the kerosene for this damn thing? It's dark in here." He swung the lamp over her face.

The little flicker of light was weak, but allowed her to catch just a quick glimpse of his face. Another pain moved up, and she held her breath. The lamp thudded to the bedside stand. The old bed dipped down as Alex knelt beside her.

"You got to breathe, Laura."

Her eyes cranked open under Alex's rough hands as he grabbed her shoulder and pulled her up in the bed. His hands beat back the offending blankets and suddenly stopped. She watched him bring his hands to his face. Alex played his fingers under the light from the kerosene lamp. Both hands were smeared with her blood.

The floodwaters signaled their retreat, and Annalaura watched Alex stare at one hand and then the other before he grabbed at the lamp and swung it back to her face. Her neck and shoulders pressed against the bed headboard. She let the air make its way back into her. She had just enough strength to lay an arm against Alex's chest as he knelt over her.

"Go. You gotta leave. John...my husband...he back...pistol. Please go."

"Your face? What the hell happened to your face?" Alex leaped from the bed.

Didn't he hear her? Annalaura bit down on her bloody lip as she tried to make her ribs move to take in more air. She had to get Alex out of the cabin.

He scrambled around the room, filling the lamp. She heard Alex's boots stomp back across the floor to her.

"Now." She moaned. "You gotta go now."

The mattress sagged under Alex's weight as the bright light from the refilled lamp swung across her face. She closed her eyes all the tighter until the lamp moved away.

"That bastard hit you, didn't he?" The sound of Alex's voice frightened her all the more.

A cloth dampened in cool water dabbed at her eyes, lips, and all her cut and sore places. She sensed his hot breath of anger fan her face. She couldn't have Alex mad at John and John mad at Alex. If only God would take these awful pains away so she could sort out this mess.

The first shock wave of cannon thunder rolled into position in her belly. She heard Alex lift the pitcher on the nightstand and pour fresh water over the cloth.

"Where's yo' aunt? Where's Rebecca Thornton?" He turned back to her and laid the coolness against her cheek.

The cloth's comfort was no match against the cannonballs ripping her insides to shreds. The battle caught her breath and held it as she grabbed for life wherever she could touch. She dug her fingers into her lifeline as the battle raged.

"Take a breath, Laura. My God, take a breath."

Hands grabbed her shoulder and pulled her body out of a deep hole just as the cannons ran out of powder. Her eyes opened as air poured back into her body. She stared directly at Alex. Annalaura remembered that his eyes were a milky blue, but she had never seen such worry in them. He sat next to her on the bed, cradling her head against his chest.

"Mmm." Her voice wouldn't work.

She pointed to Becky's food safe. She clutched at Alex's arm, and her fingernails dug in hard, drawing his blood.

"Where's the doctor?" Alex's arms wrapped tighter around her.

His mind must be unraveling. He knew as well as she that no colored in Lawnover could afford a doctor. With one hand, he grabbed Becky's feather pillow and plopped it under her back. He swept the other down to the blood-speckled cover, gave her a puzzled look, and jumped off the bed.

"I'm goin' for help." He started for the door.

"Nooo." Her scream rushed out as a new rumbling started in her belly. "No time. Baby comin'." If she could, she would have swallowed back her words. She couldn't bring this baby into the world by herself. She needed help, and right now. But if Alex stayed...and John...

"Oh, my Lord...uhhh...Go. Go, now."

Almost at the door, Alex stopped, turned, and stared at her. "Becky.

I'm goin' for your aunt Becky." Panic spurted out of his mouth. His hand reached for the door latch. "Where's your aunt?"

"No good...took...Doug...Lot...Hettie..." The scream jumped out of her mouth before she could corral it.

He ran to her side just as the pain took all of her air. "Hettie? You sayin' Rebecca's gone to Hettie's?" Alex shook his head. "No good. She's with Ben Roy. Don't know if he's done with her yet."

As her insides fought to burst through her body, Annalaura tried to lift an arm to reach out to him.

"Safe. Becky's safe. Knife. Then go." The pain eased up just enough to get the words out.

Alex didn't move.

"Knife to cut the pain. Put it under the bed." She nodded toward Becky's safe before she sank back into the pillow.

Alex started to pace the floor, ignoring her. He threw more kindling into the fireplace and poked the embers into flame as he stared at her.

"Birthin' a calf." The frown lightened as he muttered.

A moan pushed out of Annalaura's bruised lips. Becky was beyond her reach and could offer no help. And now Alex was talking out of his head. She watched him drop the poker to the brick hearth. He rushed to the other side of the cabin. Alex fetched a bucket of water, Becky's dish-drying towels, and a knife. Annalaura turned her head to see him set the pail in the flames of the fireplace and splash the knife into the water before he returned to her side. She inched a hand out to him again.

"My grandma said...knife under bed...cut pain." She let her hand fall against the bedcovers as she felt the winds of a tornado begin their slow wind-up in her stomach.

She said her last prayers. Lord, take pity on my motherless children. Let Alex get out of this cabin before John gets here. She closed her eyes, the better to see death coming.

"If it's like birthin' a calf, the hot knife will cut the cord clean." Alex climbed on the bed, knelt behind her, and shifted her weight from the pillow to his chest.

Calves. Cows. What was the man talking? The winds of the tornado picked up.

"Ahm…my God. It's stuck. Can't get it out." The iron bar moved to her belly and held her fast. If only dying didn't hurt this much.

"It'll come." Alex, his voice carrying a sudden strength, took her hand, and with his, slid it down her belly.

"When a mother cow gets stuck like this, I rub on her belly to get her to moo it out. Each time we rub yo' belly, you grunt it out." Alex shifted her body higher up against his chest.

Annalaura let her head rest under his chin, her knees bent. Alex was right. This was a less hurting way to die. She felt the strength of Alex's grip on her hand as he rubbed down her belly.

"It's comin'." The scream burst out of her as the old cabin splintered into a hundred logs all rushing to get out of her at once. "Alex, help me. It's comin'." Her eyes clamped shut, she felt him shift her weight from his chest back onto Becky's wadded-up pillow and bedcovers.

Alex moved down to her spread legs. She opened her eyes enough to see the top of his head. Dozens of logs jammed inside her as each one turned and twisted to find the opening.

"Help *meeee*."

With death tightening the baling wire, somewhere Annalaura felt the hand of Joshua reach in to free the jam just like he fought the battle of Jericho in the Good Book.

"Laurie, another grunt. One more good grunt." It was a saint's hand she felt, but Alex's voice she heard.

"Can't Alex. I got no mo'." Though the sainted hand wrestled with the devil, the logjam started its move back up into her belly. Even the saints couldn't save her.

"I can touch the head. Laurie, you got to grunt one mo' time." Alex laid one hand on her belly and pushed down. "Give me yo' hand."

Did he want to hold her hand when he bid her good-bye? She hadn't asked for another man in her bed, or in her life, but if one had to come, she would thank the Good Lord when they soon met, that he had sent Alex. She stretched her hand toward him. The tips of her fingers brushed his. At first touch, Alex grabbed her fingers and squeezed hard. A new pain shot sharp jabs up her arm, and her yelp exploded into the room. The logs, stuck together in her womb, splintered apart and began to slide out of her belly, one by one. Annalaura dropped her head back onto the pillow, her hands limp in the muddled bed-coverings.

"Um." The broken bits of timber eased out, and air oozed back into her chest.

"I've got it," Alex shouted.

Annalaura lolled her head to the side in time to see a little smile cross Alex's lips.

"I've got…Laurie, I think it's a girl." He lifted his surprised eyes to her face.

She felt the muscles in her cheeks work themselves into a weak half smile. Alex's eyes flitted from her back to the baby. She saw the light in them move from bright surprise to fear. He moved a bloody bundle into Annalaura's line of sight. She turned away.

"Ain't she supposed to cry?" He whispered the question like he knew he was sitting in a death room.

"Hold her up by the feet. Smack her bottom." Annalaura's eyes closed when she heard the light tap on the baby's backside. Sensible thoughts started to make their way back into her head. Maybe the Lord wasn't going to take her tonight after all.

"Harder." She felt her voice growing stronger.

As she let her chest take in more welcome air, she smiled at Alex. Even if he didn't know how to do this part, she wished she had the strength to tell him her thank-yous for all the rest. The second, louder slap brought a familiar wail that filled the cabin.

"She's just fine," Annalaura announced. She hadn't wanted this baby. She knew that nothing but a lifetime of trouble awaited them both. Still, she needed to know if the child had all her fingers and toes. Annalaura would do what she had to soon enough. But for now, it was best if she didn't set eyes on Alex's baby.

Alex stared at the infant as though he'd never seen a newborn human before. Then Annalaura remembered. Hadn't his own firstborn come out dead? If Tennessee allowed, she would take Alex in her arms. Another gulp of air, and Annalaura pushed that thought away faster than any of the others about the man who had just saved her life.

"Let her down and clean her up." Those were the only safe words to say to him.

Alex's face showed confusion as he looked for a place to lay the baby. He reached toward her stomach.

"No. Lay it on the bed. I don't need to see it." The words came out too late. As he wiped a damp rag over the baby's chest, Annalaura caught a glimpse. A new fright washed over her.

"Alex, what do she look like?" Please, Lord, let the herbs have done their job.

"She's good. She's better'n good. She's perfect." The sound of wonder seeped into Alex's voice.

"But what do she look like?" Annalaura pressed.

Alex lifted the baby toward her, and she wrenched her head toward the far cabin wall. She knew she had to care for this child, but if she could put off looking at it just a bit longer...

"No. I don't need to see it. Just tell me." She caught his confusion in that little shrug of his shoulders.

"I reckon all her parts are here, but is she supposed to be this little?" That edge of fright hadn't left his voice.

"What color?" Annalaura knew what Alex didn't.

Colored babies often came out with pale, alabaster-looking skin, only to darken nicely in the first few months. But that first quick glance she'd had of Alexander's child told her that this baby's color was all wrong. The infant's skin showed a creamy pink, not a golden, fried-potato brown. Maybe there was still hope. Aunt Becky told her that a colored child's true color stayed hidden behind the ear. Many a light-skinned newborn child was lucky enough to turn a beautiful chocolate brown within three months. And, if Aunt Becky was right, the truth lay there all the time, just behind the ear.

"They're blue." Alex shouted the word in the room.

"Blue?"

"Yeah, her eyes are blue...like...like mine." His voice dripped pride.

He cradled the baby in his arms, stroking a finger over the child's foot. He caught Annalaura's eyes staring at him and returned a funny little half smile like he couldn't quite trust the good fortune his own eyes were telling him.

Her heart spiraled down into her chest. Only fair-skinned colored children started out with blue eyes. The darkest they ever showed was a greenish-brown.

"Behind her ear. Look behind her ear." Annalaura tried to keep the fright out of her voice, but the worry crept in.

She held her breath while Alex shot a puzzled look at her.

"What's behind her ear?" She shielded her eyes as he turned the baby, but not in time to escape a glance of the top of the child's head.

"She don't look no different there than anywhere else." Alex shot her a puzzled look.

Of all the pains that had ravaged her body in the last hours, this one hurt her heart the most. Alex reached around for one of the unused cloths and wrapped it around the squirming, squalling infant.

"This ain't warm enough." He looked for another clean drying rag, found none, and grabbed an unsoiled corner of the bedcover.

The baby's cries settled into soft little mews. Carefully, Alex stood, holding the infant in his arms, trailing the end of the cover behind him. He walked to the safe and returned with a Mason jar. He handed it to Annalaura.

"Take a sip of this." He sat down beside her with the baby still in his arms.

Annalaura turned her head as she brought the cup to her lips. She sipped Aunt Becky's best medicinal whiskey. Its smooth warmth soothed her all the way down to her ravaged insides.

With one hand Alex took another cloth and finished wiping away the blood between Annalaura's legs. As he held the baby in the crook of one arm, the color of the infant's still damp hair registered in Annalaura's mind. Tufts of burnished gold covered the child's head, and Annalaura fought against a fresh wave of misery.

"She don't look real. She looks just like a doll." Tossing the rag on the pile of tousled covers, Alex turned the baby toward her.

Before she could shout her no, it was too late. Annalaura got a full look at the child she had just delivered.

"What're we gonna call her?" Alex looked down at the drowsy infant girl.

Even with her head still clouded over worse than a winter's day, Annalaura already knew that she had asked too much of Alex. If he'd been in Becky's cabin, even little Henry would have seen that the mind tottering on the edge of some unreal place was not Aunt Becky's. It was Alex's. What are we going to call her? There could be no "we" when it came to this baby. Not in all of Tennessee had there ever been a "we" for a colored woman's baby with a white man. Annalaura looked at Alex as he held the child like she was that newborn calf he talked about. He touched her soft and gentle like she was the pure gold the wise men brought to the Baby Jesus. Didn't he know, didn't he understand that he could have no parts of this baby? No white man could claim a colored child as his own. What little bit of a mind she had coming back to her, Annalaura had to spend on easing Alex out of the cabin and away from all thoughts of a "we."

"Becky say it might be best to let the child stay down county fo' a bit." She said it gently.

"Down county? Why would I want her there? No, my daughter ain't leavin' Lawnover." He snapped his head toward her, his voice carrying a sting in it.

Annalaura squinted over Alex's shoulder. How much time 'til daybreak? How much time until John found her and the child? How many more hours until he found Alex and pulled out that pistol? She shut her eyes tight as turmoil roiled up in her chest again. She had to find the strength to bring Alex back to his senses and wash his mind clean of any thought of a "daughter."

"Baby's mine, Alex." She let the words flow as soft as she could. "It can't be none of yours."

His eyes stayed on her face, and a slow smile started at the corners of his mouth. All the words went out of her head when Alex leaned forward to brush his lips over her cut and bruised mouth.

"I know she's mine, Laurie." The words were low and sure. "I know it more sure than anything. I can feel it all the way here." He tapped the pocket of his gray shirt. "She's the most beautiful thing I've ever seen, and she belongs to me."

Annalaura sat stiff on the bed, afraid to move, wondering if he was going to kiss her again.

"Dolly, I want to call her Dolly."

Alex reached to stroke the side of Annalaura's cheek. She closed her eyes and for a moment breathed in the healing touch of his hand. The soreness melted away quicker than any Cherokee salve Rebecca ever rubbed on her. Annalaura forced her eyes open to stare at the glare from the lamp. She needed its hot, yellow light to remind her that the devil had lost one battle to take her away with him this night. If she couldn't set her mind on the right road and put away those other thoughts, that red-coated fella was there to remind her that he wasn't through with her yet. Why else did things that could never be keep planting themselves in her head?

She looked toward the curtained window. There was no time for foolishness. Every second in this too-short night had to be spent making a plan to put a considerable distance between Alex and John, between herself and the devil.

"It might be best for the baby if John don't see it right off. Down coun..." She started off as slow as she could.

Alex laid the sleeping baby on the bed. He took Annalaura's hand in his.

"If John Welles lays one hand on her or you, it won't take me no time to kill him." Alex's voice came out strong and certain.

A chill ran from her shoulder blade to her tailbone. Alex released her hand and reached over to pick up the child.

"I'm takin' her to my place."

A horde of white-robed night riders rode into Annalaura's head. Carrying the baby, Alex moved toward Becky's safe.

"Yo' place?" The words rushed out on a shriek. "You can't take her to yo' house. She just born. You'll kill her if you take her out in the night air." Another wave of fear welled up in her stomach.

Alex pulled an old shawl, brown from age, from a drawer and wrapped it around the baby.

"Give her to me. She needs to be fed." Annalaura pushed herself up farther in the bed.

Alex stood by the safe, patting the shawl-wrapped baby. Annalaura leaned forward in the bed, her arms stretched out as far as they would go, willing Alex to come to her. With his face set hard like she'd never seen before, he started slow steps back to the bed. As though she knew the trouble she was in, the baby started wailing. If John came to Becky's cabin and heard the cries of a newborn... When Alex stepped close enough, Annalaura tugged at his pant leg.

"Alex, please. I needs to feed her."

He stood looking down at Annalaura, his jaw muscles locked tight, the baby's wails coming louder.

"When can she leave?" Alex looked at the child, then back at Annalaura. "When can the two of you leave?"

Annalaura tried to walk her hands up his leg to the shawl-wrapped baby. Alex took a step backward. Behind him, she watched the darkness that had covered Becky's window curtain all these hours show signs of letting go. Daybreak was about to cut through the night. Annalaura let her hand drop from the edge of the shawl to Alex's leg. She laid on gentle strokes just above his knee.

"We both got to get a little stronger, but that ain't gonna happen if you don't let me feed her." She succeeded in keeping her voice steady.

Alex turned away and walked to Becky's wood-block table, where the light from the kerosene lamp couldn't reach him.

"If you say you and Dolly can't leave now, I'll just wait here 'til you can." He kicked Becky's shored-up chair from the table, but he didn't sit in it. "Where's the old woman's shotgun?"

Annalaura fell back on the bed, her tired arms dropping on the mussed covers. The baby thrashed in Alex's arms as he scanned the walls. She watched him step around the broken chair and walk to the fireplace wall, where Annalaura remembered Becky kept the shotgun. Thank the Lord, that rusty old blunderbuss was gone. Alex peered at the empty hook, turned, and stomped at every floorboard in the cabin that looked loose. The newborn filled the room with her cries, and her mother's heart with fear.

"Listen to me, Alex." She had no more time. "Give me Dolly, now." She pushed herself upright in the bed, struggled her feet out of the covers, and stretched them to the cold floor. She felt the old floorboards skidding out from under her as she started to slip off the bed.

Alex blocked her slide with his free arm and chest. He laid the baby on the bed and took her in his arms.

"I ain't gonna leave you here for that nig…for that bastard to come after you again." His lips brushed her ear with a softness, but his voice sounded hard.

Alex shifted her back in bed to the squalls of little Dolly. Annalaura reached a hand toward the shawl-covered bundle, her eyes away from him. She gathered the baby in her arms, unbuttoned her chemise and set the little mouth to her breast while Alex smoothed the blankets around her. Dolly lost the nipple, screamed, and Annalaura put her breast back into the infant's mouth.

"I want the two of you to come to the farm." Alex settled in on the bed next to her. His voice carried the sound of wonder.

As the baby suckled, Annalaura let his hands lean her body against his chest. Alex stroked a finger from her breast to the baby's puffing cheeks.

"He ain't comin' after me again, Alex. He already took out his mad on me." She was too tired to move her body, and there was no time to push back how good she felt. "He won't hurt the baby either. When he takes one look at her, he ain't never gonna want to see me or her again." Maybe telling Alex the truth would get him out of the cabin.

If John Welles wanted her dead, he would have done the job when he first saw her fat belly in their barn. Cleveland, and his pitchfork, would never have stopped John if he'd really meant business.

"How long before he gets outta Lawnover?" Alex's hand played between the baby and her breast. His voice carried more calm, but Annalaura heard the purpose in it.

"If he don't meet up with you, no more'n a week." She tried to make her own voice sound casual as though his life and John's life weren't hanging on a thread thinner than any ever sold by Mr. Bobby Lee to colored.

"Can't wait no week." Alex raised up on the bed and fumbled through the blankets until he uncovered her shirtwaist and skirt. Shaking them free, he pushed them toward her. "Soon as you feed the baby, put these on. Night air or not, I'm gonna take you both to my place."

"We can't ride out nowhere, tonight." Annalaura shuddered, and the baby's tiny mouth fell away from her breast. "In a week, maybe…"

Alex bounced off the bed. Scouring the floor, he reached down to retrieve her boots.

"Let's get some clothes on you." He grabbed at her foot and tried to jam on the shoe. She pulled her leg back.

"Can't do it, Alex."

"I know it's soon. You can rest when I get you home."

"That ain't it." She swallowed trying to bring up more strength.

He stopped buttoning the boot to look up at her face.

"Me bein' there might not set too well with yo' wi…with…Miz McNaughton."

Clutching the sleeping baby, Annalaura watched Alex's brow furrow into a scowl. He looked like he'd just heard her say that roses bloomed straight out of snow for all the sense her words made.

"What in the name of the Good Lord has Eula got to do with this?" He stared at her. "I left her asleep." Alex slipped another button through its loop, looked up at her, his eyes bright. "Matter of fact, once you finish yo' layin' in, you can help her with the cookin' and cleanin' and such. She'll thank me for bringin' in help." He went back to the shoe.

"Thank you?" Annalaura fought back a tremble that shook her shoulders. "Alex, it's just that I fear that the surprise of me and little Dolly livin' there without much notice might…well…it could be a might sudden."

He fastened the last button.

"Besides, I got fo' other children." She kept her voice as gentle as she could.

He slipped on the second shoe.

"I don't reckon yo' missus wants fo' children and a baby clutterin' up her house."

His hand on the bottom shoe button, Alex looked up at her at last. Annalaura took in a breath as she watched the slow nod of his head. He had heard her after all.

"Well, a house full of youngsters might take her a spell of gettin' used to." Alex tapped at the shawl-wrapped baby as his eyes settled on the far wall of the cabin. "I reckon Cleveland and them can stay here with yo' aunt."

"This cabin sets on Thornton land, Alex. Mr. Ben Roy might not take too kindly to..." Annalaura bent her head toward the baby, but she let her eyes search out Alex's face.

"Could be so... if Ben Roy... Thornton land..." When he raised his eyes to her, they blazed. "Then I... we... me, you... the baby... we can all go off to Chicago." His hands dropped from the button as he moved back to the bed. "I've got enough money to get us a start up north. Eula can have the farm. I can find a place for Cleveland. Maybe for Doug, too, if I pay a farmer for his keep."

Annalaura couldn't tell if the pile of words spilling out of his mouth were for himself or for her. Either way, his second plan was worse than the first.

"I reckon the colored preacher will take Lottie and Henry if I give him some money."

A second tremor shook Annalaura. Did he really believe what he was saying? Chicago. She'd never seen any colored person who'd ever come back from that northern paradise once they'd left Lawnover, but she'd heard all the glory talk from their jealous kinfolk back home. But even if Chicago did sound close to God's heaven, Annalaura knew that not even in that place could a white man live in the open with a black woman.

Annalaura looked down at the sleeping baby. This little one would need more than Aunt Becky's Cherokee medicine to spare her all the misery that life was going to throw at her. How could her little girl survive when she had a mother so overcome with worry that her milk was bound to taste worse than rancid butter. And her father? Didn't Alex understand that white men who got too close to colored also got killed? There could be no running off to Chicago with a colored woman and a half-colored baby. The trees in Montgomery County had limbs on them marked for white as well as black. Dolly could be an orphan before the next sunset. Annalaura had to try again.

"Their father... John... he's gonna want the big boys. Ain't no need to pay for their keep." If she started to agree with him, he would soon see that his plan was impossible.

"I don't want Welles nowhere near any of the children." Alex shook the bed with his shout. "Not Cleveland, not Doug, not you, not Dolly.

I want his as . . . butt out of Lawnover." He laid his arms around her shoulders, pulling her closer to him. "Once spring plantin' is over . . . Let's just say Welles will be happier if he gets out of Lawnover sooner better than later."

Annalaura clenched her grip on Dolly. Her whole body started to shake. What could she do to make Alex understand that if he stirred up the white farmers over her—a colored woman, any colored woman—both he and John would be dead? John would be found hanging from the branch of an oak, and Alex would meet up with a runaway wagon.

"I don't want no trouble, Alex. Not fo' John, not fo' Miz McNaughton, not fo' you." She had tried almost everything.

Her body and mind had little more to give. Careful of little Dolly, she twisted around and slipped her free arm around his shoulders. She leaned into him and pushed her tender breast into his chest. She readied herself against the pain and kissed him hard on the lips.

"Alex, you gotta help me make this work." She released his lips and looked into his eyes. "You gotta let me handle John. He ain't gonna hurt me nor this baby. Without you doin' a thing, John gonna light out of Lawnover before the week's out." She nuzzled her nose against his ear, raised her hand to his neck and stroked the side of his face. She watched his eyes. She had him captured in her gaze. "Let me have two days, just two days, and I promises you that I will bring Dolly to yo' house. I'll stay low 'til yo' missus get used to us bein' there. Just, please, darlin', let me bring Henry and Lottie 'til I can find somebody good to take 'em in." She kissed him again.

Little Dolly, squeezed between the two, started to whimper.

Alex leaned away from her and laid a hand on his daughter.

"You'll move in with me in two days and bring the baby?"

Her time had run out. If this was the price she had to pay to keep alive the two men who had captured every feeling her heart ever held, then she would make the bargain.

"I'll do whatever you say and be glad for it. Just, no more talk 'bout John. He don't want my face to give him reminders. He'll be outta Lawnover in two days' time." She let her head fall back on his chest. She could fight no more.

Alex kissed her forehead as he eased her back onto the pillow. He smoothed one of the rumpled blankets over her, bent down to kiss the top of Dolly's head, and headed for the door. As he opened it to let in the first pale gray streaks of the new day, he turned and looked back at Annalaura.

"I love you, Laurie. Thank you for the baby." He was gone.

# CHAPTER TWENTY-FIVE

The pink of the sky settled in and brushed away the last of the night. If the day was going to be hot, cool, or middling warm made no difference to John. If the newly turned earth scented the fields with wild primrose or mule dung was of little notice to him. Every thought in his head, every picture before his eyes, every sound in his ears, and every touch upon his skin was fixed upon the barn and the mid-forty. Last night had been no different.

He had passed most of it sitting upright in the little lean-to right behind the colored Baptist church in Lawnover, hearing their voices and seeing their intertwined bodies as he waited for daylight to banish the night. His only company in that long space between sundown and sunup had been the hired man's words that kept playing in his head, and his own black-handled pistol. He had run his hands over the weapon so much that it was a wonder he hadn't rubbed off all the black.

As soon as that first gray light of dawn brushed half the sky clean of the dark, he had made his way over to the blacksmith's shop, awakened the drowsing man, and rented a horse. Now, it took all he had not to whip the animal into a frenzied pace to reach the barn and Annalaura, and whoever else she might have in there. So far, he had won the fight to keep the bile down in his stomach. The pictures plaguing his mind left him no room to plan anything beyond what he knew had to be done on the McNaughton acres.

As the horse trotted down the lane, John struggled to keep enough of his wits about him until he could get the job done. Then he would welcome mindlessness. Riding as sedate as a preacher's wife was the first thing he had to get right. A colored man riding a horse at breakneck pace down a country lane would stir up the suspicions of any early rising cracker farmer. By the time he turned off the lane and onto the path leading to the mid-forty barn, the hired man's words were beating a tattoo in his head.

He'd had no real trouble finding the man, despite Annalaura's claim that her so-called lover had hied himself off to Kentucky. A few well-asked questions at the colored juke joint down near the river bottom let him know that Isaiah Harris, along with his new wife, was staying the night in the colored principal's back room. The two planned to catch a ride the next morning for a tenant farm ten miles south of Lawnover. As John's horse made its slow way to the hitching post, his head swirled with bigger-than-life pictures of the hired hand, Annalaura, and her lie. When he first set eyes upon the rather flabby-looking man, John knew something wasn't right.

Harris couldn't look him in the eye, only at some place between his eyebrows and the middle of his forehead like John bore the mark of Satan. As soon as he spotted that look on Isaiah's face, he knew the truth. John knew that look. It was the same one black men all over the South gave one another when a man's woman had just been made some white man's whore. Even before the first real words passed between the two, John read all the answers in that look of Isaiah Harris. No black man had fathered Annalaura's baby.

John's heart beat faster as he climbed down and tied up the rented animal. Walking through the barn, he neither heard the cows grazing at their hay, nor smelled the pigs, happy in their swill. He put one hand on the side of the ladder and stopped. He had clamped down so hard on his lip that the taste of his own blood stained his teeth and flowed into his mouth.

John began the slow climb up the ladder. A trembling Cleveland greeted him. The boy still held the pitchfork.

"You promised you ain't gonna hurt her no mo'." The boy raised his voice.

"I ain't come here to hurt yo' momma. I come to talk." Halfway up the ladder, John tried to peer around his son.

Cleveland shoved the sharp tines two feet from his father's chest. John's ears finally cleared of the hired man's voice, only to have it replaced by a strange silence in the loft. Where were the sounds of a crying Henry and a pouting Lottie?

"Where's yo' momma?" John moved to climb up one more rung, but neither Cleveland nor the pitchfork budged.

"She don't wanna see you."

The fork brushed his shirt.

"I wanna see her." John grabbed a corner of the metal tines and jabbed the wooden handle into Cleveland's little chest, knocking the child onto his bottom.

Before the boy could regain his feet, and control of the pitchfork, John bounded up the last few rungs of the ladder. Throwing the fork to the opposite side of the room, he reached down with one arm and scooped up Cleveland.

"You a good boy to look after yo' momma. It's just what I wanted you to do when I had to be away." John watched his son's face dissolve into tears.

He wrapped his arms around the child as the wetness dampened his shirt. John kept his hand firmly around Cleveland's head, grateful that the boy could not see his own throat swallow back the beginnings of unmanly tears. What had he done? Had he left a child to do a man's job?

"Stop yo' blubberin' now." He meant his voice to sound gruff.

"Papa. It ain't her fault, it ain't..." Cleveland spoke on great gulps of air.

John released his boy and looked around the empty loft.

"I ain't meant to hurt yo' momma. Just lost my head fo' a minute. Won't do it again." He laid both hands on the boy's shoulders. "Cleveland, you done the job of a man whilst I was away, and I'm feelin' good 'bout that." He heard the sobs simmer down in his son's throat. "But, I'm back home now, and it's the business of a full-growed man to look after you all. Where's yo' momma and the others?"

Cleveland rubbed his eyes with his shirtsleeve. "I reckon I can't tell you that."

"I can understand that, son." John sorted it out. Rebecca Thornton. "When you next see yo' momma, tell her to find me in Lawnover when she get ready. I'll wait fo' her there." He patted Cleveland on the shoulder and walked down the ladder.

⤬

The pile of white man's junk lay at the end of the path as John galloped the horse toward the lane. The bile lurched up from his gut, and he yanked on the horse's reins. He stared down at the debris. The feel of two red-hot pokers burning deep into his eyes would have soothed him more than looking at what was on that stack of goods. John couldn't recall how he got off the horse, nor how long he'd stood shredding, stomping, tearing, and kicking at the pile when he remembered the matches he carried in his pocket. He struck one and threw it at the mass. When the flames licked too slow at a new patchwork quilt, he tossed in another match. He remounted the horse. The low-burning fire made slow work of the stack. To his satisfaction, the little flames lapped at a woman's blue serge coat.

Dawn came into full bloom. He kicked at the horse's sides to quicken the pace and get away from the nightmare misery of knowing what had happened to his wife in his very own bed. The sour taste of bile rested just at the back of his tongue.

How could she let this happen? He dug his heels into the horse's side and snapped the reins. Why didn't she just kill that cracker herself when he came near her?

His throat felt sore, raw, from the cries flying out of his mouth. The labored snorts from the horse made their way from his ears into his brain. He eased up on the animal when it nearly stumbled at the fast pace John had urged upon it. His head throbbed with the furnace-red heat of his anger. Something gnawed at him.

When Alexander McNaughton set his sights on his Annalaura, there was no way for her to win. If she killed him, it wouldn't be no time before her neck was stretched to a tree. And, if she didn't, if she let him have his way, well, she was still lost to John.

As Becky's cabin loomed into view, the gall puffed out his cheeks until he won the battle to push it back down into his chest. Turning onto Becky's path, he laid his hand on the pistol tucked under his shirt and in the waistband of his pants. It was the only thing that could keep him calm enough to talk to Annalaura.

"Becky, open up this damn do'. I know she in there." John pounded on the old door until he felt it on the verge of splintering.

"Get yo'self gone from here, John Welles. You ain't nowise welcome." The old voice sounded like it came out of a deep well behind the closed door of the cabin.

"Becky, it won't take me no time to kick this damn thing in." John delivered the first blow with his foot and the old wood showed the beginnings of a faint, jagged crack. He readied his boot for the second assault when the door creaked open no more than two inches. He couldn't make out Becky's face in the gloomy slit of an opening.

"I'm tellin' you fo' yo' own good. Get gone from this place."

A bandanna still wrapped around Becky's head made John believe that she had just come in from the outside. He had no time to deal with a crazy old woman. He leaned his shoulder into the door and shoved it open. Once inside, he stopped an instant to adjust his eyes to the dimness, only to feel cold, hard metal pointed between his shoulder blades. Slowly, the specter of Becky coiled around in front of him, holding a blunderbuss of a gun older than John himself. As she moved, she pressed the weapon into his skin, ending up with it dug deep into his chest, right over the heart.

"I ain't of a mind to tell you again. You ain't layin' another hand on my niece."

"It ain't Annalaura who needs a hand laid to her." He didn't know if Becky's old gun could still shoot or not, but he couldn't afford to have it go off at him before he had a chance to do what he must. His eyes began adjusting to the duskiness.

The old woman's mouth was set in a thin line, and those Cherokee eyes stared up at him like she was trying to put out a hex.

"Ain't gonna be no layin' on of hands. You gonna leave." Becky had stationed herself between John and the sleeping area of the cabin. With his eyes rapidly taking in the dark, John had no trouble looking over the old woman's shoulder.

There, in the bed, lay a tousle of blankets and pillows. Under it had to be Annalaura. He tried to take a step toward the worked-iron bed, but Becky's old arms held the strength of a mule team driver as she blocked the way with her gun.

"I means no harm to Annalaura. I just needs to talk to her." He reached into his waistband.

Before he could pull out the pistol, he heard the squeeze of Becky's finger on the trigger. He lifted both hands in the air.

"Hold on there, Rebecca. I'm gonna let you hold my pistol to show you that I mean no wrong to Annalaura."

"I got this gun offen Old Ben Thornton, and I knows how to use it, too. And I ain't too old to push yo' worthless body outside once I done killed you." The steely eyes were still on his face when the cry went out into the room from the mound of covers on the bed. The sound was quickly stifled by the now moving jumble of blankets.

"What the hell's that?" John asked in the direction of the sound. "Annalaura?" He started toward the bed, saw Becky's mouth move but caught only a bit of the sense of her words.

"Pistol…my hand…real slow…" The blunderbuss lowered and circled his waist.

He felt the old gun lift the pistol out of his waistband and knock it to the floor.

"Knives. You got knives?"

"Knife? Ain't got no knife. Annalaura? What is that?" He brushed past Becky and reached the bedside in no more than three strides.

"Mind my words. Touch her and I'll kill you." Becky was right behind him, though his eyes were fixed on the bundle in the bed.

Something that sounded strangely like the wail of a tiny infant started again as Annalaura pushed herself up onto the pillow. The sight of her battered face sent waves of surprise through his body.

"Annalaura, I ain't meant to…my mind…seein' you…" He reached out a hand toward her swollen cheek, but she turned her head away. He followed her eyes down to the sound of the cries.

They came from something wrapped in one of Becky's old shawls. He pointed to the bundle. The rancor threatened to reappear in his mouth at any moment.

"What's that you holdin'?" Could it be a cat?

Annalaura turned her face back to him.

"John. There's a way we can make this right."

How could her voice come out so quiet, so settled, so steady, when the whole world had just spun itself down to hell?

"Right? Make it right?" His whole body shook as he pointed a finger at the bundle. "That's a baby. You done birthed another man's bastard chile, that's what you done." His shouts shook his own body.

Waves of heat swarmed over him. Annalaura dropped her head and clutched the bundle all the tighter.

"We can try to make it right." Her words were low like she didn't believe them herself.

"Right don't come easy to yo' mind, do it?" His words burned his own mouth as he spit them out. "You layin' there talkin' to me 'bout makin' it right? Was it 'right' fo' you to lie to me?" He wasn't sure she could hear his scorched words. "Was it 'right' to tell me it was Isaiah Harris who done give you a baby last fall?" Without warning, even to himself, he jabbed a hand at the shawl only to feel the quick stab of cold metal under his armpit, jamming his arm upward and away from the bundle.

Annalaura's body shook even under the blankets. He saw her eyes puddle with tears.

"John, ain't nothin' I can say to make it like it used to be. I can't take back what I done. But fo' the children, we got to try to make it better, not worse." She sucked in her lower lip as she brought the still-covered bundle to her chest.

He could hear the squalls of a newborn.

"Worse? Hell, woman, how could it get mo' worse? You laid with a man who wasn't yo' husband like a common whore. And, now, you done give him a baby." He tried to clear his throat of the snake's venom that threatened his voice, but it was no use. "And it weren't no hired man you laid with, either." He jerked his head toward the covered bundle. "Let me see that."

The blunderbuss raked across his back. Annalaura swallowed hard as she tried to quiet the wailing bundle.

"She ain't to blame. It's all on me. I done it." The tears finally began their trail down his wife's cheeks as she glanced down at the shawl.

"She? You done give that damn white man a girl?" He wanted to double over right then and there, and let the bile finally make its way out of his mouth.

"I knows I'm no mo' good to you. I ain't askin' fo' me or fo' this chile, I'm askin' fo' Cleveland and Doug." She kept her eyes on him like she was willing him to look her in the eye like she was a decent woman again. "I'm askin' that you pay mind to them. You got money now. You take all of it and go back to wherever you was this last year and make a place fo' the boys." She stopped and swallowed.

Annalaura tried more than once to clear her throat. When she finally had sound, John heard the hoarseness in her voice.

"I'll send Cleveland and Doug to you as soon as it gets quiet 'round here." She looked at him like he was supposed to understand her words.

"You damn right I got money now. I earned it all for us, for you and Cleveland and little Henry..." All those long, big city months at Miz Zeola's came flooding back to him. "Right now, I wouldn't give a bent penny to a tart like you."

He had lain with Sally and the colored schoolmarm, but neither of them had meant a thing to him. God knew he had done the deed for Annalaura, and not against her, so why was the Lord visiting this awful punishment upon him? If he'd been wrong to do it, still, wasn't no need for God and Annalaura to get back at him this way. Not this way.

"You damn right I'm gonna take Cleveland and Doug with me, but not befo' my business here is finished."

Annalaura's head started a frantic shake.

"Them boys need you. They needs their daddy." Her voice rose as she pulled the blanket half across her chest.

Though he couldn't see, he knew she was suckling that new thing in the bundle. A white man's bastard with its mouth on his wife's tit. The bile strangled his tongue.

"Take them with you now if you feel like you have to." Her voice trailed away.

She looked up at him. Did he see a plea for mercy in those beautiful eyes?

"Cleveland's of some size and can help. So is Doug, but he still young. I'd like to come visit my boys sometimes if that be all right with you?" Trembling joined the pleading in her voice.

"Come visit? You wanna look in on Doug and Cleveland? Why in hell would I let my boys see a mother who ain't nothin' but a worthless everyday slut? I bet this wasn't the first white man you done slept with either, now was he?"

The blunderbuss almost poked a hole right through his back. Annalaura's lip crumpled, and she looked as though he had slung the rope around her neck himself. He couldn't help it. Somebody caused this mess. And if it wasn't Annalaura—as much as he wanted to believe otherwise, something told him it wasn't all McNaughton—then who was it? John's head felt ready to explode.

"What you think you know 'bout it?" Becky's voice dripped with menace and the blunderbuss held steady.

"Rebecca, I ain't gonna lay a finger on Annalaura, but I can't say as much fo' you." He drew back a hand.

Becky's mouth set as hard as blacksmith iron. The sound shot out of the tousle of covers.

"John. It ain't Becky. It ain't the baby. It's me who done wronged you. You don't never have to look on my face again after this night." She swallowed hard. "Doug's at Hettie's. Becky will fetch him here, and the three of you can leave Lawnover this mornin'."

"Well, I reckon you got it all worked out. And yo' white man will take care of Lottie and Henry as soon as you let him back in yo' bed? He'll buy Lottie some mo' pretty-haired dolls? Maybe a drum for Henry?" He whipped around toward her, crashing his fist down so hard on the table stand that the water pitcher skittered to the floor and shattered into a dozen pieces.

The surprise of his action startled Becky and she almost dropped the old gun.

"I ain't havin' it, woman. That cracker ain't gonna have a chance to get his hands on my children. Not none of 'em." He let his eyes narrow to slits. "I will have all of 'em with me." Cleveland, Doug, Lottie, and Henry. I'll have 'em all."

Annalaura looked worse than when he'd punched her.

"But Lottie and Henry is too young to leave me." Her eyes tried and failed to blink back more tears.

"Ain't no chile too young to leave a ten-cent whore." His words banged around the cabin. He saw Annalaura fall back onto the pillow.

She looked like she was dying. He knew his words slammed hard, crushing every hope, every prayer, and every plea for mercy in its path. He could see the light go out of his wife's eyes, but what she couldn't see was what that sledgehammer was doing to his own heart. Every mean word, every cruel name he let spark out of his mouth and aimed straight at her, did not belong on those brave shoulders. Even a fool, crazy mad like he was, could see that it wasn't Annalaura who deserved the stoning. But he couldn't fix his heart, nor his mouth, to tell her that. None of this was the blame of Annalaura. Even with all of his hurt, he knew that none of the fault lay at her feet. That damn cracker.

"He's right, Aunt Becky," Annalaura's reed-thin voice pronounced. "I'll send all yo' chil'ren with you as soon as you is ready for them."

"I'll be ready as soon as I kill that cracker."

"You ain't killin' nobody." Becky stroked his earlobe with the gun. "You gettin' yo' ass out of Montgomery County as fast as yo' legs can carry you."

He turned around to face her. "You think I'm gonna run? You think I'm gonna let some cracker ruin my wife?"

This was not the time to let the women know that Annalaura was everything to him. She always had been ever since he picked her out from the edges of that crowd of silly women always around him.

"Night riders will kill you dead." Becky's voice was low, soft, and certain.

"Old woman, do you think I is afraid to die?" He couldn't tell Rebecca

Thornton that he was already dead. That he'd died yesterday afternoon when he first saw Annalaura's swollen belly.

"Ain't no white man gonna put his thing in my woman and get away with it even if she ain't nothin' but a worthless slut." He couldn't bear the pain.

What he really wanted more than anything was to wrap his arms around that wounded body, that shattered mind, and tell her that all of her misery, all of her shame, all of her pain belonged on his soul, not hers. He longed to tell her that he loved her far more than anything else in this world. But he couldn't say the words. When he spoke, all that came out were devil thoughts and low-down names that he could never mean against her in a thousand years. John knew the real whore, and it wasn't Annalaura.

"You is a bigger fool than I took you fo'." Becky waved the gun around. "First, you run yo' ass off to God-knows-where for most parts of a year. Then you hauls yo' sorry self back home and bad-mouths Anna-laura when she ain't had no mo' say in what happened than little Henry. Next, you talk 'bout killin' a white man." The gun steadied again and its aim was dead at his heart. "You think you can kill a white man and that's gonna be that? He dead and you dead and you got no mo' worries?"

Although he hadn't seen the telltale wad in her cheek, the old woman spit out a stream of weak tobacco juice.

"Ain't you got sense 'nough to know that it ain't just you who gonna die?"

"Crazy old woman." He muttered in Becky's face. Why didn't she shut up? His heart couldn't take the truth right now.

"They'll string up my Annalaura as sure as you standin' there. They'll say she stole all that stuff offen that white man." Becky waved the gun in the air.

"Woman, you is crazy as a bedbug. Why would they put a hand to Annalaura? She ain't the one gonna kill that cracker. Lawd knows, she sho' should have." He turned a glare on his wife that he did not feel or believe.

"You know as well as me that every white farmer in these parts know that Alexander McNaughton been layin' with my girl. If you kills him, they gonna kill you and her, too, so she won't be 'round to tell the truth of the tale." Becky's words only chunked away a part of the fuzz in his head.

That he had to die was all right with him as long as Alexander McNaughton got dead first. But Annalaura couldn't be part of the killing.

"Hettie ain't dead, and she's laid up with Ben Roy fo' most six years."

"If a black man kill Ben Roy, white men will kill Hettie, all them children, and the nigger who done the first killin', all in one night." Rebecca confused him with her Cherokee reasoning.

"And it won't jest be you and Annalaura. It'll be Cleveland. They'll call him a 'complice. He's twelve, big enough for white mens to hang. My Johnny wasn't but seventeen when they hung him high."

The barrel of the gun suddenly pointed to the floor, and the old woman's arm dropped to her side like soft butter. So this was why Becky was talking crazy. It was all about her Johnny. A wave of relief swept over John.

"Aunt Becky." Annalaura jerked up in the bed, her voice full of alarm. As she sat up, the covering blanket fell away from her breast and the suckling baby.

John turned to look at that face, the color of cream from a fresh-milked cow. A blue eye flitted open and then closed. Ignoring Becky and her dragging shotgun, John walked over to the chair and plopped down in it. His legs turned weaker than water corn bread. It was true. Alexander McNaughton had fathered Annalaura's baby. To see the living proof with his own eyes left him searching for the cabin window to let in more air. He turned to the old woman.

"Rebecca," he kept his voice steady. "I ain't seventeen, and I ain't yo' Johnny."

She turned a rheumy eye toward him. "Seventeen. Weren't but seventeen." She stared at John as though she thought he should have been there on that long-ago night. That he should have been the one to pull Johnny out of harm's way all by himself.

"They killed yo' boy over nothin'. They'll kill me over somethin'." John kept his eyes on the Cherokee.

"Old Ben Thornton said he could have that hoss." Rebecca lifted her head toward Annalaura. "Old Ben was my Johnny's daddy, you know."

"Becky, I've known that story since before I married Annalaura… yo' niece."

"Like that one." Rebecca ignored him as she pointed to the baby in his wife's arms. "My boy had the light skin like that one, and that ain't never set well with Ben Roy."

"Aunt Becky, fetch me a quilt, I'm cold." Fear leaped out of Annalaura's voice.

"Old Ben said he wanted our Johnny to have that hoss." Rebecca looked at an empty space between herself and the front door. "Was the only decent thing he ever did fo' me. I think he done it to spite Charity. He had her first, you know. He took my momma, and when he got tired of her, he lay on top of me even when he knew I was his own flesh and blood. That's a God's sin, you know." Becky nodded to the vacant space.

"Auntie," Annalaura called out.

"It was a big roan and the best thing on this here farm." The woman ignored the girl she had raised. "I was there when Old Ben said Johnny could have it, but his oldest boy, Ben Roy, took a jealous streak."

John struggled to his feet. Becky, lost in her world, ignored him.

"Ben Roy always hated Johnny 'cause Old Ben took a shine to the boy. To vex Ben Roy, he would give Johnny little things like a toy that Ben Roy wanted first. A pair of shiny high-button boots fo' Johnny, even if they wasn't new, whilst Ben Roy got new, but they was only work shoes. Then, he sold my boy the best, that big, pretty roan—worth 'most five hundred dollars. He give it to him fo' one silver dollar. And he give him the God's truth legal sale papers." There was pride in Rebecca's voice. She took a step toward John, the barrel of the blunderbuss bumping along as she crossed the cabin floor.

"Aunt Becky, my quilt, please. The baby's gettin' cold," Annalaura pleaded.

"When Old Ben took sick and died, Ben Roy told everybody that Johnny had come by the hoss illegal. Ben Roy riled up them night riders, and they came fo' my boy. Johnny tried to tell 'em he had the papers, but Ben Roy tole them riders not to listen, that my Johnny was a liar." The old neck swiveled from the empty space to her kitchen safe, slow like a gate on a rusty hinge. She looked at the floor beneath the bottom drawer.

"I had them papers hid all along. Right there in the false bottom of that drawer." The eyes blinked. "But I ain't had no time to get them papers to the night riders. They killed my boy befo' they had a chance to see that everthin' was all legal. It was Ben Roy, Johnny's own half brother, who threw that rope over the tree, and he knew all the time that his daddy done give Johnny them papers." The gun thudded to the floor.

John rushed to her side and guided Becky to the rickety chair.

He tried to give her a sip of her medicinal whiskey when he heard Annalaura shudder. He couldn't bear to look upon his wife's face as he walked across the floor, stooped, retrieved his pistol, and headed for the door.

"Have my children ready to go as soon as I tell you." At the doorway, his back to her, he paused and called over his shoulder. "I may not kill that cracker tonight. Maybe not even tomorrow, but I'm gonna get black man's justice offen Alexander McNaughton. You just worry on it." He let the door slam behind him.

Mounting his horse, he nudged the animal in the side and let him have his head. He knew what he had to do, but he hadn't yet sorted out the when and the how of it. The sun was well placed in the sky, and it was halfway to noon already. When the horse reached the lane, the animal slowed its pace, waiting for the tug on the reins. Right would take him across Ben Roy's acres and onto the roundabout back road to Lawnover. Left would take him back to the mid-forty and right past Alexander McNaughton's house to the direct road to Lawnover. Whichever way he went, some farmer would see him. And no colored man not working on some white farmer's land had any business astride a horse riding up and down country lanes during prime work hours. He jerked the reins to the left and let the animal find his own pace.

As the horse clopped closer to the McNaughton place, the nightmare vision that hung in his head of another man touching Annalaura churned at his stomach. When he blinked his eyes to make the spectacle go away, he could see the leaves on the trees move slightly. He supposed the day had brought a slight breeze with it, but his outside body had gone numb. There was no feel of the warming sun on his arms, or the touch of the light breeze brushing his cheek.

John neared the McNaughton back-forty, where a little stand of cherry trees stood twenty yards distant. The buds on the limbs were full, but they hadn't yet blossomed. Normally, he could taste their promise, even this early in the season, but now, all that was on his mind, all that he could see, was Alexander McNaughton putting the sweet honeyed lips of his Annalaura into his own white mouth. And worse.

He couldn't recall how he got to the ground, nor when the horse stopped near the little stand of cherry trees. He only knew that his knees barely made it to the grass when all the bile stored up over the past twenty-four hours spilled out over the fresh green carpet. He couldn't stop. His woman had been dirtied, defiled, and his world told him he had to take it like it never happened. To hear Becky tell it, half the colored population of Lawnover would be murdered before tomorrow sunrise if he touched one hair on McNaughton's pale head.

Cleveland, and even ten-year-old Doug, might swing from a tree. Little Henry would wind up in some workhouse, while Lottie would become another white man's plaything before her seventh birthday, all because he dared do what any husband on this earth had a right to do.

Becky and Annalaura would have him believe there was justice in running away. Justice and honor in swallowing his words and tipping his cap to every white skin, if that's what it took to keep his children safe. Didn't those women know there were worse things than not being safe? The bile grabbed him, and he retched to get it out. All of it had to be gone to cleanse his own soul before he died of the pain.

Ten minutes, maybe twenty, passed before he had the strength to try to stand. He reached for a branch to help pull himself to his feet, but its suppleness reminded him of Annalaura. In time, the cherry buds would deliver rich, ripe, dark fruit. But time would never deliver Annalaura to him the way she was. How could he ever mount her again when all he would ever see and smell when he came near her from this day on would be Alexander McNaughton? He rolled to his side, pulled his knees to his chest and squeezed his eyes so tight that the pain came to his head.

"Forgive me, Lord Jesus. I can't let this go, though I knows a good man should. It's best fo' me to keep on livin' so I can take care of my children. I knows it's best to stay breathin' so I can look after Annalaura though she can't never know it." At the sound of his voice, the horse quieted. "But, Lord, I is weak. A better man than me would say nothin', do nothin' that would bring harm to his family. But I ain't got that kind of strength in me, Lord. I can't stand by and let another man soil my wife." His throat felt raw, scratchy, from the retching and the shouting to the sky. "I know that because I can't turn it loose, more'n me will die. Lord, I prays you will show me a way to make it all go quick for them." The tears broke through like a spring river flowing over an earthen dam. "I don't care nothin' 'bout myself. Ain't no torture night riders can put on me worse than this hell I'm already in. To spare them the fright, if it be Yo' will, Lord, I will shoot Annalaura and all of my children in the head befo' the night riders get to them to spare them the fright. I just prays fo' you to show me the way."

John didn't know how long he lay there, but when he finally came to himself, he saw the sun filtering through the tree branches. It was directly overhead. As slow as though he carried the chains of his slave ancestors, he moved to his feet. He knew what he had to do and when. He was finally ready.

# CHAPTER TWENTY-SIX

Alex dug his heels into the gray. Only when the horse failed to quicken his pace did he come back from his daydream. Out of habit, he looked at the sky. The sun told him he had one more hour of daylight. No wonder the animal could not give him what he'd asked. Since dawn's light, when he left Rebecca's cabin, and Laura, he'd ridden the animal hard. Alex eased up on the reins. He patted the horse's neck. There was no rush to hurry home. Nothing there but Eula Mae. And the cradle, of course. That rush of pleasure that thundered up from his gut when he least expected it washed over him again. Ever since she, his Laura, had agreed to move in with him, Alex had struggled not to let his excitement show in front of the other farmers. That's why he'd ridden the gray all the way to Clarksville. To Clarksville, where he could buy clothes and toys for his baby, and even pretties for his Laura. All without too many prying eyes. His smile broadened. It felt good to his soul to realize that he had just become a father. And, better yet, his baby and her mother would soon be under his roof. Now, away from peeping eyes, it was all right to grin all the way home. He felt like he had just brought in the biggest cash crop in the whole of Tennessee.

&#x269C;

John mounted his rented horse, his stomach still cramping from all the retching he'd done. He aimed the animal up the road. As he settled himself on the broke-down horse he'd paid a pretty penny just to rent, he shook his head. What foolishness was he thinking? What did money mean to him or to any of them now? He had to do what must be done. He slid his hand into his overalls pocket—his old work clothes. He'd dug them out special this morning. It wouldn't do for a Lawnover colored man to be seen wearing city slicker clothes. Not this day. John reached a hand inside the pocket and pulled out the pistol. He held it by the hard metal butt. He'd wanted a fancier revolver, of course, but at the time he bought it, there had been no money for such frippery. Back then, in his early days in Nashville, he'd scraped together every penny just to get by. Twenty acres. That's what he'd wanted. Twenty acres to make a stab at earning a decent living for his children and . . . Annalaura. The gall clambered to his throat again, and he pulled up on the horse. He looked at the sun as he dismounted. There was still time. He just had to sort out the order. With planting season just starting, McNaughton wouldn't be getting home 'til dusk. And here it was, a good hour left of daylight.

⸺❧⸺

The gray bobbed his head like he was saying his thanks to Alex for slowing the pace. Alex rubbed the horse's flank. No harm in letting the animal take a break. He spotted the old oak that told him home was just two miles away. He led the animal off the road. There was a tiny spring just a quarter mile distant. Let the horse have a drink. Besides, he supposed he could use the time to figure out the sleeping accommodations once he got Laura to his house. There was that spare room, upstairs next to where he and Eula slept. His wife used that one as her sewing place. No problem there. He'd just find another spot for Eula's sewing stuff. Maybe outside in the storehouse. Alex trotted by the oak tree and turned the animal's head to the left. His mind drifted back to the sleeping arrangements. Having two women with beds on the same floor just might not work. He sensed that Laura's passion for him was growing. It wouldn't be too much longer before she screamed out her nighttime feelings for him. Suppose her moans woke Eula Mae? No, he'd have to come up with something different. He spotted the little stream, dismounted, and led the horse to water's edge. As the stallion drank his fill, Alex reached into the saddlebag. He fumbled underneath the wrapped doll he'd just bought for his new daughter and pulled out the pistol he always carried to Clarksville. Laura had asked him to let her handle things. Two days, she said. Two days and she'd make sure the husband wouldn't bother them anymore. One week and he'd be out of Lawnover for good. Alex snapped open the bullet chamber. One. Two. Three, four, five, six. Fully loaded. "Just in case." He spoke out loud with only a horse to hear him.

⸙

John leaned against the tree trunk, mindful of the time. The sun said about forty-five minutes 'til setting. His pocket watch told him it was half past six. There was enough time to sort it all out, but when he started to put his mind to it, his belly told him no. He rechecked the bullets in his pistol. Yep. Six of them. One for McNaughton, that was for damn sure. One for Annala...His head started its throbbing again, and his eyes their watering. He knew what had to be done. The Good Lord knew what was right. No man could be expected to have his wife dirtied the way McNaughton had ruined his Annalaura. It would pleasure John no end to put a bullet straight between McNaughton's eyes. Yes, he wanted the man to look him full in the face when he pulled the trigger—so the farmer could see just who was delivering what he had coming. But when it came to the part about who he had to shoot next, John couldn't order it out in his head. Another mile and he'd be at the turnoff to the barn where his family lived. Cleveland would still be there. His boy was only twelve. He'd acted like a man, and for that John was proud, but in God's truth, the boy was not yet full grown. Should he shoot Cleveland after McNaughton? Or before? The cramp rolled up so fast from his belly that it forced John to his knees, and four minutes of dry heaving. When the knives in his gut finally quit their chopping, he took out his handkerchief and wiped his face. "Let me just lean here for a minute." He guessed he was asking permission of the Lord. The Lord's answer must have been yes, because he still had thirty minutes before he had to lay in wait in the shadows of McNaughton's barn.

⸙

The gray whinnied his satisfaction, but Alex was in no hurry to remount. He was still mulling over who was going to sleep where. Eula's pantry might make a tidy little room for Laura and his baby. "My baby. My very own little girl." Alex rummaged the saddlebag again and pulled out the parcel the shopkeeper had wrapped. He lowered himself under the tree and undid the string. He looked at the porcelain-faced doll he'd bought for his daughter. Daughter. Every time the word entered his head, he felt like pinching himself. Yes, he was going to have his child and Laura with him tomorrow. He closed his eyes and let his mind drift into that place of happiness that only came when he thought of her—Laura. "One more week and it's all over." The words whispered out of his mouth so soft, even the horse couldn't have heard. "One more..." The caw of a hawk jolted him out of his dreaminess. One more week was a hell of a long time to wait for John Welles to get his ass out of town. Alex felt around the ground for the pistol he'd laid under the tree. Laura had pleaded so. Yes, she was certain she could get John Welles gone from Lawnover. Alex picked up his pistol and stroked the barrel. "That would surely be best." His words were loud enough for horse or man to hear if they cared to listen. "That would keep Ben Roy, and all those other afraid-to-spit farmers, quiet." He tucked the gun into the back of his britches' waistband. "But if that nigger lays another hand to Laura, he's dead, planting season or no." Alex placed the doll back into her wrapping paper. It had been a hard night and a long day. "Twenty minutes 'til dark. No need to face Eula tonight. Tomorrow will do. I'll just take myself a little nap."

~~⁂~~

The rented horse was not the swiftest thing on four legs, John declared. No matter, there was McNaughton's barn just up ahead. There were still ten more minutes 'til nightfall. John slowed the horse, raised a hand to his eyes and scanned the fields on either side of the road. No sign of McNaughton. Good. He reined in the horse to a slow trot as the side path off the main road came into view. Nobody else on the road. McNaughton wouldn't be home before dark. John's only worry was Miz McNaughton. At dusk, she would be heading to bed. It wasn't like she was a colored woman who had to tend the fields from sunup to sundown, then fetch and carry for her own family 'til the moon was good and set in the sky. McNaughton's wife was a white woman, and every white, churchgoing farmer's wife took their night's rest right after sunset. The last red-orange glow of the sun had just sunk below the horizon. There it was. The path to the McNaughton house. John slipped off the horse and tied the animal to the sycamore tree. If Miz McNaughton happened to be about, it would not do for her to see a colored man riding to her back door on a horse. John walked slowly up the path toward the red barn, his head down the way a respectful colored man should hold it. He sidled his eyes toward the main house. A lone lamp shone from one bottom floor window. Upstairs, another lamp lit a window. Good. Miz McNaughton was readying herself for bed. The downstairs light meant McNaughton hadn't made it back yet. John moved to the far side of the barn. It was a risk. The pistol shot was sure to wake the wife. She'd most likely look out the window, but darkness would cover him. He would have to scoot around the back of the barn and get out of the woman's sight if she took a notion to come outside for a look around. If he made a clean shot, McNaughton ought to drop without a sound. That way, maybe the missus would let everything be 'til morning. In the twilight, he spotted Miz McNaughton's buckboard at the far side of the barn. He eased toward it and squatted behind one of the big-rimmed wheels. He was good and out of sight now to anybody coming up the night-dark path toward him. John pulled out his pistol, scooped up a handful of dirt and rubbed it across the barrel just in case the moonlight reflected off the gun barrel. He was giving McNaughton no warning.

Something furry scurried across Alex's foot. A squirrel? A chipmunk? Whatever it was took him out of his dream about Laura. He ran a hand over his eyes and looked around for the horse. Night had fallen hard, but he spotted the animal where he'd left him, loosely tethered to a branch near the stream. Alex gathered his belongings and returned them to the saddlebag. He reached behind him. The gun was still there. Nice and snug. Not that he expected any trouble. But it was getting harder to keep his promise to Laura.

He mounted the horse. Hell. He had only one more day, and she'd be there in his kitchen. In the pantry he'd fix up for her. He'd have to get a good-size bed, of course. No need for the two of them to be cramped like back at the barn. He felt a warming in his pants as he headed the gray back onto the path to home. The gentle rocking of the animal as it trotted up the road to his house jostled the gun into his back with each step. He laid a hand on the handle and readjusted the revolver. "Just in case," he muttered.

⁂

Full dark had fallen and tonight's moon was not quite a quarter in the sky. The upstairs light had gone out. John's leg cramped from staying too long in his crouch. He sat long-legged to give his muscles a rest. He kept his eyes on the road.

⁂

The clop-clop of the horse did little to help Alex keep awake. That nap had not really done the job. But no matter. There was his house. He couldn't make out the color in this sliver of a moon, but he knew every inch of the silhouette. Clop-clop, the gray approached the path leading to his property. "Damn, if this horse don't know the way home better than I do." Clop-clop. One more day and he'd have his Laurie in his arms.

⁂

The cramp eased in John's leg. He shifted to his knees and crouched behind the back wheel of the buggy. As he settled himself, his shoulder brushed the spoke and sent up a little creak. "Damn thing needs oiling. Can't McNaughton do nothing right?" John mumbled. He peered up the lane for what seemed the one hundredth time. Nothing except the occasional rabbit. He clicked open the gun. Six bullets. The first for McNaughton. The second for...for... No. It couldn't be Cleveland. Maybe, he'd go after Annalaura next. Bullet number two for her. But what about the baby? Sure, it looked as white as snow, but could he really put a bullet into a baby? The gall swarmed up again. Aunt Becky had told him to swallow seven times quick when that happened. It hadn't worked at all today, but now was not the time to start retching again. He swallowed four times before he let loose with one dry heave. He snapped the gun closed. Snap.

Clop. Clop. The gray made the turn with very little direction from Alex. "Where the hell am I going to sleep tonight?" Alex asked out loud. Not with Eula, for sure. He'd make himself a pallet in the pantry. He smiled. Tonight was as good a time as any to get used to sleeping in the little room. Clop. Clop.

The sound came at him like a heavy hailstorm. Clop. Clop. A horse. McNaughton's horse. John leaned into the buckboard wheel. Creak. In the darkness, he could see no figure, but sure as hell, that was McNaughton. As his eyes strained at the pathway to the house, his brain told him to sort out the order. Quick. First, McNaughton. Second, Annalaura. Third. Who? Which of his children would he have to kill next? Not Cleveland. Not that soon. Doug? Oh Lord, no. That was his smartest child. He'd make something of himself one day. Then it had to be Lottie. Lottie? His only girl? Sweat poured off John. His hand felt slippery against the butt of the gun. Clop. Clop. Good God. There he was. A horse and rider. Decide. Decide now. John leaned into the buckboard wheel. Creak.

What was that? Alex pulled up on the red reins. Sounded just like that creaky wheel on Eula's buggy. What was she doing out here at this time of night? Clop. He slowed the horse 'til the gray barely moved. Alex pulled out his gun.

Sweat swarmed down John's forehead and into his eyes. He had McNaughton in his sights. His finger stroked the trigger. One little squeeze. That's all there was to it. One little squeeze, and good-bye to a no-good, rapin' white man. She rose up out of nowhere, or from that place Becky called "beyond the pale." Annalaura's face with a body that looked made of rippling water popped into his head and would not move away. Worse. The figure blocked his view of McNaughton. John's gun hand shook as he stared at the thing Becky called a 'Parition. Some Cherokee women possessed it—a force as strong as fury that took hold of them and pushed their will into another person's head. Once it was in there, no power on God's earth could make it leave until the thing got good and ready. Even so, John jerked his head hard. Anything to make the Annalaura 'Parition go away. The vision that chilled to the soul wouldn't budge. John took his gun hand and swiped at the image clouding his eyes. No good. He leaned into the wheel to look around the thing. Creak. There was just no getting away from Annalaura.

⚬

Creak. Alex pulled up on the horse. He leaned forward and rested the side of his face against the gray's neck. Something was out there. Could be a coon. Could be a possum. Best not to take any chances. He let up on the reins, and the horse ambled toward the barn.

⚬

Annalaura's mouth moved, but she still shrouded John's target. Then, like a hammer dropping down on a nail, the 'Parition called out the names of their children. Cleveland. Doug. Lottie. Henry. Oh, Lord. He'd forgotten all about little Henry. He'd have to kill him too. The 'Parition moved closer, and with a force stronger than he'd ever known the flesh-and-blood Annalaura to show, the thing that looked like the woman he loved pushed him back from the wheel. Whatever it was dropped to the ground and got hold of his feet. It pushed them backward and away from the wagon wheel. To keep from falling, John reached out a hand to steady himself against the side of the barn. Two more steps and he'd be behind it, out of sight of anyone coming to inspect the buckboard. He dug in his heels, flailing out. The 'Parition grabbed his gun hand and jammed it against his chest. A force stronger than a mule pushed him behind the barn, and out of sight of curious eyes.

⚭

The gray halted at the barn door. Alex slid off. He peered over at the buckboard as he flattened himself against the barn wall. He sidestepped toward Eula's buggy. The black silhouette, with its high seat, stood undisturbed. Alex stared inside the open contraption. Empty. He let his eyes travel to the undercarriage. Nothing. With the gun cocked and pointed, he checked the front wheels. No change. He slipped along the side of the buckboard toward the back wheels. Snap. His shoulders tensed and his breath stuck in his chest. Gurgling for air, he raised his gun arm, sighted as best he could, and aimed his pistol into the darkness. A raccoon ran from the back of the barn and straight into the wheel of Eula's buggy. Creak. The stunned animal staggered away. Alex sucked in a mouthful of air. "Hell, ain't nothin' but a coon." He laughed out on a burst of air as he lowered his gun arm. Alex shrugged that burst of fear off his shoulders as he moved to the front of the barn. "All right, horse. Let's get you settled in for the night."

⚭

John's breath came in spurts. McNaughton ought to be dead. But the way Annalaura had lied to protect the man...All that new stuff McNaughton had given her? White men didn't give black women fancy presents unless they were sweet on them. More gall coated his throat. He did Rebecca's seven-swallow routine in quick succession. Was that the truth of it? That fool white man had real feelings for Annalaura? He leaned against the back of the barn, both hands clutching his stomach. John's chest burned, and his head felt like it was going to cave in on itself. Annalaura, Henry, Lottie. Cleveland. Doug. Even McNaughton. They all did a whirly-twirly dance before his eyes, each one reaching out a hand to squeeze his heart. That farmer ought to be for sure dead, but McNaughton was nowhere near good enough to take the lives of John's four children to the grave with him. His breathing slowed, and his head quit throbbing. The

Annalaura 'Parition began a slow fade into the blackness of the night. If white men found out McNaughton had honest-to-God feelings for a colored woman and was careless enough to let them show, they'd kill him themselves. One thing to use her, another thing to love her. The sweat began to dry on John's forehead. His lips forced themselves into a crooked smile. "This white man ain't worth it. Not if the price is my four babies. Not even An..." Leave his killing to those of his own kind.

⚭

With the gray settled for the night, Alex walked through the porch door, a half smile twitching across his lips. He'd pull out that old cradle and get it set up for his new baby. His own Dolly. Tomorrow he'd have both mother and child with him. The porch door snapped closed behind him.

⚭

John waited fifteen minutes by his pocket watch. He never did see a lamp relit upstairs. But at least two more lamps blazed on the first floor. Whatever that white man was up to would do him no good come morning. McNaughton could live. For now. But he'd never have Annalaura. John bent double and made his way back to the main road and the rental horse.

# CHAPTER TWENTY-SEVEN

That first pale light of dawn filtered under Eula's eyelids and gently brought her out of sleep. But before she let the day come into focus, she stretched out an arm to search for that unfamiliar tingling "down there." It had been with her through all of yesterday with its heart-pumping exhilaration. She wanted to revel in that strange pleasure as long as she could. Disappointed that it was fading, she re-created the memory of that amazing night in her mind. Even though Alex hadn't been beside her in the morning when she rolled over to greet the new day, she had been filled with such pleasure and awe that she paid little mind that he hadn't shown up back at the house 'til close to suppertime.

He had gone to Clarksville, he announced upon his return, and brought back a package wrapped in brown butcher paper. She didn't believe she'd shown him impatience when he said she'd have to wait until tomorrow for its opening, but everything she thought she knew about her husband had turned inside out. The man who had been as predictable as the planting and prayer dinner had become a man of surprises.

Eula let her eyes drift to the first light at the window as she remembered her excited anticipation of last night. That she had to wait until morning to open a package was of small matter. Her husband had gotten her a gift, and it wasn't even her birthday. But it wasn't a butcher-paper-wrapped surprise that she'd wanted from Alex when she climbed into bed beside him last night.

First, she had debated if she should take off her own clothes or let him do it. When she saw him with his eyes closed and the kerosene lamp dampened, she shakily stood beside the shut chifforobe and, in full view in the dimly lit room, shed her corset and donned her summer gown. Alex never opened his eyes that she could tell, and by the time she crawled next to him, she was certain he was asleep. Perplexed, she finally eased off to dream, wondering what he would expect when he came after her in the morning. Whatever it was, she was determined to give it to him.

When she finally let this morning float into her mind, she quietly turned her head to the pillow beside her. Like yesterday, Alex's side of the bed was empty. A quick frown furrowed her brow, but she banished it just as fast as it came. She climbed out of bed and hurried to dress. She had to ready breakfast for her husband, and she had to be ready for whatever else Alex might want this morning. Eula giggled. Not since their first year together had he ever wanted her in the morning. She said a quick prayer for forgiveness at the sinful thought as she hurried to the kitchen.

Her mind lingered on the faded warmth of two nights past as she swung into her kitchen to find Alex sitting in a chair, a dusting rag in his hand, his head bent down over something sitting on the floor. Her gasp brought Alex's head up sharply from the object he was polishing. It was the smile on his face and the pure delight shining from his eyes that first made her feel like she'd fallen through February ice. She didn't know how long she stood there staring at her husband and the baby rocker. But she supposed it was long enough for him to speak first.

"Look what else I've got." He pointed to the butcher-wrapped package he'd brought in yesterday, completely ignoring the pain that spilled out all over her at the sight of the twenty-year-old cradle.

Alex had cut down the trees and sawn and hammered the cedar and elm pieces into place with his own hands. It had taken him over three months to ready it for their child. With their own precious baby so long gone, how could he bear to touch that wood again? Before she could nod her head nay or yea, Alex tore into the paper wrappings. Inside, the painted porcelain face of a child's doll emerged, its blonde hair made of bleached horsetail. As the paper came off, arms, legs, and a body stuffed of excelsior stared up at her.

"Alex, it's a baby doll?" Her question matched the look on her face. One heavy arm pointed to the toy and then swung to the cradle Alex stroked with such tenderness. What had brought his thoughts to their dead daughter after all these years? "What am I to do with a baby doll?"

"I'm gonna set you down, Eula Mae." Alex walked over to her and slipped both hands to her shoulders.

When he looked at her, Eula could have sworn that she saw real caring in those blue eyes. Was this the way of men after twenty-one years of marriage? Did they blow hot one minute and then act like you mattered no more than a gnat the next? Before she could blink again, he pulled her close, hugged her to his chest, and propelled her toward the kitchen chair he always took.

"I couldn't have done no better." He brushed her forehead with his lips.

"I got to get your breakfast on." It was the only thing that came to mind.

"That's just it, Eula. You been cookin' and cleanin' this place since the day I brought you here. Brought you to a po' piece of land." He took her big work-reddened hands in his.

Of all the things that she hated about her large, awkward body, her hands came high on the list. She felt Alex's strong fingers stroke them like they were made of the most delicate Nashville silk. He brought them to his chest.

"You're a Thornton girl, you deserve better."

"I been a McNaughton woman for twenty-one years." What was he saying to her?

"You're the best wife I could have picked. You keep this place spotless and your cookin' is more than tolerable." He lowered her hands to his knees. "I ain't never heard a mean, nor a nay-sayin', word out of you."

"Work ain't nothin' to me, Alex. I'm the oldest girl with four brothers." She stumbled out the words. "Of course, I can work."

"You're a Thornton woman. Fedora has lots of help. Even Wiley George got Hettie for Tillie. It ain't right that you don't have help after all these years." He bobbed his head.

"Fedora needs all the help she can get, and Hettie's wet-nursin' Tillie's baby until my niece can get the hang of it." In her confusion, she put a half smile on her face. "Wiley George can't afford to keep a real hired girl."

"I wouldn't be much of a man if I couldn't give you what you should have, now would I?"

"That's not the way of it," Eula managed.

"I'm gonna set you down."

"Alex, I'm not near 'bout worn out yet." Was that what he thought of her after their astonishing night? That she was too old and too broken down to do what a good wife ought? That she needed help?

"I ain't gonna let you get worn out, Eula. I'm gonna get you a hired girl." Alex couldn't have sounded more jubilant.

"A hired girl?" She couldn't remember who let go first, but her rough hands were suddenly free. "I don't need no hired girl."

Alex moved back to the rocker. "Of course you do. She'll do a lot of the heavy work around here. She's a strong woman." He laid a hand on the cradle and set it into gentle motion.

"Nooo." Eula felt her stomach sway in time to the old baby crib. He knew how she hated to look upon it.

"Best part about the hired woman is that she's comin' with company." He ignored her cry as he gave the cradle another gentle push. He laid the baby doll in it. "I know it's been hard on you all these years since you lost the baby."

"Please, Alex, put it back," she whispered as she turned her eyes away.

"It took me hard, too . . . the loss . . . God's will, I said. Best not to dwell on it." He left the cradle and walked back to her. "But that ain't helped much. I still felt mostly like I was missin' something. A baby will put life back into the both of us." He stood there looking down on her as though every word he said didn't feel like he was slicing the skin right off her bones.

"I can't talk about no baby." The tears were on the verge of falling, but Eula knew her husband would never tolerate them. She blinked them back.

"The hired woman will be here in a few days and she's bringin' a baby with her." He grinned at her.

Eula swallowed her spit so many times she was sure it would dry up in her throat before she could get the words out.

"Baby? What baby?" Missing preserves, disappeared hams, lost nights, her silver-trimmed wedding bowl, all popped out of their hiding places and marched before her eyes. She knew the answer.

"Looks like I've got to let out the mid-forty to some new tenants." His eyes didn't meet hers.

"The mid-forty?" Her voice croaked.

"Yeah. The new hired man's family is movin' in. The other one...the woman on the mid-forty...she'll make a very good hired girl for this place." His eyes slid back to the cradle.

"The...the woman on the mid-forty..." The words coming out of her mouth belonged to another voice. "Her name. What is the name?" If Alex could speak the syllables, could she bear to hear the sound? She glanced out the kitchen window.

The sun, the new leaves on the trees, the blue on the jaybirds, all looked painted, unreal, frozen into their places like on a play-acting stage. Only her breath, trying to fight its way out of her tight chest, made any noise in the room. Hours passed as Eula watched Alex's mouth set itself to form the words.

"It's Laurie...ah...Annalaura. I mean Welles...Annalaura Welles." Alex's face showed that he had no idea that the stops and starts of his voice made her ears hurt.

"Laurie..." She let her mouth form the noises into order and push them back out into the dull, dead world that was now hers.

She heard the pantry clock tick five times. Each click felt like a knife cutting the inside of her throat. Her one night of perfect, exquisite pleasure banged against those sounds coming out of her husband's mouth. The only laurel wreath Alex had called out in ecstasy that night belonged to Annalaura...Laura Welles.

"The ba...baby?" She coughed each syllable through the heavy grit of sandpaper.

Alex finally turned his eyes back to her. "Dolly. Her name is Dolly." He said it like he was praying soft in church. He said it like it was the most precious name ever to fall off the tongue of humankind in the history of the world.

Only the sudden knock on the porch door at ten o'clock in the morning stopped her body from sliding from the kitchen chair to the floor.

"Mornin'." Ben Roy stepped through the back door. Eula barely noticed her brother.

"Mmm." Alex's curt response glanced off her ears without leaving a dent.

She paid no mind when Ben Roy walked right past her husband and stared at the black cast-iron skillet she had set out last night. Only the rustle of a go-to-town skirt hovering near her told her that Fedora must have followed Ben Roy through the open door. Why these two were paying the McNaughtons a morning visit took second place to Eula trying to stop her kitchen from spinning. She put a hand to her head and leaned heavily on the table. She felt Fedora lay a hand on her shoulder. She had no strength left to even wonder at the why of it from a sister-in-law who had never expressed any fondness for her.

"Took Fedora over to Bobby Lee's this mornin'." Ben Roy kept staring at her skillet.

She thought she had cleaned it good last night, but for the first time in her life, she didn't care. Fedora's hand dug into her shoulder. Even though she sensed anger coming out of Alex, she didn't have the will to look at him.

"Uh huh." There was an unmistakable eagerness in Alex's voice to get his brother-in-law out of the house.

Ben Roy still stared at her skillet like it might throw itself off the stove if he didn't keep a close eye on it.

"Talked to Bobby Lee this mornin'." Ben Roy kept his eyes gripped on the skillet.

"What can I help you with, Ben Roy?" Alex sounded impatient.

Ben Roy turned halfway around to face her husband.

"Bobby Lee knows of some niggers that can help you out."

Fedora's fingers dug deep into Eula's arm. She lifted her head toward the two men standing on her kitchen floor.

"I already got enough hired hands to work my acres." Now Alex sounded annoyed as well as impatient.

"You gonna need mo' help on the mid-forty." Ben Roy puffed out his cheeks.

"Don't you worry none, the mid-forty will be all right." Alex stared down Ben Roy.

Fedora's fingers dug in so hard that Eula squirmed away from the pain.

"Bobby Lee was up in Clarksville way early this mornin'. He saw that nigger, John Welles, pull out on the eight ten train." Ben Roy slipped a hand into a pant pocket.

"What?" Alex drew the word out long, but at its end, Eula was sure she heard pleasure and relief.

Ben Roy gave a fast and strong shake of his head after shooting a quick glance toward Eula.

"Welles is gone? He left town?" A smile started across her husband's face.

"Took the whole damn family with him."

"What?" Alex shook his head, his eyes blazing. "What you say?"

"Even his wife." Fedora aimed each word at Alex.

"Took every damn one of 'em." Ben Roy spat out an imaginary stream of tobacco juice on her kitchen floor. "Took his fo' kids and even that old woman—Rebecca."

Eula's eyes moved to Alex. Nothing on her husband moved. Even his eyes had gone numb.

"Nigger must have stole him some money 'cause he bought one-way tickets fo' all eight of 'em," Ben Roy added.

"Eight of them," Fedora repeated.

"Welles, his fo' kids, Becky, the woman, and..." Ben Roy stopped.

"Bobby Lee's wife said the woman just had a baby two days ago. Colored," Fedora clucked. "Don't even allow their women proper lyin' in time, but then, I don't reckon a colored woman needs much of that anyway. Birthin' comes easy to them."

Ben Roy turned toward Eula. "I reckon Welles wanted to get his woman and the new baby away from that man who was botherin' her...that hired hand from last harvest. What was his name, Fedora?" Ben Roy kept his eyes on Alex.

"Harris somethin' or other." Fedora's hand stroked Eula's shoulder.

Eula stared at the floor and wondered why it was fast rising to meet her face. Ben Roy's arms wrapped around her. Her brother sat her back on the chair. Fedora laid a wet cloth across her forehead.

"It's a damn lie." Alex exploded across the kitchen.

Eula's closed eyes flew open as the tornado that was her husband brushed roughly past Ben Roy and stormed onto the porch. Through the open door, she heard kettles, supply barrels, water pails, metal-tipped leather straps, and nails, all crashing and clanging together. Alex thundered back into the kitchen, the shotgun in his hands.

"You comin' or not?" He looked like a man possessed as he shouted at Ben Roy.

"Comin'? Where the hell you think we goin'?" Pushing Fedora almost onto Eula's lap, Ben Roy rushed up to Alex.

"For God's sake. To catch that damn train, of course." Alex turned toward the back door, but Ben Roy wrapped an arm around his shoulder.

"You ain't doin' no such a thing. Hell, man, that train done left Clarksville two hours ago. It's clean to Kentucky by now. You can't catch no train." Ben Roy took a quick look at the women, jerked his head toward Fedora, and slammed the porch door.

"Let's get you to the parlor and onto yo' settee," Fedora soothed. "You can rest better there. You been workin' too hard, Eula Mae." Fedora kept her eyes on the shut door.

"If you ain't goin', then get out of my damn way," Alex's voice shouted through the closed door. "I'm after me a train."

"And what the hell you gonna do if you catch that train?" Ben Roy yelled at Alex.

"The same damn thing you and Wiley George should have helped me with two nights ago," Alex roared. "Run that nigger out of town."

"He is outta town, damn it. He left this mornin'. Ain't that what you wanted? Bobby Lee says the tickets was one way fo' Chicago. They ain't comin' back. Not none of 'em." Ben Roy's barked words were lost in the moan of pain Eula heard erupt from Alex's throat.

She felt Fedora try to pick her up bodily from the chair. Eula clung to the edge of the table.

"That's a damn lie. He's takin' her against her will. I'm gettin' her off that train. She promised me…" Alex's shouts could be heard straight to Lawnover, Eula was certain.

"She's the man's wife. She can't make you no damn promises," Ben Roy shouted right back at her husband. "Alex, it's over."

"I can't. I can't do it. Don't you know that I can't let her go? I love h…" The sound of fist against flesh and bone rocketed through the porch door.

Eula felt Fedora lurch at the commotion.

"Are you fo' sure crazy? You can love her all you want, you just can't say it out loud, and you sure as hell can't keep her." The sounds of a scuffle bumped out of the porch as Ben Roy shouted.

"I'm gonna bring her back. Her and my baby. It's me she wants, not that n…" Bone against muscle shook the room.

From the crashes on her porch, Eula was certain that not even one wall would be left standing.

"You ain't doin' no such a thing, and you ain't sayin' them words to another livin' soul. Talk like that will get us all killed," Ben Roy panted.

"I don't give a damn." The sounds of wrestling on the floor blasted into the kitchen.

"You'd better give a lot of damns, 'cause ain't no power on earth gonna let you love a colored woman and live it out loud. Not here in Tennessee." Ben Roy struggled for breath as the most soul-rattling sound Eula ever heard emerged in a keen, low moan from her husband.

Eula lurched in a haze. Fedora shouted words at her that she couldn't understand. Her sister-in-law may as well have been speaking Geechee.

"What you gonna do with that?" Fedora's face went white. Eula followed the short woman's horrified gaze to her own hand, where she held her just-sharpened chicken-butchering knife. Fedora pushed on Eula's hand with all her might, but it felt like a ladybug crawling up her arm.

"I'm goin' to kill Alex." She said it just like she asked her husband if he wanted a third cup of breakfast coffee.

Eula started toward the door, each foot feeling like a dozen horseshoes were nailed to it. Almost at the porch, she didn't see it coming. The chair crashed across her back and she stumbled. The knife skittered from her hand. Fedora kicked it to the other side of the kitchen.

"Get on into this parlor, now." Fedora put her full weight behind her sister-in-law and shoved her into the front parlor.

Eula's feet stumbled out from under her but not before Fedora gave her a final push in the direction of the settee. Eula fell more on than off of the horsehair-stuffed sofa. Fedora grabbed both of Eula's feet and draped them over one arm. She pushed a pillow under her head and started for the sideboard after she first slammed the door to the kitchen.

"Where the hell does Alex keep the key to the whiskey?" Fedora began pulling out drawers and opening doors.

The sounds coming from the back porch were now too muffled for Eula to hear.

"Bottom drawer. In the matchbox." Eula watched her sister-in-law fill two glasses almost to the top. Did Fedora actually think strong drink was going to help?

Walking to the settee, Fedora pushed a glass in Eula's hand as she pulled up a straight-backed chair.

"Where's yo' rose water? You got fresh lavender soap?" Fedora's words grated on Eula's ears. "Day after tomorrow, me and Ben Roy will take you over to Clarksville. I know a place where all the wives go. Yo' momma told me 'bout it."

Eula pushed herself up on the settee, looked at the amber liquid in the glass, and brought it to her nose. It smelled like tar paper.

"You can get black bloomers there." Fedora kept up her chatter. "They just cover the crotch and got pretty lace 'round the edges." Fedora stopped and took a deep swig from her glass. She coughed once and leaned in closer. "Eula, you payin' me any mind?"

"Alex don't think a lady should take a drink unless she's close to dyin'." Eula looked at the full glass in her hand and sloshed it around, not caring that some of the liquid dripped onto her tapestried settee. She raised the glass, nodded her head toward Fedora, and took in one large gulp of whiskey. Eula held it in her puffed-out cheeks until it burned, and then let it flow down the back of her throat.

"You listen to me," Fedora commanded, but her words held about as much force as Tillie's new baby, Little Ben. "It's a woman's job to keep her husband happy, especially if he's off with a colored woman. Now, this place has got short corsets that are made to push 'em up and out just like when you was twenty."

How many of these whore clothes did her sister-in-law own? Eula wondered.

"Fedora, to keep my husband in my bed, you want me to dress and act like a Clarksville trollop." She took another deep swallow of the whiskey. "I'd rather kill myself." She spit the word out at her sister-in-law.

The slap across her face came sudden and hard. In fact, if she hadn't seen Fedora standing there with her open palm, Eula would have sworn that the woman struck her with a closed fist loaded with buckshot. The whiskey spilled all over the settee.

"Who do you think you are, Eula Mae McNaughton?" Fedora was a head shorter, but right now, she towered over Eula. "Who told you that you could kill yo'self?" Fedora tossed her head back and started to laugh.

"Fedora?" Had her sister-in-law gone as mad as Alex? "Take yo'self another drink." Eula sat up straight as she tugged at Fedora to sit beside her.

Fedora could not be moved. "You sit there like you are the queen of all the Thorntons. Always did think you was the best of us. The best cook, the best canner, the best at managin' the farm. Yo' husband never had to say a harsh word to you, nor lay a hand to yo' head 'cause you never did nothin' wrong. Everthin' you touched was right." Fedora screamed every word at her.

"Fedora...no...I never thought I was better. I..."

"You're a damn liar, Eula Mae. And now you think you're too good to have yo' man climb into bed with a nigger woman. You think your stuff is so special, so pure, that you should be the only white woman in Lawnover, hell, in all of Montgomery County, whose man ain't never laid with a black woman." Fedora finished off the whiskey and threw the glass to the floor.

Eula watched it roll, unbroken, on her carpet. "Alex wouldn't...he couldn't."

"Why is that? Because he loves you so much? 'Cause he can't live without you?" The steam suddenly shot out of Fedora, and she flopped, deflated, on the settee beside Eula. "Don't you know it ain't got nothin' to do with how well you cook, or how many preserves you put up for the winter, or how clean your kitchen, or how many times you write in that damn journal of yours?"

"What...what was it then?" She welcomed Fedora's arms around her shoulders.

"Eula, you ain't the first in this family whose man took up with a colored woman." Fedora swallowed hard. "You know as well as me that Ben Roy got three yella bastards by Hettie. And, yo' own daddy, Old Ben, had that yella Johnny by that old woman all you Thorntons is scared of—Rebecca."

"That can't be right. My pa never..."

"And, if you're thinkin' Rebecca was the only woman yo' pa put it to, you're wrong. I hear tell there were at least a dozen others." Fedora sounded so sure.

"I don't believe you." Eula stared at her sister-in-law.

"Everybody in Lawnover knows the truth of it 'cept you. Kept it from you because you're his daughter." Fedora took her arms from around Eula's shoulders and folded them in her lap. She stared down at her hands.

"Does Tillie know about Hettie?" Eula managed.

"Not yet, praise the Lord." Fedora looked at Eula with hollow eyes. "Wiley George don't have a colored woman yet, far as I know, but he will. It's the way of a Southern man. It ain't nothin' you done or didn't do, Eula. You just been lucky that it took Alex twenty years to find him a black woman. All men do it, and it's just somethin' we wives have to get used to." Fedora let the words come out of her mouth like there was no other way to it in all this world.

"Well, I'll be damned if I'll ever get used to it. I'm leavin' Alex. I'll go to Kentucky with Bessie." She shrieked her agony when Fedora's second blow landed across her mouth.

"Ain't gonna be none of that. You ain't disgracin' this family. There'll be no Kentucky. Ben Roy won't have it no other way." Fedora was on her feet, towering over her again.

"No." The sobs finally came. "I'll never stay with Al...that bastard."

"Oh, yes, you will. You'll bear it the way the rest of us have. You ain't never gonna mention none of this to Alex. Is that clear?"

"The hell I won't. I'm gonna kill Alex if I don't leave him first. I'm…" Her sobs drowned the words in her throat.

"No, you ain't. Now, this woman's gone, but if he takes up with another, you're gonna pretend that you don't know nothin' 'bout it. Like you barely know her name. You ain't never gonna say a cross word to Alex, because if you do, it's only you who bears the hurt, not him." Fedora stooped and picked up the empty glass from the carpeted floor. She marched to the sideboard and refilled it. Wiping the rim, she handed it to Eula.

"I don't think I can do that…"

"Oh, you'll learn how quick enough."

"And if I can't? If I complain out loud to everybody who will listen about what… what Alex done to me?"

"They'll take you for a fool and a disgrace." Fedora reached for the whiskey bottle. "Not a disgrace to your husband, but a disgrace to every married white woman who ever lived in Lawnover. If you complain that yo' husband is cheatin' on you with a nigger, then you're telling everybody in all of Montgomery County that a colored woman is the same as you. That she's as good as you. That she's even better, because she's got yo' man. Eula, you can't do that." Fedora took a deep swig from the bottle.

"He said he lo… loves… her. He said he loves a… nig…" The word ripped out of her gut.

Fedora lowered herself slowly to the settee. "That's why you can't never complain. Not to nobody. Different if it was another white woman he said he loved. Then you been wronged fair and square, and you can cry and scream all you want. Everybody will come to yo' side. But when it's a colored woman, it just can't be the same. White men ain't supposed to love black women over us. My Lord, if we acted like that was true, there wouldn't be no sense to this world." Fedora patted her shoulder. "Never you mind, Ben Roy will make sure you never have to hear those words out of Alex's mouth again." She slipped an arm back around her shoulders. Her voice was full of sorrow.

Eula saw the floor coming up to greet her again. Fedora grabbed her hand and squeezed her wedding ring.

"You feel that?" She dug the too-tight ring into Eula's skin.

"Umm."

"You feel that ring on yo' finger, Eula Mae? You'll always be the wife. Let that be yo' comfort. He may be in her bed every night. He may even tell her that he loves her and mean it with all his heart. But that black woman will never have what you have. God and the law will always see to it that you come first."

"What do I care 'bout bein' first with the damn law. What do I care 'bout bein' first with God? What about his heart? I need to be first in his heart." The tears ran down her cheeks and onto her shirtwaist. "I want Alex to say those words to me...that he loves me. Only me." Eula pulled at the bodice of her dress so hard that one button clung to the garment only by a single thread.

Fedora slipped both arms around her, pulled her close, rocked her in her arms.

"Men think different than we do. They have their colored women for their own silly fun, but they have us fo' wives." She smoothed Eula's hair. "Alex is just talkin' crazy right now. He don't really want to trade." She sounded so certain.

"If he could, he would." The words rode out of Eula on a great wave of sobs.

All the years with Alex slipped in front of her eyes. Everything about her husband played over and over in her mind. She knew when he was happy, when the planting had gone wrong, when the harvest had gone right, when he was pleased, what he disliked, what and when he wanted more, when he wanted quiet, when he wanted laughter. Even the passion he'd shown two nights ago fit into the picture puzzle that was her husband. It had been there, simmering deep down in the center of him always. Now, somewhere, at her own core, she knew she'd never been, and would never be, the one to unlock that power. The woman...Laura...held the only key.

"If he could, he would." Eula repeated the sounds that squeezed out of her mouth in a soft moan.

Fedora squeezed the ring again. "You will get through this, Eula. Every other white wife in Lawnover married more than five years has lived through the exact same thing you're goin' through right now."

"Belle? Cora Lee? Jenny?" Their pain couldn't be as great as hers. She loved Alex. Did Fedora love Ben Roy like that?

"All of us. Don't worry none. Alex won't never talk about this day again. You ain't seen Ben Roy bring Hettie anywhere near my house, now have you? Why do you think I wouldn't allow Ben Roy's woman to come serve the plantin' dinner and prayer when Tillie carried on so? Your brother knows better than to parade his colored whore in front of me. All the men do." Fedora put her hand on Eula's whiskey glass and tipped it to her sister-in-law's lips.

As she drank, Eula knew the woman, Alex's woman, was no whore. He couldn't love her if she was.

The parlor door slammed open, and Ben Roy walked through, Alex's shotgun crooked in one arm.

"Let's get outta here, Fedora." Ben Roy spoke through blood-oozing lips. One eye was swollen almost shut, and a large lump had made an appearance on his forehead. He turned to his sister. "Eula Mae, get on in to yo' husband. He's in the kitchen." Ben Roy crossed the floor of the front parlor in two strides, grabbed Fedora by the wrist, turned, and stalked out of the house.

Eula eased up from the settee as though she'd been the one in the fistfight. Her feet wanted to take her into her bedroom, away from Alex, away from Fedora telling her to be brave, away from Ben Roy ordering her to comfort her scoundrel of a husband, away from all thoughts of... She wanted to climb into her bed and pull the coverlet over her head, go so deep into sleep that she would dream only of darkness. She turned her head toward the parlor door leading to the kitchen. Alex was in there. What could she say to him? Nothing, as Fedora suggested? Everything, like she wanted? Cajoling like Ben Roy ordered?

First one foot, then a second, led her toward the bedroom. She stopped in her tracks. Through the partially cracked door, she spotted the neatly made bed. The bed where Alex had lain with her two nights ago. Lain on top of her, and lied with his body. Her stomach churned as she moved like a stiff scarecrow into the kitchen.

Alex's usual chair was empty. Whether Ben Roy put him in there or he just didn't care, her husband sat at her regular place at the table.

Alex must have heard her walk into the room, for he turned a cheek reddened with an upcoming bruise toward her. His moist eyes looked right through her. He turned his head back to the crib and rocked it slowly. Moving like a rusted plow badly in need of oiling, he picked up the porcelain-headed doll and brought it to his lips. He rubbed his face against the horsetail hair.

On the kitchen table within easy reach was the knife. Ben Roy must have picked it up from the floor. Slow like she was treeing a possum, Eula inched her hand toward it. Alex paid her no mind. Grabbing the handle, she slid the knife as quiet as she could toward her. Even though it was twenty-one years old, the rocker was so well made that only the whisper of a sound came from its runners as Alex let his hand keep it in motion.

Eula brought the knife before her eyes, both of her hands clasped tight on the handle. She twisted it in the almost-noonday sun. A ray caught on Alex's yellow hair. She wondered what it would look like drenched in the red of his blood. She switched the knife to her right hand and took a step toward him. Her feet moved like great lead weights had been sewn into her cotton stockings. Standing right behind him and breathing hard, she wondered that he hadn't noticed her. She brought the knife handle to her right shoulder, the blade pointed straight at the middle of Alex's back. She closed her eyes and willed her arm to move with the strength of David in his battle with Goliath. The heavy image brought up a laugh from her gut that she failed to keep down.

Alexander McNaughton was no giant, and she wasn't a weak but brave child. She was poised to plunge a knife into the back of a cheating, lying husband, and he hadn't even noticed. That was just it, Alexander McNaughton had never noticed her—not to hurt her, not to hate her, not to love her. And if she did plunge the knife into his back, straight through to his front, and he turned his dying eyes toward her, he would only wonder at her actions. In his mind, he had done nothing to cause her harm, nothing to inflict pain. He had done nothing wrong. He would go to his Maker with innocence in his mind, pondering what had driven his wife to such an uncalled-for act. The laugh came out of her mouth garbled. It sounded more like she was wishing him a cheery good morning.

"Sorry 'bout breakfast. I'll get yo' pork chops on fo' dinner." She lowered the knife and laid it on the table as she moved her lead feet to the stove. Behind her, she heard Alex pick up the cradle and carry it back to her pantry.

As he walked past her like a dead man, she caught a glimpse of the porcelain-faced doll inside. For long seconds, she stood at the stove, a lit match in her hands, wondering what to do. Should she follow him into the pantry? Should she tell him something about loving him so much that she could die of it right now? That she could forgive him? The match singed her fingers, and she blew it out. She took a step toward the pantry. Trying to find a way to put her feelings into words, she moved almost to the door. Alex turned those vacant eyes on her. She reached out a hand to him just as the pantry door closed. She felt for the knob. She heard the pin drop into the latch. He had locked her out.

No thoughts came into her head. All the pain had left her heart. All the agony had fled her soul. She felt nothing as she walked to her safe. She didn't need her calendar to tell her that it was time to check her journal records. She reached up on the shelf and pulled out her account book. She laid it open on her table, reached for her pencil, and sat down. Thumbing through the book until she reached the Planting page, she began to write in her careful hand. Under Vegetables she wrote String Beans: 10 rows. Corn: 20 rows. Lima Beans: 5 rows. She had always been proud of her nicely squared-off handwriting, and though she couldn't feel the pencil in her hand this day, she saw that her writing was still meticulous. She turned to a fresh page to continue her accounts, but without her hand or her head willing it, the pencil lead began to write.

*May 21, 1914*
*Township of Lawnover*
*State of Tennessee*

*I, Eula Mae Thornton McNaughton, testify that this day I have become, at last, the perfect Southern wife. I now know to the core of my being the lessons my mother, and her mother before her, tried to teach me. On the eve of my wedding, when my mother told me that my husband would give me pain beyond all imaginings, I foolishly thought she spoke of his breaking of my flowering. Now, I know that it was my soul that would be shattered. It is because I am condemned to never be a whole woman that I make this testimony.*

*This day I resolve to do what I must. If, and when, my husband ever returns to my bed, I will hold him harmless for any pain he has ever put on me. I vow to never bring shame to him, or his name, by letting the world know that my entire life, and my love for him, have been betrayed in the most foul way. None will hear the cries that I dare not release from my chest even though each one strangles the breath out of me. If my husband ever again puts an arm around me, or even his lips to my forehead, I will never allow him to see the sorrow that drowns my heart. Beyond all doubt, I know that he wishes with all his being that the shoulder he touches, the lips he brushes, the breast he caresses, belonged to the body of another.*

*When he looks through me, I will always pretend that I don't know that it is the love of another he prays to possess. To my God, this day, I resolve to be the best white wife who has ever lived in Tennessee.*

## Eula McNaughton

In letters no more than tiny pricks of the pencil to the middle-aged eye, the words printed on the right-hand edge of the paper running up and down read *Alexander McNaughton loves Annalaura Welles.*

She punched the pencil deep into the book, but she could not bring herself to scratch out the words. Carefully closing the journal, Eula pulled her usual kitchen chair to the safe, climbed on it, and reached to the top shelf to move her silver-tipped wedding tureen. She tucked her journal behind it knowing she would never touch it again.

Not bothering to look where she landed, Eula stepped off the chair, grabbed a pitcher, and walked into the shambles that had been her porch. She pushed aside her now dented tin bathtub to reach the pump handle. Filling the pitcher with water, she returned through her kitchen, picked up the butcher knife, and retreated to her bedroom. She barely glanced at the bed as she poured water into the blue-flowered basin she had bought for herself as a remembrance of her tenth wedding anniversary. Bending over the water-filled basin, she saw the reflection of an old woman, her face lined with care, staring back at her.

Eula dipped her hand into the clean depths, her fingers splashing away the image. She would never look upon that woman again. She patted her face with the water, but she couldn't tell if it was hot or cold. With her hands dripping, she smoothed her hair and tugged the wrinkles out of her dress. She gave the knife only a quick glimpse as she picked it up and aimed its point at the fleshy part of her finger. Like everything else this morning, and for the rest of her life, she couldn't feel its sharp point jab into her flesh. When the bubble of blood oozed to the surface, she took her finger and smeared its redness across both cheeks. She pulled open a bottom drawer of her dresser and took out the hand mirror that she'd never seen a need to use. She checked her cheeks in its glassiness. The care-lined old woman was gone.

She was ready. With the lead from her shoes and stockings quickly disappearing, Eula walked through her parlor, past her kitchen, out through the now unhinged porch door, and over to her buckboard in the barn. She must ready her horse for a trip to Fedora's. It was time to redeem herself. When she sees her sister-in-law, of course she will act as though nothing unusual has happened in the McNaughton household these last ten months. Her fainting spell was just part of the change of life. Fedora will recognize that. All in all, she will act as though she understands everything and knows nothing. She will be the good Southern wife. After all, that is the only thing that Alex has ever loved about her.

# CHAPTER TWENTY-EIGHT

"Cairo. Cairo, Illinois." The porter walked down the aisle of the just-stopped Illinois Central train.

Sitting bolt upright in the stiff seat, Annalaura looked out the window into the lamp-lit dimness of the railway station. A few passengers from the car just ahead walked the platform, puffing on cigarettes. The letters, C-A-I-R-O, stared back at her from the sign suspended just over the round-faced brass clock that showed ten p.m. But she read only A-L-E-X.

Annalaura, the children, Becky, and John had boarded that first train fourteen hours earlier in Clarksville, and after a three-hour layover in some Kentucky town too small to notice, John had settled them all into the dusty, colored-only coach bound for Chicago. Now, the porter, in his white railroad jacket, paused briefly beside Annalaura's seat. The man give her a slight wink as he pretended to check the destination on her ticket.

"Chicago, Illinois. Land o' Lincoln." He let a slight smile play across his lips.

Did he want to make sure she understood that he had just announced her arrival into the Promised Land? But was it?

Annalaura looked down at Lottie, lying beside her, deep in sleep, her little girl braids resting heavy against Annalaura's shoulder. Doug and Henry sat across the aisle with Aunt Becky, all three sound asleep—Henry curled up on his great aunt's lap. Annalaura started to whisper to Rebecca that freedom had come to her for real this time. They had crossed over into the North, where a colored person's life was supposed to be easier.

She glanced over at Henry's still body, dreaming his little boy dreams of drums and shoes that didn't pinch. Staring out of the window into the darkness as the colored porter pulled up the portable steps, Annalaura spotted Becky's wavering reflection in the window. Her aunt sat with her chin on her chest, her mouth open. The unfamiliar feel of a tiny smile played across Annalaura's lips, the first since the beating John had laid on her three days ago. If she could believe her husband, Alex still lived. She let her shoulders sink into the seat back.

It would be midnight in another two hours, and the porter had already announced that Chicago was eight hours away. Dawn would come soon enough, and with it, her own uncertain future. As the railroad man made his way out of their train car, clanging the heavy connecting door behind him, Annalaura caught a glimpse of movement across the aisle and four rows up. John carefully stepped over a sleeping Cleveland and swayed his way down the aisle. Before she could let the full fright take hold in her brain, her hand clutched the shawl covering baby Dolly. The husband she had betrayed walked toward her, his face bathed in flickering shadows.

Since this morning, when he had stormed into the cabin before Becky could get the old blunderbuss off its fireplace hook and ordered them all to come with him "right now," he hadn't said more than a dozen words to her. When she scooted her sore and bruised body to the far side of the bed, fierce to let no harm fall on Dolly, John let his words come out as cold as the barrel of Becky's gun.

"McNaughton ain't dead...yet," he'd said, "and if you wants to keep it that way, you'd best be comin' with me."

Annalaura couldn't recall much more about that moment, only that her brain felt like it was pushing her eyes right out of her head. Little Dolly's gasping cries brought her back to herself in the train car. She searched in her head for the right questions to ask her husband the why of it. She remembered that John had swooped her and the baby into his arms and shoved past a flabbergasted Becky. Even when he laid the two into the back of a wagon already crowded with Doug, Henry, and a shivering Lottie, Annalaura couldn't muster a word. Was it a trick? Had John killed Alex after all? Before the night riders caught up with them, was John driving her to take a final tormenting look at Alex's lifeless body? In the wagon, she'd clung tight to the swaddled infant as she grabbed at Doug's foot, resting in the small of her back.

"Cleveland?" had been the only sound she'd managed.

"Up top with Papa and Aunt Becky," Doug had whispered back.

Annalaura felt Lottie wrap her skinny arms around her mother's drawn-up knees. Why was John taking his own children to see a dead man? Even in her bewilderment, she'd felt the tears forming.

Under Becky's blanket, and a pile of old burlap that still smelled of last year's husked corn, Annalaura stretched out a hand to pat Henry's face, now snuggled hard into Doug's chest. Where was her husband taking them?

With Doug's shoe putting fresh bruises on her back, Annalaura tried to reach for Lottie as the wagon jolted to a wobbly stop. She remembered croaking something at Doug about covering little Henry's eyes. She would stop her children from seeing the sight of white-hooded men, stout trees, and knotted ropes, for as long as she could. She shut her own eyes tight, clung to Dolly, and let the darkness swoon over her.

The feel of rough hands, and the sharp stab of daylight, brought her back to the world. Annalaura worked Lottie's head and shoulders between her knees and rolled her own body on top of the newborn. But, it was too late. Before she could blink, she felt John take her in his arms and toss the blanket over her and the baby. She turned her face into his chest. She didn't want to see where he was carrying her as the hissing sound of steam beat into her ears.

By the time John set her down on the second of the train steps, Annalaura already knew that her husband had taken her to the Clarksville railroad station. With her eyes blinking in the light of the early morning sun, she looked up to see a stern-faced Cleveland grab for her arm. As John gave her a final boost, she turned to look out at the platform. Other than her husband handing up Lottie and Henry, and the four or five other people boarding the cars ahead, she didn't see a white-hooded man in sight.

As John boarded the train with Becky and Doug trailing him, Anna-laura tried to pick the words out of her confusion. Her husband brushed right past her and settled next to Cleveland. As the porter called out his "All Aboard," John turned his face to the window. Only Aunt Becky had words for her, "Gal, thank yo' Jesus fo' this day."

$$\infty$$

"They say Chicago's three or fo' times bigger'n Nashville." John stood over her in the train aisle of the Chicago-bound train. "Lots mo' people livin' up there, but ain't many of them colored." John looked through the window at the Cairo sign.

Annalaura's shoulders shuddered at her husband's ramblings. What did he want?

"Too easy to find colored folks livin' up there." In the semi-darkness of the train car, John shifted his eyes to the top of her head. "Like pickin' out raisins in a bowl of milk."

Annalaura tucked the shawl tighter around Dolly, and eased the baby closer to her breast.

"This night air . . . " John pushed a hand toward the shawl.

The move sent a sudden shiver through Annalaura. Could she believe her husband? Did Alex still live?

John lifted his head. His eyes roamed across Annalaura's face. She watched him work his lips but no words came. A burst of steam belched from underneath the train. She heard her husband's loud swallow as he aimed a finger toward the shawl-wrapped bundle.

"Night air can be bad for little ones, the old folks say." He stared down at his spread fingers.

Annalaura stared back at him. Something was new. His eyes swam with eye-brimming water. "Night air bad," she managed.

The bulge in her husband's throat bobbed up and down. His nostrils flared.

"She all right?" He whispered as he looked at the shawl.

"She's just fine." Annalaura kept her eyes on John.

"Annalaura . . . " The word came out strained. "I'm not wantin'

Chicago… but…if you think it be best fo' you and the chil'ren, then I…I mean that I won't…be stoppin' you." He parceled out the words.

Annalaura shook her head. What was her husband saying?

"McNaughton…If he took a notion to look…" John thrust a hand in his pant pocket and pulled out a balled-up wad of paper. "If you got yo' mind set on Chicago, then I reckon there's not much I can do to change it. But I wants you to know…" he shifted his eyes to Dolly, "I ain't never leavin' you again, Annalaura…not lessen you tells me to go." John pushed the wad between the folds of the shawl. "Becky says you callin' her Dolly." He settled his eyes somewhere between Annalaura's neck and chin. "Dolly…Dolly Welles. That'll set just fine with me."

John jumped to his feet, took a step toward the front of the railcar, stopped, and stared straight ahead just as the train began its slow roll forward. "Becky let me know who pulled that baby out of you. Snatched you away from the angel of death, she say." He started to move. John called over his shoulder as he began his way up the aisle. "You pick."

Annalaura shook her head in confusion. Her hand fingered the wadded paper. She smoothed out the wrinkles just as Lottie stirred. Annalaura stroked her daughter's shoulder, and offered up a silent prayer that her firstborn girl would never know the pain of being torn in two. She let her eyes rest on the paper. Her husband had given her the train schedule.

She stared at the printed sheet now lying open on Dolly's shawl. On the creased paper, she saw the names of unfamiliar towns lining the state of Illinois from Cairo, all the way north, to the end of the line. John had put a pencil scratch through the black letters of the last town listed—Chicago. Annalaura raised the shawl and stared down at her sleeping youngest daughter. Had her ears heard right? Did John tell her to pick? Pick what?

"Raisins in a bowl of milk." She murmured out loud. She let her fingers travel down the schedule, touching the strange names—Carbondale, Urbana, Danville, Bloomington. Strange towns. Unknown towns. If Alex cared to search for her, and in her heart, she knew he would, he would think only Chicago. Never Urbana. Never Bloomington.

Annalaura patted Lottie's shoulder. Across the aisle, deep in sleep, his head resting against the hard armrest of his seat, Doug smiled like he was dreaming of books with more words than pictures. Henry, Lottie, Cleveland—she owed them all the promise of hopes and dreams. And John. Annalaura let her eyes travel to the front of the train car. The head of the man she'd promised to love and obey 'til death parted them rose over the seat back, rigid, like he was waiting for something important to be settled before he could allow himself to drift into his own sleep.

The Land of Lincoln, the Land of Promise. She had promised John. She had promised Alex. Choose one, and all her yesterdays with the other had to be forgotten. Pick an anonymous name on a train schedule, and her new life would begin with John. He would take Dolly as his own. Love her as his own flesh. Choose Chicago, and become a raisin in a bowl of milk. A raisin waiting for Alex to knock on her door—to claim her as his own.

The train lurched around a curve. Annalaura felt Lottie's body burrow deeper into her aching chest.

Choose, John had made it clear. Two towns. Two men. A shaft of moonlight so bright that it won out against the gaslight in the coach car, bounced off the window, and settled across the back of her hand. She stared at her arm. A slight movement across the aisle drew her gaze to Becky. She shifted her eyes from the old Cherokee to her hand cradling Dolly, and back again. For the first time, she realized that she didn't need her aunt's second sight to give her the strength to do what she must.

Annalaura bent down to the shawl-covered bundle and kissed her daughter on the forehead. Choose, John had said. I forgive you. She took in a gulp of air and stared long seconds at her hand. Though the errant speckles from the moonlight had slipped away, her fingers still showed golden. The ache in her head that had held fast since John left her a year back slipped away. The velvet night outside the train window sparkled as clear as her mind felt. Annalaura looked up the aisle toward her husband. Bless him for his love, but she needed no forgiving. She had done no sin against her God, nor against her husband. Her eyes went to Dolly. Alex, trapped in his own world, had tried, like John, to force a choice out of her. But how could any choice belong to her when it was someone else who laid out only what he knew?

Her free hand fumbled over the thin, wrinkled slip of paper John had thrust at her. She caught her breath as she let the fingers that had caressed Alex's naked body play across each of the lined-over letters that spelled out CHICAGO. She lowered her eyelids to block out all sight in the train car. Her hand began a slow slide down the page. A raised place in the paper stopped her. She eased her eyes open and stared down at the letters on the smoothed-out sheet. Her answer had lain there all along. The pain in her heart eased. Choose, said John. Choose me, Alex had said. But she had a choice neither man had given her. Annalaura stroked Lottie's arm and settled into sleep, her mind at peace. All of her tomorrows belonged in only one set of hands—her own.

# ACKNOWLEDGMENTS

*Page* has been a story demanding a voice for almost one hundred years. Thank you, Teresa LeYung Ryan (author of *Love Made of Heart*), for being the first to let me believe that I could pull one woman's story out of the dark world of secrecy and into the light. Bless the California Writers' Club—Berkeley Critique Group/ David Baker and Anne Fox—for forcing the tools into my hands to allow me to put thoughts to paper. To that first batch of brave readers—Tootsie, Gloria, Juanda, Dora Jean, and Rozelle—I owe you an immeasurable debt of gratitude for wading through my sea of words. And just when I thought *Page* would forever languish on a bookshelf for my eyes only, Gilles, you gave me the encouragement for that final push.

I will always be in awe of whatever force directed Terry Goodman, senior acquisitions editor at Amazon-Encore, to choose *Page* from all the manuscripts submitted to him. Terry's input, encouragement, and support have been incredible. A special acknowledgment to the entire team at AmazonEncore. They've made this an experience of a lifetime.

I thank my family, Hank and Doug, for putting up with me these past six years. But, most of all, Grandma, this is for you.